Applications of Laser–Plasma Interactions

Series in Plasma Physics

Series Editor:
Steve Cowley, Imperial College, UK and UCLA, USA

Series in Plasma Physics

Applications of Laser–Plasma Interactions

Edited by

Shalom Eliezer
Soreq Nuclear Research Center
Israel

Kunioki Mima
Osaka University
Japan

CRC Press
Taylor & Francis Group
Boca Raton   London   New York

CRC Press is an imprint of the
Taylor & Francis Group, an **informa** business

A TAYLOR & FRANCIS BOOK

CRC Press
Taylor & Francis Group
6000 Broken Sound Parkway NW, Suite 300
Boca Raton, FL 33487-2742

© 2009 by Taylor & Francis Group, LLC
CRC Press is an imprint of Taylor & Francis Group, an Informa business

First issued in paperback 2019

No claim to original U.S. Government works

ISBN-13: 978-0-367-45247-6 (pbk)
ISBN-13: 978-0-8493-7604-7 (hbk)

**Visit the Taylor & Francis Web site at
http://www.taylorandfrancis.com**

**and the CRC Press Web site at
http://www.crcpress.com**

Library of Congress Cataloging-in-Publication Data

Applications of laser plasma interactions / Shalom Eliezer, Kunioki Mima.
    p. cm. -- (Series in plasma physics)
    Includes bibliographical references and index.
    ISBN 978-0-8493-7604-7 (hardcover : alk. paper)
    1. High power lasers--Industrial applications. 2. X-ray lasers--Industrial applications. 3. Laser-plasma interactions. 4. Laser fusion. I. Eliezer, Shalom. II. Mima, Kunioki. III. Title. IV. Series.

TA1677.A67 2009
621.36'6--dc22   2008044198

It is an honor for us to dedicate this book to the pioneers of laser fusion

and high-power laser applications: Professors Chiyoe Yamanaka, Yoshiaki Kato,

Toshiki Tajima, Guillermo Velarde, Jose M. Martinez Val, and Manuel Perlado.

Contents

Foreword

This book is about the practical use of laser interactions with the fourth state of matter: plasma. This is the story of the coherent interactions of photons with electrons and ions. Both the laser and the plasma have unique properties that offer a multitude of applications through their interactions.

The most important match between laser and plasma would be the solution of the energy problem for mankind. If this could be realized even by the middle of this century, then all the intellectual effort on this project would be one of the greatest achievements of laser–plasma applications. Without energy, the human society cannot be sustained. Energy sources can be divided into two categories: the good and the bad. The good, that is, the use of economic solar energy on a large scale; nuclear fusion energy from inertial fusion and magnetic fusion; and a safe, secure nuclear fuel cycle for fission reactors are still not available. The bad, that is, the present major source, fossil fuels, may cause irreversible damage to our planet.

Laser has the ability to deliver a wide range of energy densities during various time durations. For example, present laser technology can provide a mega-joule of energy into a space of about a millimeter cube during a time duration of nanoseconds, while, with femtosecond lasers, the power of a petta-watt can be achieved on a scale with dimensions of microns.

Plasma can cover very wide density and temperature ranges that extend over many orders of magnitudes. Plasma has the advantage over other matter because it does not break or disintegrate when heated to extreme temperatures or when applied to strong electromagnetic fields. This makes plasma a good and interesting companion for the high-power laser in multiple applications.

Other applications in this book are diverse enough to justify multiple laser–plasma relationships. Due to the "unbreakable" plasma in very high electric fields, this relationship may lead us to the ultimate accelerator. The laser is capable of stripping every atom from any electron number it desires, including stripping off the 92 electrons from uranium. This makes the "laser–plasma couple" an extremely useful candidate for x-ray sources. Even a coherent x-ray laser has been achieved under laser–plasma interactions. The equation of state of matter is also analyzed at very high pressures and high energy densities with the help of the laser. Today, a very popular research field in applications of high-power lasers is the nano particle matter produced by lasers. The nano particles have been created recently through laser–matter interactions.

As is well known, basic research is used for application purposes. In our book, the application of the laser–plasma couple is used for carrying out basic research. This includes nuclear physics, elementary particle physics, and basic quantum mechanics.

This book is all about the "laser–plasma couple" that has the potential to solve the energy problem of mankind, to be used for many applications in industry, and to carry out new basic research.

Shalom Eliezer
Kunioki Mima

Preface

As early as 1917, Albert Einstein suggested the theory of stimulated emission of light that led to the development of laser. The first laser, based on Einstein's theory, was demonstrated by Maiman in 1960 and was believed to be the solution to an unknown problem. As a solution to one of the foremost problems, it was suggested by Professors Edward Teller and Nicolai Basov to use nuclear fusion induced by lasers to solve the energy problem. This was proven correct after Maiman's demonstration in 1960. Since then, intensive international research has been carried out to attain this goal and nuclear fusion has been developed successfully by compressing a laser pellet. In a few years, ignition and burn of imploded fusion fuel will be demonstrated at the National Ignition Facility at Lawrence Livermore National Laboratory, Livermore, California and at Laser Megajoule, Bordeaux, France. However, it will take more than two decades for it to be used commercially. Chapter 1, by Mima et al., deals in detail with the science and technology of this important subject that could be the solution to the energy crisis for generations to come.

The recent advances in developing lasers with more energy, power, and brightness have opened up new possibilities for exciting applications. The electric fields that can be achieved with intense lasers have paved the way to a new technology for particle accelerators as described in Chapter 2 by Kogan. Lasers, while interacting with plasma, can produce efficient high-power x-ray sources. The applications of these sources are used with high-power, short-pulse durations and small dimensions obtained in laser–plasma interactions. These topics are discussed in Chapter 3 by Nishimura, while the development of x-ray lasers used in biological sciences and solid-state physics is described in Chapter 4 by Daido et al.

Ultrahigh power lasers generate high fluxes of energetic photons, electrons, protons, and ions that are used for the study of applications of nuclear and elementary particle physics. Chapter 5 by Mendonça and Eliezer deals with this topic. The very high power density that can be achieved in laser–plasma interactions can be used for the study of matter under extreme conditions as described by the high pressure and high temperature equations of state. This topic is summarized in Chapter 6 by Eliezer and Henis. The very short femtosecond laser pulses used in material processing are described in Chapter 7 by Fujita. And finally, the popular subject of nanoparticles in the context of laser–plasma interactions is put forward in Chapter 8 by Eliezer.

This book reviews the current status of the above topics related to high-power laser applications. We hope that it will attract readers to the high-power laser world.

Shalom Eliezer
Rehovot, Israel

Kunioki Mima
Osaka, Japan

Acknowledgments

We are grateful to Yaffa Eliezer and Kazuko Mima for their lifelong support of our scientific activities and also express our thanks to Professor Masakatsu Murakami and Kaoru Nishimura for their patient efforts in arranging and collecting manuscripts.

Kunioki Mima
Shalom Eliezer

Editors

Shalom Eliezer studied physics at Technion University, Israel, where he received a DSc in 1971. He has published about 300 papers in scientific journals; written or co-authored 6 earlier titles, of which 2 have appeared as second editions; contributed chapters to other books; and given over 80 invited lectures at international conferences. His previous book *The Interaction of High Power Lasers with Plasma* is a thorough self-contained presentation of the physical processes occurring in laser–plasma interactions.

Professor Eliezer is a fellow of the American Physical Society, an honorary fellow of the Institute of Nuclear Physics at the Polytechnic University of Madrid, a recipient of the Israeli Landau award for achievement in physics on the subject of laser–plasma interaction, a member of many international advisory boards, and an editorial board member of several scientific journals. His areas of research include laser–plasma interactions, nuclear fusion, high-pressure and high-energy density physics, nuclear physics, and elementary particle physics.

Kunioki Mima has been serving as the director of the Institute of Laser Engineering (ILE), Osaka University, since April 2005. He received his PhD in physics at the Graduate School of Science, Kyoto University. He was appointed as an assistant professor at the Faculty of Science, Hiroshima University, and then moved on to Osaka University in 1975. Since then, he has worked at Bell Telephone Laboratory, Murray Hill, New Jersey and at UCLA from 1975 to 1977. In 1984, he was appointed as a full professor at ILE. From 1995 to 1999, he served as the director of ILE to initiate the fact ignition research. He is the general co-chair of the International Conference on Inertial Fusion Sciences and Applications (IFSA). The 5th IFSA was held from September 9 to 14 at Kobe, Japan. He is also the vice president of the Japan Society of Plasma Science and Nuclear Fusion Research. He is a fellow of the American Physical Society (APS) and has received the Award for Excellence in Plasma Physics Research from the APS and also the Edward Teller Award for his contribution to laser fusion research from the American Nuclear Society at Kobe on September 13, 2007.

Contributors

Hiroyuki Daido   Advanced Photon Research Center, Japan Atomic Energy Agency, Kyoto, Japan

Shalom Eliezer   Soreq Nuclear Research Center, Yavne, Israel

Masayuki Fujita   Institute for Laser Technology, Osaka, Japan

Zohar Henis   Soreq Nuclear Research Center, Yavne, Israel

Tetsuya Kawachi   Advanced Photon Research Center, Japan Atomic Energy Agency, Kyoto, Japan

James Koga   Advanced Photon Research Center, Japan Atomic Energy Agency, Kyoto, Japan

José Tito Mendonça   Grupo de Lasers e Plasma, Instituto Superior Técnico, Lisboa, Portugal

Kunioki Mima   Institute of Laser Engineering, Osaka University, Osaka, Japan

Kengo Moribayashi   Advanced Photon Research Center, Japan Atomic Energy Agency, Kyoto, Japan

Masakatsu Murakami   Institute of Laser Engineering, Osaka University, Osaka, Japan

Sadao Nakai   The Graduate School for the Creation of New Photonics Industries, Shizuoka, Japan

Hiroaki Nishimura   Institute of Laser Engineering, Osaka University, Osaka, Japan

Alexander Pirozhkov   Advanced Photon Research Center, Japan Atomic Energy Agency, Kyoto, Japan

1

Inertial Fusion Energy

Kunioki Mima, Masakatsu Murakami, Sadao Nakai, and Shalom Eliezer

CONTENTS

1.1   Introduction

In this chapter the production of inertial fusion energy (IFE) is discussed. The final goal of the research described here is the production of "clean" energy for commercial applications by igniting fusion fuel pellets. Besides the deuterium–tritium (DT) fuel, other fusion fuels are considered. "Direct drive" describes the scenario where many high intensity laser beams irradiate uniformly a spherical shell with the nuclear fuel. At stagnation the nuclear fuel is highly compressed and heated inducing nuclear fusion.

The "indirect-drive" scenario starts with the conversion of the laser beams into soft x-rays within a cylindrical cavity of a few mm size. The compression and heating of the spherical shell with the nuclear fuel is done by the x-ray photons in the cavity.

In both schemes described above, the direct and indirect drive, the shell must be stable to hydrodynamic instabilities, such as Rayleigh–Taylor (RT), in order to obtain the high compression necessary for ignition. The hydrodynamic instabilities are initiated by all spherical symmetry imperfections in the shell surface and the nonuniformity of the pellet irradiation.

The severe requirement of hydrodynamic stability has been relaxed in the scheme of "fast ignition (FI)." In this case the compression is achieved like in the direct drive or indirect-drive models; however, the necessary heating for the nuclear ignition is derived by another laser heating a segment of the fuel. The "impact-FI" possibility is also described.

The physics involved in all scenarios is described and analyzed. The scientific and technological gaps in reaching an IFE reactor are considered. The final goal is to obtain an economical electric power plant based on IFE.

1.2   Nuclear Fusion Reactions

Nuclear fusion energy is the energy source of our universe. It is the origin of energy in our sun (Atkinson and Houtermans, 1929) and in the stars (Bethe and Peierls, 1934; Bethe, 1937, 1939; Gamow and Teller, 1938; Landau, 1938). The main energy production mechanism in the sun is due to the combination of four protons into an alpha particle, as described by the following nuclear fusion reactions:

$$
\begin{aligned}
p + p &\to D + e^+ + \nu_e + 1.2 \text{ MeV}, \\
p + D &\to {}^3He + \gamma + 5.5 \text{ MeV}, \\
{}^3He + {}^3He &\to \alpha + 2p + 12.9 \text{ MeV}, \\
\Rightarrow 4p &\to \alpha + 2e^+ + 2\nu_e + 2\gamma + 26.3 \text{ MeV},
\end{aligned}
\tag{1.1}
$$

where

p, D, ^{3}He, and α are proton, deuterium, helium-3, helium-4

e$^+$, γ, ν_e are the positron, a gamma, and a neutrino (of the electron type), respectively

The nuclear fusion of hydrogen is a weak interaction and it is possible due to the gravitational force of the large mass of the sun.

In 1939, Hans Bethe suggested (and for this he received the Nobel Prize in 1967) that the energy source of stars more massive than our sun is obtained by the carbon cycle (Bethe, 1939). This cycle is described by the following nuclear reactions:

$$
\begin{aligned}
p + {}^{12}C &\rightarrow {}^{13}N + \gamma + 1.9\,\text{MeV}, \\
{}^{13}N &\rightarrow {}^{13}C + e^+ + \nu_e + 1.5\,\text{MeV}, \\
p + {}^{13}C &\rightarrow {}^{14}N + \gamma + 7.6\,\text{MeV}, \\
p + {}^{14}N &\rightarrow {}^{15}O + \gamma + 7.3\,\text{MeV}, \\
{}^{15}C &\rightarrow {}^{15}N + e^+ + \nu_e + 1.8\,\text{MeV}, \\
p + {}^{15}N &\rightarrow {}^{12}C + \gamma + \acute{a} + 5.0\,\text{MeV},
\end{aligned}
\tag{1.2}
$$

where C, N, and O are the nuclei of carbon, nitrogen, and oxygen accordingly. The result of this cycle can be represented by

$$
4p \rightarrow \alpha + 2e^+ + 2\nu_e + 3\gamma + 25.1\,\text{MeV}. \tag{1.3}
$$

In stars with core temperatures greater than 15 million kelvin the carbon fusion cycle seems to be the dominant effect rather than the hydrogen fusion described in Equation 1.1.

Furthermore, the formation of the elements up to iron occurs by nuclear fusion (Burbidge et al., 1957). For example, the following reactions describe the production of the light elements:

$$
\begin{aligned}
\alpha + D &\rightarrow {}^6Li + \gamma + 1.5\,\text{MeV}, \\
{}^3He + {}^3He &\rightarrow {}^6Be + \gamma + 11.5\,\text{MeV}, \\
\alpha + {}^3He &\rightarrow {}^7Be + \gamma + 1.6\,\text{MeV}, \\
{}^6Li + {}^6Li &\rightarrow {}^{11}B + p + 12.2\,\text{MeV}, \\
{}^6Li + {}^6Li &\rightarrow {}^{11}B + D + 3.0\,\text{MeV},
\end{aligned}
\tag{1.4}
$$

where Li, Be, and B are lithium, beryllium, and boron, respectively. One of the advantages of using thermonuclear fusion of light nuclei as the source of energy lies in the fact that a small investment in the kinetic energy of the ions (~10 keV per ion) can induce an output of large energy (~10 MeV per reaction).

Between the possible reactions the first candidate for fusion development relies on DT interaction creating an α and a neutron (n)

$$D + T \rightarrow \alpha + n + 17.6\,\text{MeV}. \tag{1.5}$$

Other interesting candidates for fusion reactions for potential reactor applications are

$$D + D \rightarrow \begin{cases} p + T + 4.1\,\text{MeV}, \\ n + {}^3\text{He} + 3.2\,\text{MeV}, \end{cases}$$

$$D + {}^3\text{He} \rightarrow p + \alpha + 18.3\,\text{MeV},$$

$$p + {}^{11}\text{B} \rightarrow 3\alpha + 8.7\,\text{MeV},$$

$$D + {}^6\text{Li} \rightarrow \begin{cases} {}^7\text{Be} + n + 3.4\,\text{MeV}, \\ {}^7\text{Li} + p + 5.0\,\text{MeV}, \\ 2\alpha + 22.3\,\text{MeV}, \\ {}^3\text{He} + \alpha + n + 1.8\,\text{MeV}, \\ T + \alpha + p + 2.6\,\text{MeV}, \end{cases} \tag{1.6}$$

$$p + {}^6\text{Li} \rightarrow {}^3\text{He} + \alpha + 4.0\,\text{MeV}.$$

It is worthwhile to point out that in considering the DT nuclear fuel one gets the interactions of DD and TT besides the DT described in Equation 1.5. Furthermore, all possible reactions of the products occur (Eliezer et al., 1997, 2000). For example, for a pure deuterium fuel one has to consider the following reactions:

$$D + D \rightarrow n + {}^3\text{He} + 3269\,\text{keV},$$

$$D + D \rightarrow p + T + 4033\,\text{keV},$$

$$T + p \rightarrow n + {}^3\text{He} - 764\,\text{keV},$$

$$D + T \rightarrow n + {}^4\text{He} + 17{,}589\,\text{keV},$$

$$T + T \rightarrow n + n + {}^4\text{He} + 11{,}332\,\text{keV}, \tag{1.7}$$

$${}^3\text{He} + D \rightarrow p + {}^4\text{He} + 18{,}353\,\text{keV},$$

$${}^3\text{He} + {}^3\text{He} \rightarrow p + p + {}^4\text{He} + 12{,}860\,\text{keV},$$

$${}^3\text{He} + T \rightarrow n + p + {}^4\text{He} + 12{,}096\,\text{keV},$$

$${}^3\text{He} + T \rightarrow D + {}^4\text{He} + 14{,}329\,\text{keV}.$$

Not all these reactions have the same probability and it turns out that only a few of them contribute significantly to DD fusion (e.g., the D³He and DT have important contributions).

The DT fusion cross section σ and rate σv are the largest for the low energy range (~10 keV) as can be seen from Figures 1.1 and 1.2. Therefore, the DT fusion

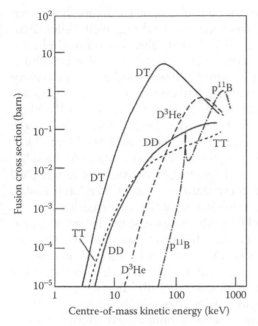

FIGURE 1.1
Fusion cross sections for various reactions.

process requires the smallest investment of energy per ion and it provides the greatest energy amplification. The nuclear fusion reactions of DT yields, per unit mass, eight times more energy than nuclear fission of uranium and more than a million times in comparison with fossil fuels. The main question is how to use the nuclear fusion energy here on our planet in a controllable way.

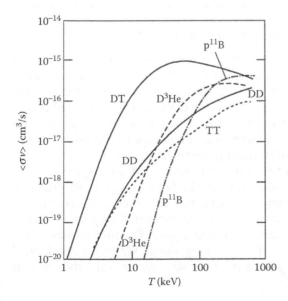

FIGURE 1.2
Values of $<\sigma v>$ for various reactions.

Two different distinctive schemes for controlled nuclear fusion have been investigated in the past 50 years (Glasstone and Lovberg, 1960; Teller, 1981; Dolan, 1982; Herman, 1991; Fowler, 1997; Harms, 2000; Eliezer and Eliezer, 2001). (1) Magnetic confinement fusion (Post, 1956; Stacey, 1984; Hazeltine, 1992; Kadomtsev, 1992; Berk, 1993; Rosenbruth, 1994) based on high intensity magnetic fields (~several teslas) confining low-density (~10^{14} cm^{-3}) plasmas for long times (~10 s). (2) Inertial confinement fusion (Basov and Krokhin, 1964; Nuckolls, 1972; Deutsch, 1986; Yamanaka, 1991; Eliezer, 1992, 2002; Velarde et al., 1993; Hogan, 1995; Lindl, 1998; Liberman, 1999; Hora, 2000) based on rapid heating and compressing the fusion fuel contained in a spherical target of several millimeters in diameter. At densities of about 1000 times liquid DT (~200 g/cm^3) and temperatures around 5 keV the fusion reaction rate occurs efficiently before the plasma pellet disassembles.

The idea of using lasers to fuse a DT mixture was first published by Basov and Krokhin (1964). This idea was incredible taking into account that Maiman constructed the first laser only in 1960 (Maiman, 1960). As indicated in the first publications on this topic (Basov and Krokhin, 1964; Dawson, 1964; Kastler, 1964), it appeared that it was possible to build the necessary short laser pulses needed to obtain high-gain fusion. After the first fusion neutrons had been detected by laser irradiation of targets containing deuterium (Basov et al., 1968) and were confirmed in a sufficiently convincing number (Floux, 1971), the biggest center for laser fusion was established in the United States at the Lawrence Livermore National Laboratory (LLNL) followed by other centers. In 1972, the fact that it is necessary to compress the fusion fuel to very high densities was published for the first time (Nuckolls, 1972).

1.3   Background for Implosion and Ignition Schemes

Thermonuclear ignition has been a long-awaited goal in the research and development of IFE. The core physics of IFE is implosion of a fusion fuel pellet, which has been well investigated since 1972 (Nuckolls, 1972). The megajoule laser facilities that are now under construction in the United States and France are expected to demonstrate fusion energy gain within several years. This will be an epoch-making achievement, which will give us the real means to solve the future energy and environmental problems of the world.

When an intense laser light is uniformly impinged on a spherical fuel pellet with an intensity of 10^{14}–10^{15} W/cm^2, the laser energy is absorbed on the surface to generate a high-temperature plasma of 2–3 keV, and an extremely high pressure of a few hundred megabars is generated. This pressure accelerates the outer shell of the target toward its center. The mechanism of the acceleration is the same as that of rocket propulsion. When the accelerated fuel collides at the center, compression and heating occur. If the dynamics is sufficiently spherically symmetric, the central area is heated up to 5–10 keV (called the "central spark"), and a fusion reaction starts.

In reactor-scale implosions, a fuel pellet with an initial radius of about 3 mm should be compressed to a radius of about 100 μm. For this purpose, highly precise uniformity is required for irradiation intensity distribution over the fuel pellet as well as high-quality sphericity and uniformity of the pellet. The implosion velocity for achieving the fusion ignition temperature of 5–10 keV at the center is required to be $3-4 \times 10^7$ cm/s. Therefore, the pulse length of the laser should be 10–20 ns, and in this timescale the directed megajoule energy has to be delivered to a fuel pellet of radius about 3 mm. The directed energy to drive the implosion is simply called the "driver."

In addition to laser there are such other candidate drivers as particle beams and pulse power as shown in Table 1.1. High-power lasers have been used for physics investigation of implosions in the past because advanced technology is needed to deliver a relatively large energy in a short pulse and to focus the energy into a small spot. Inertial fusion by a laser is called "laser fusion." An intense particle beam or pulse power could also be a driver if the energy could be compressed in time and space for the implosion of a tiny fuel pellet.

There are two main schemes for pellet implosion, i.e., direct and indirect drives. Their concepts are schematically shown in Figure 1.3. In "direct-drive implosions," the surface of the fuel pellet is directly irradiated by the driver beams. In contrast, in "indirect-drive implosions," the driver energy is converted into soft x-rays, which are confined in a cavity. The confined soft x-rays are then absorbed on the surface of the pellet to generate ablation pressure to drive the implosion. There is another scheme called "hybrid" (Nishimura et al., 1998), which combines the direct and indirect concepts of driving an implosion.

As for ignition of the imploded fuel, there are basically three concepts, i.e., central ignition (Nuckolls, 1972), FI (Harrison, 1963; Maisonnier, 1966; Yamanaka, 1983; Basov et al., 1992; Tabak et al., 1994), and volume ignition. The corresponding distribution of density, temperature, and pressure is illustrated in Figure 1.4. When the implosion is spherically symmetric and there is no significant instability to cause mixing of the boundary, the central hot spark is formed as shown in Figure 1.4a for the isobaric model (Meyer-ter-Vehn, 1982).

TABLE 1.1

Different Types of Driver, Implosion Scheme, and Ignition Scheme can be Combined to Achieve Energy Gain

Driver	Implosion Scheme	Ignition Scheme
Laser	Direct	Central ignition
Particle beams	Indirect	Fast ignition
Pulse power	Hybrid	Volume ignition

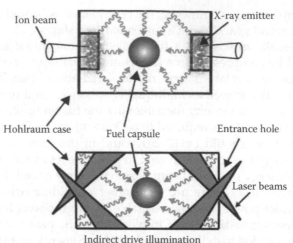

Ion beam   X-ray emitter

Hohlraum case   Fuel capsule   Entrance hole

Laser beams

Indirect drive illumination
X-rays from the hohlraum or emitter rapidly heat the surface of the fuel capsule

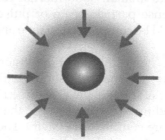

Direct drive illumination
Laser beams rapidly heat the surface of the fuel capsule

FIGURE 1.3
Basic implosion schemes, direct and indirect drive. (From Nakai, S. and Mima, K., *Rep. Prog. Phys.*, 67, 321, 2004, figure 2. With permission.)

If the implosion produces a compressed fuel of uniform density, we can ignite it with local heating by introducing an intense laser of pulse length much shorter than the expansion time of the compressed core. It is called FI, the core configuration of which has been depicted in Figure 1.4b. When the whole compressed core is of high density and the temperature is high enough to ignite the entire volume, it is called volume ignition as shown in Figure 1.4c.

As mentioned above, the concept of the FI is to separate fuel compression from fuel ignition and to ignite precompressed fuel by a separate external trigger (Harrison, 1963; Maisonnier, 1966). The advantage of the FI over the conventional spark ignition is that it can achieve higher energy gains at lower invested driver energies. In an orthodox FI scheme, high energetic particles such as electrons (Puhkov and Meyer-ter-Vehn, 1997; Key et al., 1998; Atzeni, 1999; Norreys et al., 2000; Kodama et al., 2001, 2002), protons (Roth et al., 2001), and macroparticles (Harrison, 1963; Caruso and Pais, 1996) are expected

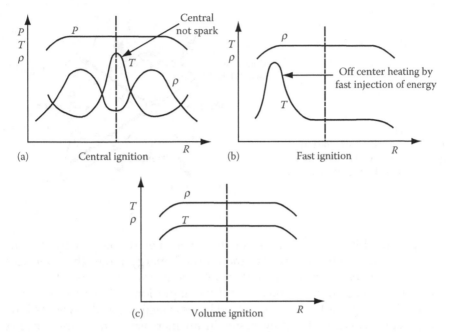

FIGURE 1.4

Ignition schemes of compressed fuel. The central ignition scheme (a) analyzed with an isobaric model, and the fast ignition (b) is analyzed with an isobaric model. Uniform distribution of density and temperature is assumed for volume ignition (c). (From Nakai, S. and Mima, K., *Rep. Prog. Phys.*, 67, 321, 2004, figure 3. With permission.)

to be transported into the core of the compressed DT fuel and to heat it successfully enough beyond the ignition temperature. On one hand, however, this scenario still bears many unknown physics such as the interaction between relativistic electrons and matter and the resultant energy transport. In contrast with the above-mentioned schemes using electrons and ions, different ideas for FI using plasma momentum have been proposed by other authors (Velarde et al., 1997; Gus'kov, 2001). However, in these schemes, enough amount of thrust moment and thus enough heating of the fuel cannot be expected because only a small amount of driver energy will be converted into kinetic energy.

Recently, a new ignition scheme, impact ignition (Murakami and Nagatomo, 2005; Murakami et al., 2006), has been proposed. Figure 1.5 shows the initial target structure of the impact ignition overlapped with the compressed fuel image at maximum compression. The target is composed of two portions: a spherical pellet made of a DT shell coated with an ablator and a hollow conical target, which is stuck to the spherical pellet. The conical component has a fragmental spherical shell (the impactor) also made of DT and an ablator. The key idea is to accelerate the impactor shell to collide against the precompressed main fuel. On the collision, shock waves generated at the contact surface transmit in two opposite directions heating the fuels to produce an igniting

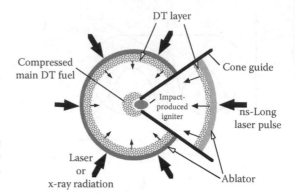

FIGURE 1.5
Initial target structure of the impact ignition target overlapped with the compressed fuel image at maximum compression.

hot spot. In this case, the impactor itself becomes the ignitor by directly converting its kinetic energy into the thermal energy rather than boosting the main fuel heating as in the particle (electrons or ions)-driven FI.

When we focus our scope on direct-drive implosion by lasers, we find that remarkable progress has been achieved by new facilities and new data have been accumulated, which may result in an increase in the confidence of achieving high-gain fusion with megajoule class lasers. The most important and remarkable of these achievements are (1) the demonstration of a hot spark at the center of the compressed fuel with the OMEGA-Up Grade of 60 beams, 30 kJ, and 351 nm at the Laboratory for Laser Energetics (LLE, University of Rochester). The improvement in the irradiation uniformity on the fuel pellet and the development of sophisticated diagnostics on the implosion physics and compressed core have clarified the conditions required for the spherically symmetric implosion to form a hot core at the center (McCrory et al., 2002). (2) Detailed studies on the growth and nonlinear behavior of the various kinds of instabilities and on the energy transport in laser–matter interactions with plane target experiments using the HYPER system at Institute of Laser Engineering (ILE), Osaka University, and with KrF laser at NRL have given us backup data about implosions for reliable numerical simulations and for credible gain prediction of direct-drive laser fusion. (3) The experimental demonstration of heating an imploded fuel in the GEKKO XII petawatt (PW) laser system (Kodama et al., 2001, 2002) has stimulated research on FI scheme.

In this chapter, we focus our description on laser-driven fusion with the direct implosion scheme for central ignition, FI, and impact ignition together with the equation of state (EOS) problem. Our goal is then to understand implosion physics in terms of demonstrated results and to address the physical and technical prospects of realization of IFE power plants. Note that the two world's biggest lasers with output energy of about 2 MJ, i.e., the U.S. National Ignition Facility (NIF) (Haynam et al., 2007) and the French Laser Megajoule (LMJ) (Bigot, 2006), are mostly configured for the indirect drive. The details of the physics related to the indirect drive were well reviewed by Lindl (1998).

1.4 Compression and Gain Scaling

1.4.1 Fusion Reaction and Energy Gain

Among all the possible fusion reactions the DT fuel seems to be the first candidate for the reactor since it has the highest cross section for low energies (few keV). The DT reaction products and energy excess are given by $D + T \rightarrow \alpha + n + 17.6\,\text{MeV}$ as described earlier, where α and n are the helium-4 nucleon and the neutron accordingly. For this reaction, the fusion energy overcomes the bremsstrahlung losses at temperature higher than 4 keV; therefore, the maximum theoretical gain is $G_{max} = 17.6\,\text{MeV}/(4 \times (3/2) \times 4)\,\text{keV} \approx 730$, where in the denominator of the ideal EOS, $E = (3/2)k_B T$, was used for the two ions and the two electrons, i.e., $D + T + 2e$ with a temperature $T = 4\,\text{keV}$ in energy units, where k_B is the Boltzmann constant.

If such a high gain can be achieved then what is the problem? The real gain is much smaller since one has to also take the following efficiencies into account: the driver absorption rate by the pellet η_a, the hydrodynamic efficiency from the absorbed energy to the heating energy η_h, and the fraction of the nuclear fuel ϕ that burns before the pellet breaks apart. Therefore, the gain G is

$$G = \frac{\eta_a \eta_h \phi q_{DT}}{q_{th}} < G_{max} = \frac{q_{DT}}{q_{th}} = \frac{3.39 \times 10^{11}\,[\text{J}/\text{g}]}{1.16 \times 10^8 T_{keV}\,[\text{J}/\text{g}]} = \frac{2922}{T_{keV}}, \qquad (1.8)$$

where q_{DT} and q_{th} are accordingly the fusion energy yield per gram and the ideal EOS for heating the fuel to a temperature T_{keV}, in units of keV.

Solving explicitly the nuclear rate equations in the domain of temperatures, 5–20 keV, one obtains the fraction of the nuclear fuel ϕ that burns before the pellet breaks apart:

$$\phi \approx \frac{\rho R}{\rho R + H_0}, \qquad (1.9)$$

where

ρ and R are the core plasma density and its radius, respectively; $H_0 = 8m_i c_s/<\sigma v> \approx 7\,\text{g/cm}^2$ for DT fusion at $T = 10\,\text{keV}$, where m_i is the ion mass, c_s is the speed of sound, σ is the nuclear cross section, and $<\sigma v>$ is the Maxwellian average. From this equation one gets $\phi = 30\%$ for $\rho R = 3\,\text{g/cm}^2$. For a realistic example $\eta_a = 0.8$, $\eta_h = 0.1$, $\phi = 0.3$ and for the temperature required to overcome the bremsstrahlung $T = 4\,\text{keV}$, one gets a gain of $G = 17.5$. This is not a very good gain taking into account that a power plant requires $G > 100$.

This problem was solved by designing the target and the driver pulse shape in such a way that only a spark (ξM_f with M_f being the fuel mass) at the center of the compressed fuel is heated and ignited (Lindl, 1998; Rosen, 1999)—the central spark ignition (CSI) scheme. The rest of the fuel is heated by

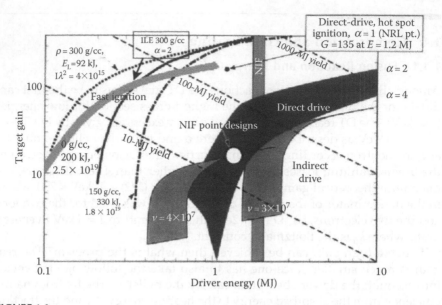

FIGURE 1.6
Target gain (fusion energy gain) as a function of the driver energy for central spark ignition with direct and indirect implosion and for fast ignition (based on LLNL chart). (From Nakai, S. and Mima, K., *Rep. Prog. Phys.*, 67, 321, 2004, figure 4. With permission.)

the α-particles produced in the DT reaction. For this to happen it is necessary that the density of the fuel (ρ) times its radius R is larger than $0.3\,\text{g/cm}^2$. For example, using the above values it is sufficient to heat only a fraction $\xi = 0.175$ of the fuel in order to increase the gain from 17.5 to 100.

Figure 1.6 shows the fusion energy gain scaling with incident laser energy. The gain scaling curve for indirect and direct drive with central ignition, and the FI gain curve with different fuel densities are those reported from LLNL in the United States. The gain curve from ILE is shown for FI with $\rho = 300\,\text{g/cm}^3$. Several point designs from different groups are also plotted on the same figure.

1.4.2 Why Compression?

In Inertial Confinement Fusion (ICF) the "name of the game" is compression. The mass M_f of the nuclear fuel is

$$M_f = \frac{4}{3}\pi R^3 \rho = \frac{4.19}{\rho^2}(\rho R)^3 \approx \frac{113}{\rho^2}[\text{g}], \tag{1.10}$$

where $\rho R = 3\,\text{g/cm}^2$ was used. Without compression ($\rho = 0.2\,\text{g/cm}^3$) the above equation implies a nuclear DT mass of about 2.8 kg, an undesired value for the controlled fusion reactor. However, a compressed fuel with density $\rho = 600\,\text{g/cm}^3$ needs a pellet with only 1.0 mg DT fuel ($\phi = 0.3$ was taken, consistently with $\rho R = 3\,\text{g/cm}^2$).

The laser energy E_L that heats the DT fuel within a sphere with radius R and particle density n is given by

$$\eta_a \eta_h E_L = 3nT \frac{4}{3} \pi R^3, \tag{1.11}$$

$$\Rightarrow E_L = \left(\frac{1}{\eta_a \eta_h}\right)\left(\frac{4\pi}{2.5 m_p \rho_0^2}\right)(\rho R)^3 T \left(\frac{\rho_0}{\rho}\right)^2, \tag{1.12}$$

where the ideal gas EOS has been used, m_p is the proton mass, $\rho_0 = 0.2\,\text{g/cm}^3$ is the initial DT density, and the temperature T is in energy units. Taking a laser energy that heats the plasma to the threshold temperature of $4\,\text{keV}$ (fusion energy equals the bremsstrahlung losses) and inducing a 30% burn of the nuclear fuel, we get

$$E_L\,[\text{J}] = \left(\frac{1.3 \times 10^{12}}{\eta_a \eta_h}\right)\left(\frac{\rho_0}{\rho}\right)^2. \tag{1.13}$$

Taking typical values of $\eta_h = 0.1$ and $\eta_a = 0.8$, the ICF ignition and nuclear burn require a pulsed (few nanoseconds) laser system with an energy

$$E_L\,[\text{J}] = 1.6 \times 10^{13} \left(\frac{\rho_0}{\rho}\right)^2. \tag{1.14}$$

It is evident from this result that without compression one requires an unrealistic pulse laser with energy of 16 TJ, while a 3000 compression (i.e., 600 g/cm³) reduces the laser energy to 1.75 MJ. Compressing a plastic shell with an initial density of about 1 g/cm³, a density of 1000 g/cm³ has already been achieved experimentally at Osaka University in Japan (Azechi et al., 1991). Therefore, it is not surprising that the two largest laser systems under construction at Livermore (the National Ignition Facility) in United States and the MJ laser in France will have energy of about 2 MJ!

From the famous report of Nuckolls et al. (1972) in *Nature* the ICF philosophy is based on high compression. The reasoning is that (1) it is cheaper (energetically) to compress than to heat and (2) nuclear reactions are proportional to density square; therefore, the more you compress the better you are in ICF. Of course, the only limitations of compression are the hydrodynamic instabilities (like RT, etc.).

In Section 1.4.3 it is shown that there is an optimum of compression, namely the gain G is maximum for a finite compression. The value of this density, for a given fuel mass and particular ICF scheme, depends on the EOS. We calculate this value for FI and compare it with the CSI. Since we do not know the EOS we shall follow the literature by taking the ideal gas EOS for the ions and the Fermi–Dirac EOS (Eliezer, 2002) with an effective multiplier (or so-called isentrope parameter) α, for the electrons. In particular, the pressure P_e

(in megabars) and the Fermi energy ε_F (in J/g) that gives the specific energy costs of cold compression are given in terms of the density ρ [g/cm^3]:

$$P_e = \alpha P_{FD}; \quad P_{FD}[\text{Mbar}] = 2\rho^{5/3}, \tag{1.15}$$

$$\varepsilon_F = \alpha \varepsilon_{FD}; \quad \varepsilon_{FD}[\text{J}/\text{g}] = 3.3 \times 10^5 \rho^{2/3}. \tag{1.16}$$

The "optimum compression" idea is easily understood from the following argument: From EOS data, one needs an infinite energy to compress to an infinite density. Since the energy output is finite and known it is clear that G is zero. On the other hand for normal density with a small fuel mass (~few mg) the gain is also zero. Therefore, a maximum should exist somewhere in between. For the DT fuel with a mass of few mg one gets an optimum at a few hundred g/cm^3. If you compress more then the gain reduces, namely there is an optimum compression (maximum gain) fixed by EOS. In our example the EOS is represented by the various values of α.

1.4.3   Gain Scaling for FI

The gain is calculated for a given final state defined by the nuclear fuel, its temperature, density, and mass. For this final state it seems reasonable (but only yet proven experimentally!) to use the simple phenomenological one parameter, α. However, in order to reach this final state from a given initial state one has to solve the hydrodynamic systems with the appropriate extra terms (e.g., thermal transport of electrons and photons; two or three dimensions, etc.) supported by reliable EOS. The target design for a given driver is based on these calculations. Since the initial target contains the nuclear fuel in a shell of one or more materials, it is necessary to know the EOS of all the materials involved in the process of ICF. Furthermore, since the initial shell of the pellet is in the solid state and the final state is in the plasma form, the EOS should be able to describe the transitions from solid to liquid, gas, and finally to plasma state of matter. At present, the desired EOS for the wide range change in temperature, density, pressure, etc., cannot be derived from basic principle; therefore, phenomenological EOS based on experiments are required.

We consider here a simple model for optimization of target gain in the FI scheme. The total driver energy E_d splits into two, i.e., the compression energy for the main fuel E_{com} and the ignition energy E_{ign}, which are given respectively in terms of the density ρ [g/cm^3] by

$$E_{ign}[\text{J}] = 1.4 \times 10^4 \eta_i^{-1} \rho^{-\nu}, \tag{1.17}$$

$$E_{com}[\text{J}] = 3.3 \times 10^4 \eta_c^{-1} \alpha M_f \rho^{2/3}, \tag{1.18}$$

$$E_d = E_{ign} + E_{com}, \tag{1.19}$$

where isochoric fuel profile is assumed; η_c and η_i are the efficiencies from driver (e.g., laser) to thermal energy for the compressed fuel and the ignitor (i.e., hot spot) respectively. Here it should be noted that in an idealized case with no energy dissipation, one can easily estimate $E_{\mathrm{ign}} = (4\pi/3)\rho R^3 T \propto \rho^{-2}$ (for a constant ρR), which corresponds to $v = 2$; however, a detailed numerical simulation (Atzeni, 1999) yields $v = 1.85$. The ideal EOS for the DT yields a pressure P_c of the compressed fuel: $P_c = n_i T + n_e T = 2n_e T$, where n_i and n_e are the ion and the electron densities, respectively. However, the electrons are degenerate and therefore the ideal gas EOS for them is not justified. The parameter α in Equation 1.12 is defined by the ratio $\alpha = P_c/P_{\mathrm{deg}} = 5T/\varepsilon_F$. In the literature (Rosen, 1999) α is taken to be ~3. The compressed mass of the fuel M_f is expressed in the form

$$M_f = \frac{4\pi}{3}\frac{(\rho R)^3}{\rho^2}.$$ (1.20)

Once the hot spot is successfully ignited, a burn wave is expected to propagate through the main fuel causing a burn fraction estimated by ϕ (see Equation 1.9). The energy gain is finally evaluated to be

$$G = 3.4\times10^{11}[\mathrm{J/g}]\cdot\phi M_f/E_d,$$ (1.21)

where the small fraction of the thermonuclear energy released from the hot spot is ignored for simplicity.

Figure 1.7 shows the driver energies versus the fuel density for $\alpha = 3$. Since the compression energy and the ignition energy have opposite behavior with respect to the density, one can easily find that the total driver energy has a minimum at a certain value of the density. In order to minimize the total driver energy we require

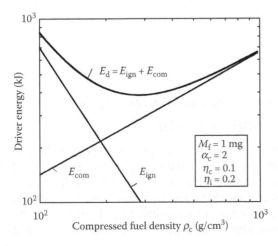

FIGURE 1.7
Driver energy versus compressed fuel density. (From Eliezer, S. et al., *Laser Part. Beams*, 25, 585, 2007. With permission.)

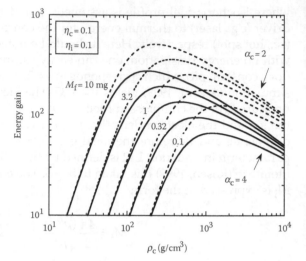

FIGURE 1.8
Energy gain versus compressed fuel density for different fuel mass for fast ignition with $\alpha_c = 2$ and 4. (From Eliezer, S., et al., *Laser Part. Beams*, 25, 585, 2007. With permission.)

$$\left(\frac{\partial E_d}{\partial \rho}\right)_{M_f} = 0, \tag{1.22}$$

implying

$$\frac{E_{\text{com}}}{E_{\text{ign}}} = \frac{3v}{2}\left(\frac{\eta_i}{\eta_c}\right) \approx 3\left(\frac{\eta_i}{\eta_c}\right). \tag{1.23}$$

This equation reveals a crucial and simple relation: To minimize the total driver energy E_d, the energy ratio between the compression and the ignition is almost constant ~3 times the efficiency ratio, η_i/η_c. Figure 1.8 shows energy gain G versus compressed fuel density ρ for different M_f and α under the fixed values of η_c and η_i. It should be noted that each curve has a peak gain at a certain ρ value (Eliezer et al., 2007).

1.5 Central Hot-Spark Ignition by Laser Direct Drive

1.5.1 Recent Progress in Direct-Drive Implosion

In direct-drive implosion by laser, the most serious physical issue is the stability of the implosion (Bodner et al., 1998). Instability grows from initial perturbation on the target and nonuniformity of irradiation by laser and leads to breakup of the shell and quenching the hot spark at the center as schematically shown in Figure 1.9. Concerning the stability of implosion, the experimental data on various parameters of the compressed core are compared

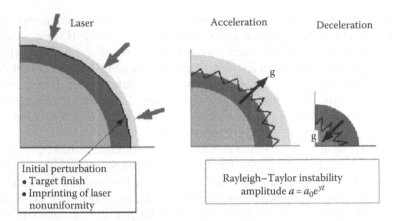

FIGURE 1.9
The most serious obstacle in direct-drive implosion is shell breakup and quenching the hot spark by hydrodynamic instability. (From Nakai, S. and Mima, K., *Rep. Prog. Phys.*, 67, 321, 2004, figure 5. With permission.)

with a one-dimensional (1D) simulation. Typical examples are shown in Figure 1.10 (Nakai, 1996), where the areal densities agree with the 1D prediction up to a convergence ratio (CR: initial target radius divided by the final radius) of about 30, when the irradiation uniformity is improved by using a random

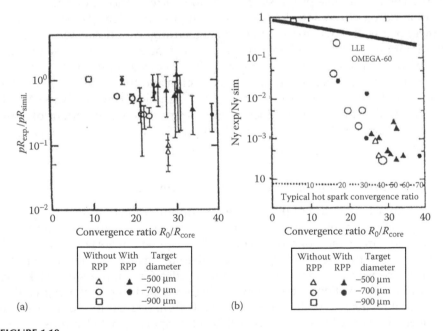

FIGURE 1.10
Normalized implosion performance as a function of the computed CR of the core area. (a) The measured areal mass density normalized by the value obtained by 1D simulation. (b) The same but for fusion neutron yield. (From Nakai, S. and Mima, K., *Rep. Prog. Phys.*, 67, 321, 2004. With permission.)

phase plate (RPP). The neutron yield, however, decreased faster as CR increased, as shown in Figure 1.10b, and about 10^{-3} of the 1D calculations at CR = 30, where the high-density compression was achieved. The estimated radius of the hot spark with the simulations at the same CR was calculated to be about $5\,\mu m$ (hot-spark CR ~50). The possible explanation for the discrepancy in the yield would be that the remaining irradiation and target nonuniformities might cause a collapse of the hot spark, from where most of the neutrons are generated in the 1D simulation.

Recently, significant progress in direct-drive implosions has been demonstrated at LLE (McCrory, 2002). The LLE experiments were conducted on the OMEGA 60 beams, 30 kJ, UV laser system. The improvements in the irradiation uniformity have significantly improved the implosion performance. A single-beam nonuniformity of 3% (averaged over 300 ps) was achieved with full implementation of 1 THz bandwidth 2D smoothing by spectral dispersion (SSD) (Skupsky, 1989), and polarization smoothing with birefringence wedges (Regan et al., 2000). This corresponds to an on-target nonuniformity of less than 1% rms due to beam overlap. The beam-to-beam power imbalance has been reduced to below 5% rms.

With these conditions, moderate CR targets (~15) produce ~30% of neutron yield predicted by 1D simulation and nearly 100% of the predicted fuel and shell areal densities. At predicted CR values close to 40, the primary neutron yield is ~20% of the 1D prediction. These results are also plotted in Figure 1.10b. The remarkable progress of implosion performance shows the validation of indirect-drive inertial fusion heading toward high-gain laser IFE for power plants. It is also reasonable to be increasingly confident about achieving ignition, burning, and energy gain at the reactor scale, because the size of the hot spark and surrounding cold fuel becomes larger with a higher laser energy. The thickness of the mixed region and the loss of the hot-spark region will become less serious, and once the ignition occurs at the center, all the fuel will be burned to produce fusion energy. Recent simulations of high-gain implosion include the mixing effect due to RT instability. The effect is included in predicting the gain curve in Figure 1.6.

1.5.2 Stabilized Implosion for Better Performance

There are still several stabilizing schemes for implosion that can be expected to be effective in achieving a high gain with a mitigated requirement for target fabrication and irradiation uniformity. They can be categorized as (1) shock preheating of the shell with a tailored pulse such as a picket-fence prepulse or a sharp-rise front edge before the main pulse for compression (McKenty et al., 2001), (2) radiation preheating with a high-Z-doped foam coat or high-Z thin layer coat on the fuel pellet (Bodner et al., 2000), and (3) multicolor irradiation (Shigemori et al., 2001) using nonlocal heat transport by hot electrons.

The physical processes of implosion related to the stability issue are schematically illustrated in Figure 1.11. Due to the irradiation nonuniformity, nonuniform ablation is driven on the target surface. At the initial phase of

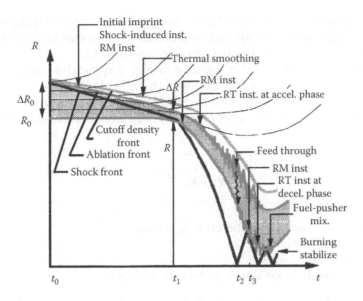

FIGURE 1.11

Implosion and physical processes related to the stability. (From Nakai, S. and Mima, K., *Rep. Prog. Phys.*, 67, 321, 2004, figure 7. With permission.)

the irradiation, separation of the absorption region (cutoff region) from the ablation front on the solid target is not increased. Therefore, the initial nonuniformity of the laser irradiation can be strongly imprinted on the target surface (initial imprint). The shock wave driven by the main pulse cab excites the Richtmyer–Meshkov (RM) instability, being coupled with the original target roughness and the initial imprint. The nonuniformity of ablation pressure is relaxed from the laser nonuniformity at later times due to thermal smoothing during the energy transport from the absorption region to the ablation surface. In the phase of main pulse acceleration, the RT instability grows. However, shorter wavelength perturbations are stabilized owing to the ablation stabilization. If the in-flight aspect ratio is too large, the accelerated shell breaks easily (shell breakup). Even when the inward acceleration is sufficiently symmetric, mixing of the hot spark and main cold fuel will take place due to an explosive growth of RT instability in the stagnation phase. In such a situation, the spark is cooled down and no fusion ignition will take place. In general, the RT instability is not generated as a single mode, but many modes grow simultaneously, finally inducing the turbulent mixing. The growth rate of the mixing layer thickness would be smaller than the linear growth rate due to the nonlinear saturation and the mode–mode coupling in the nonlinear phase. If one can manage to keep a spherically symmetric implosion, fire polishing by alpha particle heating can smooth the burning wave even with some level of nonuniform structure in the main fuel. Such elementary processes mentioned above as initial imprint, RM instability, thermal smoothing, nonlinear saturation, and turbulent mixing have been independently and separately investigated with adequately modeled experiments on planar targets of various structures and irradiation conditions of laser.

The key that governs the stability of the direct-drive implosion was identified as being the RT instability, which leads to the nonuniformity requirement of laser irradiation and the fuel pellet. The growth rate of RT instability at the ablation front (Takabe et al., 1985; Betti et al., 1998; Lindl, 1998) is given by

$$\gamma = \alpha\sqrt{\frac{kg}{1+kL}} - \beta k v_a, \tag{1.24}$$

where
  k is the wave number of the growing mode
  g is the acceleration
  v_a is the ablation velocity
  L is the density scale length at the ablation front

Empirical values of $\alpha = 0.9$ and $\beta = 3$–4 are given for comparison of numerical and experimental results for direct-drive implosions.

The concepts of stabilizing implosions aim to enlarge the characteristic scale length L, and to increase the ablation velocity v_a, with the reduced density ρ_a, keeping the ablation pressure $\rho_a v_a^2$ constant. They are (1) shock preheating, (2) radiation energy transport, and (3) nonlocal heat transport by hot electrons, as previously mentioned. The gain estimations of recent numerical simulations for direct-drive implosions include a consideration of these physical processes of the instability and stabilizing mechanism. Together with the progress in fuel pellet fabrication and laser control technology, it can be reasonably expected that fusion ignition, burn, and energy gain can be demonstrated by direct-drive implosions with multibeam megajoule lasers if the beams are properly arranged for direct drive.

1.6 FI Present Status and Future Prospects

1.6.1 Why Fast Ignition?

The FI is the new scheme for igniting a compressed fuel plasma by using an ultraintense laser pulse, called a PW laser (Tabak et al., 1994). This scheme has potential advantages over central hot-spark ignition since the compression and the ignition can be separated. As shown in Figure 1.12, the fuel ρR can be significantly higher in the FI than in the central hot-spark ignition for the same implosion laser energy because of the lack of central low-density hot spark. When the plasma is assumed isobaric, the hot-spark density is 10 times lower than the main fuel density. Therefore, the radius of a hot spark

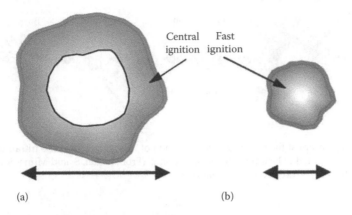

FIGURE 1.12
Compression of (a) the central ignition isobaric core plasma structure and (b) the fast ignition isochoric core plasma structure. (From Nakai, S. and Mima, K., *Rep. Prog. Phys.*, 67, 321, 2004, figure 8. With permission.)

of $(\rho R)_h \approx 0.4\,\mathrm{g/cm^2}$ is twice larger than the high-gain fuel radius, where $\rho R > 2\,\mathrm{g/cm^2}$. This means that the total mass of the fast-ignition high-gain core is about 10 times smaller than that of the central spark core. In other words, the required implosion laser energy can be 10 times smaller in the fast-ignition case. Therefore, the fast-ignition fusion reactor can be small and allows the freedom of choosing the output power and driver efficiency. Although short-pulse laser–plasma interactions have been widely investigated, imploded plasma heating mechanisms are still an open question. In this section, FI conditions, the requirements of the heating laser, and the present status of experimental and theoretical research on heating will be reviewed.

1.6.2 Ignition Condition in FI

As shown in Figure 1.13, when a fuel pellet is compressed by laser implosion, the core plasma is surrounded by a large-scale coronal plasma. In order to inject a short-pulse laser into the core, the laser pulse should penetrate into an overdense region by hole boring or cone laser guiding. When the ultraintense laser energy reaches near the core, the laser energy is converted into high-energy electron or ions that carry the energy and heat the core plasma. The required short-pulse energy for ignition depends upon the heating spot diameter, heating depth, transport and absorption efficiencies, electron and ion energy spectrum, and so on.

The FI hot spark is isobaric between the spark and main fuel. First of all, we compare the ignition condition for an isochoric spark with that for an isobaric spark. One-dimensional burning plasma simulations are carried out to determine the ignition boundaries in ρR and T space, of which the hot-spark geometry is shown in Figure 1.14a (Mahady et al, 1999).

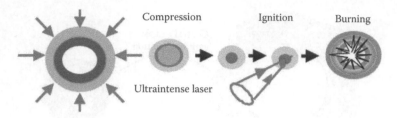

FIGURE 1.13
Fast ignition concept at the maximum compression of laser implosion: The ultraintense short pulse laser is injected to heat the dense core plasma. (From Nakai, S. and Mima, K., *Rep. Prog. Phys.*, 67, 321, 2004, figure 9. With permission.)

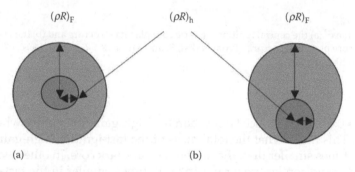

FIGURE 1.14
Structure of (a) central hot spark and (b) off-center hot spark. (From Nakai, S. and Mima, K., *Rep. Prog. Phys.*, 67, 321, 2004, figure 10. With permission.)

Figure 1.15 shows that the ignition boundary for an isochoric hot spark is higher with respect to $(\rho R)_h$ than that for an isobaric hot spark since the main fuel layer acts as a tamper layer to suppress the expansion of the hot spark. The FI is relevant to the isochoric hot-spark case. In this case, the approximated ignition condition is given by

$$T(\rho R)_h^3 = 1.0, \tag{1.25}$$

where ρR and T are in units of g/cm^2 and keV, respectively, and the above numerical factor, 1.0, is valid for temperatures around 10 keV. Since the hot-spark energy is given with the solid density ρ_0 by

$$E_g[GJ] = 10.6(\rho R)_h^3 T(\rho/\rho_0)^{-2}, \tag{1.26}$$

the required spark energy from Equations 1.25 and 1.26 is approximately given by

$$E_g[kJ] = 40\rho_{100}^{-2}, \tag{1.27}$$

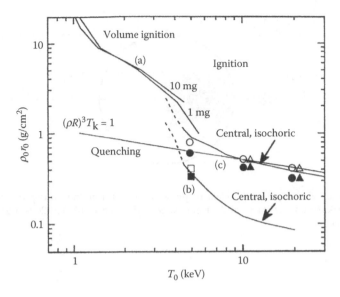

FIGURE 1.15
Ignition conditions for (a) volume ignition, (b) central isobaric spark (open and solid squares), (c) central isochoric spark (open and solid triangles), and off-center round isochoric spark (open and solid circles). (From Nakai, S. and Mima, K., *Rep. Prog. Phys.*, 67, 321, 2004, figure 11. With permission.)

where ρ_{100} denotes the hot spot density normalized by $100\,\mathrm{g/cm^3}$. The ignition condition of the off center spark has been compared with the central hot-spark ignition by Mahady (1999). For the off-center isochoric hot spark, the 2D burning simulation results in the initial geometry of Figure 1.14b. It is indicated by the triangles of Figure 1.15 that the ignition condition is the same as that of the central isochoric spark case.

However, the cylindrical heat deposition geometry and finite heating time effects change the ignition condition significantly; the hot spark is produced at the edge of the main fuel by relativistic electron heating. The heated area radius and depth are r_b and d, respectively, and the fuel density and radius are ρ and R, respectively, as shown in Figure 1.16. Note here that the stopping range of the intense electron beam is shorter than the "thin" electron beam as discussed by Deutsch (1996). Namely, when the mean distance between beam electrons is shorter than the Debye shielding distance of the plasma, the collective interactions between beam electrons and the plasma become important and enhance the stopping power. Furthermore, the self-generated magnetic field may play important roles in the heat deposition of the intense relativistic electron beam. Since the magnetic field effects on the heat deposition processes are not clarified yet, the following discussions include only the stopping range shortening (Deutsch, 1996).

In the geometry of Figure 1.16, the ignition conditions have been investigated with 2D burning simulations by Atzeni (1999). The burning process of the off-center ignition is shown in Figure 1.17, where the temperature and density

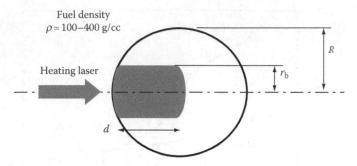

FIGURE 1.16
Hot spark-core plasma geometry. Heating pulse is injected from the left to produce a cylindrical hot spark. (From Nakai, S. and Mima, K., *Rep. Prog. Phys.*, 67, 321, 2004, figure 12. With permission.)

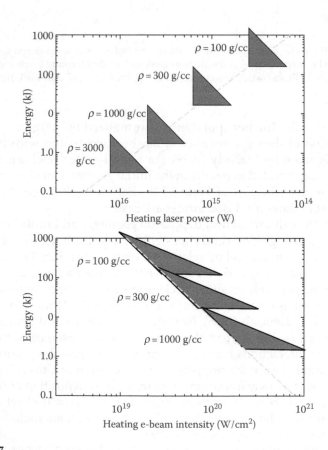

FIGURE 1.17
The iginition conditions for heating e-beam energy, heating power, and heating e-beam intensity. (From Nakai, S. and Mima, K., *Rep. Prog. Phys.*, 67, 321, 2004, figure 13. With permission.)

contours of the burning wave are illustrated. Once the ignition condition is met, the burning fraction is more or less the same as the central ignition. By the heating parameter survey of the 2D simulations, the ignition conditions for the heating energy, power and radius, the energy deposition range, and the fuel density are explored by many papers. According to the comprehensive work by Atzeni (1999), the optimized ignition conditions are given by the following relations:

$$F_g[\text{kJ}] = 140\rho_{100}^{-1.85}, \tag{1.28}$$

$$P_g[\text{PW}] = 2.6\rho_{100}^{-1.85}, \tag{1.29}$$

$$I_g[\text{W/cm}^2] = 2.4 \times 10^{19}\rho_{100}^{0.95}, \tag{1.30}$$

$$r_b[\mu\text{m}] = 60\rho_{100}^{-0.975}, \tag{1.31}$$

where the range ρd is assumed $0.6\,\text{g/cm}^2$. Note that the ignition condition of Equation 1.28 is more than three times higher than that of Equation 1.27. This is due to the electron stopping range and the heating geometry. However, the fuel density dependence of the ignition energy is essentially the same in both cases. In the actual ignition, r_b and the range ρd depend upon the PW laser–plasma interaction physics; r_b depends upon the laser propagation, the electron heat transport, and the heat deposition process. The stopping range ρd depends on the plasma resistivity and the collision processes of relativistic electrons. The experimental and theoretical works related to those issues are reviewed in the following section.

1.6.3 Imploded Plasma Heating

As shown in Equations 1.28 through 1.31, the optimum compressed plasma density decreases with the increase in hot-spark radius and the heating laser energy and the pulse width τ_p increase as $\rho_{100} = (r_b/60\,\mu\text{m})^{-1.025}$, $\tau_p = 55\rho_{100}^{-0.85}$ [ps], $E_g = 140(r_b/60\,\mu\text{m})^{1.9}$ [kJ]. For example, when $r_b = 10\,\mu\text{m}$, $\tau_p = 10\,\text{ps}$, $E_g = 4\,\text{kJ}$, and $\rho = 600\,\text{g/cm}^3$, which is 3000 times the solid density of DT. This means that 10–20 kJ heating pulse ignites the core plasma. Those conditions may be violated because the MeV electron beam generated by the laser spreads wider than $10\,\mu\text{m}$ radius or the compressed density is not as high as 3000 times the solid density. According to the experiments at Osaka, LLNL, and so on, r_b could be $30\,\mu\text{m}$ and the required laser energy will be higher than 70 kJ. Therefore, the heating spot size and coupling efficiency of the short-pulse laser energy to the core plasma are the most important critical issues in FI research.

At present, the theory and experimental research are in progress significantly at Osaka, LLNL (United States), Rutherford (United Kingdom), and so on. One of them is the relativistic electron generation, transport, and the imploded plasma heating, in particular, in the cone guide target. Another example is the efficient ion generation by the PW laser and its application to plasma heating. In the following, we discuss implosion plasma heating research at ILE Osaka University.

In the case of spherical implosion plasma heating, coupling efficiency of a PW module (PWM) laser pulse to a core plasma is sensitive to the focus position of the heating laser, because of the long-scale coronal plasma (Tanaka et al., 2000; Kodama et al., 2002). When the focus position is near the critical surface of coronal plasmas, the PWM laser pulse penetrates into the over-dense region and the neutron yield is significantly enhanced. However, the neutron energy spectra are broad and not isotropic as shown in Figure 1.18a. The width of the spectral peak around 2.45 MeV indicates that the reacting ion energy is in the order of 100 keV. Furthermore, the neutron spectral shape depends on the direction of the neutron spectrometer. These results can be interpreted by the enhanced fusion reaction around the critical surface. According to PIC simulations, electrons are expelled from laser channels and solitons by strong laser heating and ponderomotive force. As a result, ions in the laser-produced bubbles are accelerated by Coulomb explosion to high energy (Sentoku, 2000). Since the fusion reaction enhancement is dominated by those high-energy ions, the thermal neutron enhancement was not observed and the effect of core plasma heating was not clear in the PWM direct heating (Kitagawa et al., 2000). However, when the short-pulse laser

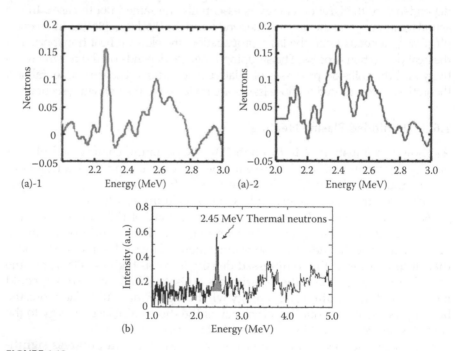

FIGURE 1.18
(a) Neutron energy spectra for direct heating an imploded CD pellet. (a)-1 and (a)-2 are the spectra in different direction with respect to the incident laser pulse. (b) Neutron spectrum for a cone-guided target. (From Nakai, S. and Mima, K., *Rep. Prog. Phys.*, 67, 321, 2004, figure 14. With permission.)

energy increases to 1 kJ level in the PW laser, the laser pulse penetrates the higher-density region since nonlinear scattering may be saturated because of the strong plasma heating. Therefore, in PW laser experiments, the thermal neutrons may also increase and core plasma heating may be possible even in the coneless target.

In a cone guide target, the neutron spectral peak at 2.45 MeV is narrow, as shown in Figure 1.18b. The neutron yield is of the order of 10^5 with heating, which is 10 times higher than that of a nonheating case. This indicates that the thermonuclear fusion is enhanced by the temperature increase in the core plasma due to the PWM laser heating. The temperature increase is estimated by the neutron yield enhancement and the observed ρ and the burning time. In the best shot, the temperature increased by 130 eV. We found from these results that 25% of input PWM laser pulse energy is transported to the core plasma (Kodama et al., 2000).

In the PW laser experiment, two kinds of cone targets are imploded and heated. The cone angles are 30° and 60°. The heating is more efficient in the 30° cone than in the 60° cone. When the 300 J/0.6 ps chirped pulse amplification (CPA) laser pulse is injected, the neutron yield reaches 10^7, while the neutron yield was 10^5 without heating as shown in Figure 1.19a (Kodama et al., 2002). This indicates that the core plasma temperature increased by 500 eV and the energy coupling between heating laser and core plasma is 20%–25%. Since the focused laser energy included in 50 μm diameter is less than 40%, a 20%–25% coupling efficiency means that actual coupling is higher than 50%. Otherwise, laser energy in the hollow of the spot is collected by the cone guide. In Figure 1.19b, a simple scaling curve is shown, where the temperature increase is assumed to be proportional to the input short-pulse laser energy and the coupling efficiency is assumed to be the same as that of the cone guide PWM experimental results. This indicates that the coupling efficiency for a 300 J case is almost the same as that for an 80 J case. This scaling law has

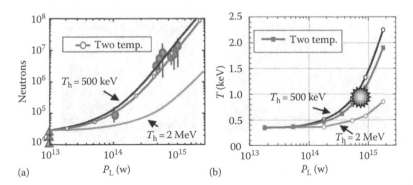

FIGURE 1.19
(a) Neutron yield dependence on the laser power (triangle, without heating; circle, with heating; curves, Fokker–Planck (FP) simulation). (b) Ion temperature dependence on the laser power (curves, FP simulation). (From Nakai, S. and Mima, K., *Rep. Prog. Phys.*, 67, 321, 2004, figure 15. With permission.)

been used for planning the FI realization experiment (FIREX). Further analysis of the experiment has been done by the Fokker–Planck (FP) simulation and the neutron yield of the simulation correlates well with the experiments.

1.6.4 Electron Heat Transport Research and Critical Issues of the FI

Recent 3D PIC simulation results for the focused ultraintense laser interaction with a solid plane target show that the relativistic electron current profile in the target is self-organized and confined to a small radius as shown in Figure 1.20. This indicates that the small hot spark could be generated by the relativistic electron heating (Sentoku et al., 2003). In the cone target, an ultraintense laser light is partially reflected on the cone surface wall and focused to the top of the cone, while the relativistic electrons are generated on the sidewall of the

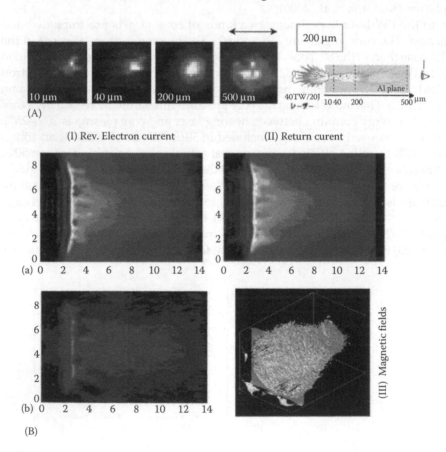

FIGURE 1.20
Relativistic electron current self-organization (self-pinch). (A) Transport experiment (B) (I)-(a) the current profile of hot electrons with energy lower than 1 MeV. (I)-(b) Current profile of hot electrons with energy higher than 1 MeV, (II) return current profile, and (III) magnetic field iso-intensity contour. (From Nakai, S. and Mima, K., *Rep. Prog. Phys.*, 67, 321, 2004, figure 16. With permission.)

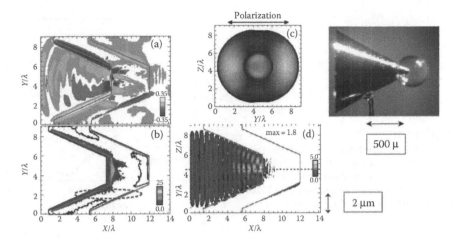

FIGURE 1.21
A photograph of a cone shell target (top right) and PIC simulation results for cone–laser inter-actions: (a) magnetic field structure; (b) electron energy density distributions on the sidewall; (c) electron energy density distribution at the top of the cone; and (d) laser intensity distribu-tion in the cone. (From Nakai, S. and Mima, K., *Rep. Prog. Phys.*, 67, 321, 2004, figure 17. With permission.)

cone and on the top wall. Since the electrons are accelerated along the laser propagation direction, a strong current is driven along the cone axis and the relativistic electron flows are pinched to the top of the cone by the magnetic fields as shown in Figure 1.21. These characters of the laser interactions with the cone target contribute to enhance the coupling efficiency of a short-pulse laser to a small core plasma as was indicated by the enhancement of neutron yield discussed in Section 1.6.3. The PIC simulations have been limited to small scale and short time durations. Therefore, carrying out longer space and timescale simulations is necessary to introduce the FP and the hybrid simulations. Then one can quantitatively compare the transport experiment results with the simulation results (Mima et al., 2002).

An enhanced coupling efficiency of the heating laser energy to a small core plasma is essential for achieving ignition with a relatively little laser energy. Because of this point, the relativistic electron transport and energy deposition in the cone guide target and in the corona of the imploded plasma are the most critical issues in the fundamental physics of FI, though extensive experi-mental and theoretical work has been conducted in an attempt to understand and control dense plasma heating by using an ultraintense laser.

1.6.5 PW Laser and Future FI Facility

A PW laser was recently completed, which delivers 500 J/0.5 ps (Izawa, 2002) at Osaka University. In PW laser experiments, imploded plasmas have been heated up to about 1 keV. The present and future PW laser target experiments at Osaka, Rutherford Appleton Laboratory, and so on (Kodama et al., 2002)

Applications of Laser Plasma Interactions

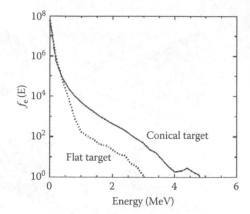

FIGURE 1.22
The number of the hot electrons for the conical target is enhanced by more than one order of magnitude above the flat target. (From Nakai S. and Mima, K., *Rep. Prog. Phys.*, 67, 321, 2004, figure 18. With permission.)

will clarify the heating scaling law that determines the relations between heating laser energy, hot-spark temperature, and neutron yield. Using the FP code, cone shell target experiments were simulated at Osaka. When relativistic electrons have the double Maxwellian energy spectrum as shown in Figure 1.22 of the cone target PIC simulation results, the predicted neutron yield and the ion temperature correlate with the experimental results as shown in Figure 1.19 (Mima et al., 2002). The results have been extended to the larger laser energy case; the gains for the multi-10 kJ heating pulse are shown in Figure 1.23. If the coupling efficiency is higher than 30%, the required heating laser energy for break even experiment (FIREX-I) is estimated to be 10 kJ for 1500 times solid density DT plasmas and 30 μm spot diameter as shown in Figure 1.23 (Mima et al., 2002; Yamanaka et al., 2002).

For ignition, the most critical parameter is the hot spot radius r_b, which strongly depends upon the relativistic electron transport. According to simulation and fundamental experiment results, the relativistic electron heat

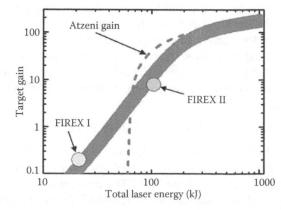

FIGURE 1.23
Gain curves for fast ignition. The core plasma is assumed $\rho = 300$ g/cc, and the isentropic factor is 2. The implosion laser energy is 10 kJ for FIREX I and 50 kJ for FIREX II. The shaded area is the route from the FIREX target to the high-gain target which is essentially the same as in Atzeni's design. (From Nakai, S. and Mima, K., *Rep. Prog. Phys.*, 67, 321, 2004, figure 19. With permission.)

flow is well confined by self-generated magnetic fields. In particular in the cone shell target, the deposited laser energy is concentrated at the top of the cone. Therefore, the hot-spark radius could be controlled by the cone top radius and the distance between the cone top and the core plasma. From the present understanding on the heating processes, the ignition will be achieved with a pulse energy less than 50 kJ/10 ps for an imploded plasma ρR higher than 1.0 g/cm². The second phase of the FIREX project at Osaka University (FIREX-II) is aiming to demonstrate such an ignition.

1.7 New Approach to ICF: Impact Ignition

1.7.1 Gain Curve for Impact Ignition

Figure 1.24 shows a schematic picture for the fuel composition at around the maximum compression; the main DT fuel coalesces with the ignitor sphere, which is transformed from the impact shell. At this time, they are supposed to be approximately in a stationary phase. For simplicity, we assume that each of the two spheres has a uniform spatial profile. Here it should be noted that the timescale of the ignitor formation is of the order of $(30-50\,\mu m)/(1-2 \times 10^8\,cm/s)$ ~20–30 ps, while the inertia time of the main fuel at stagnation with temperature of a few keV is of the order of $(100-150\,\mu m)/(3 \times 10^7\,cm/s)$ ~300–400 ps. Thus, the ignitor formation on the impact can be completed in the timescale much shorter than the hydrodynamic timescale of the main fuel.

The ignition temperature T_s is one of the crucial parameters to be specified for gain estimation and should strongly depend on the process of ignitor generation. In the CSI scheme, for example, the igniting hot spot is formed at the end of implosion of the main fuel shell surrounding the central hot spot, the plasma configuration of which is isobaric (Meyer-ter-Vehn, 1982). In this scheme, $T_s \approx 5\,keV$ has been used as a typical number (Meyer-ter-Vehn, 1982; Lindl, 1998). Meanwhile, in the FI scheme, the ignitor is formed by rapidly heating the fuel, when it is at its maximum compressed (thus almost stagnated) state. In this case, the plasma configuration is isochoric, and $T_s \approx 10-12\,keV$ is

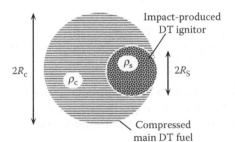

FIGURE 1.24
Schematic picture for the fuel configuration considered in the gain model, which corresponds to the maximum compression state of the main DT fuel.

predicted by systematic 2D calculations (Atzeni, 1995). However, the process of ignitor formation of an impact ignition target does not belong to either of the above two schemes. The ignitor will be formed principally via a shock-compressed process on the impact collision. However, the process is not trivial as in the planar geometry, because the compression and heating result from a rather complex hydrodynamic process through the spherically converging implosion dynamics. Therefore, the ignition temperature required for the impact ignition scheme is still an open question. In the simple gain model discussed below, such key parameters as the ignition temperature T_s and the areal mass density H_s ($\equiv \rho_s R_s$) are set to be

$$T_s \approx 5 - 10\,\text{keV}, \quad H_s \approx 0.4\,\text{g/cm}^2, \tag{1.32}$$

where ρ_s and R_s are the mass density and the radius of the igniting hot spot, respectively; hereafter, we denote the main cold fuel and the ignitor fuel (spark) with the subscripts "c" and "s," respectively. On the impact collision, the kinetic energy of the DT ions corresponding to its implosion velocity v_i is assumed to be finally converted into the thermal energy of the igniting temperature: $(1/2)m_i v_i^2 = 2(3/2)T_s$, where $m_i = (5/2)m_p$ is the averaged DT ion mass with the proton mass m_p, and the temperature is measured in CGS units. The required implosion velocity is then uniquely obtained:

$$v_i = \sqrt{6T_s/m_i} = 1.1 \times 10^8 - 1.5 \times 10^8\,(\text{cm/s}). \tag{1.33}$$

Note that this velocity requirement is somewhat optimistic because the assumption here is that all the kinetic energy is converted into the thermal energy. The total mass for the ignitor is expressed as a function of only the mass density ρ_s by

$$M_s = \frac{4\pi H_s^3}{3\rho_s^2} = 27\,\mu\text{g}\left(\frac{\rho_s}{100\,\text{g/cm}^3}\right)^{-2}. \tag{1.34}$$

The energy balance is given by $\eta_s E_{Ls} = 3T_s M_s/m_i$, where E_{Ls} and η_s are, respectively, the invested laser energy for the cone shell and its coupling efficiency to the final thermal energy. The required laser energy is then given as a function of ρ_s in the form

$$E_{Ls} = \frac{8\pi H_s^3 T_s}{5m_p \eta_s \rho_s^2} = 15\,\text{kJ}\eta_s^{-1}\left(\frac{\rho_s}{100\,\text{g/cm}^3}\right)^{-2}. \tag{1.35}$$

For the main DT fuel, the isentrope parameter $\alpha_c \equiv p/p_{\text{deg}}$ (p and p_{deg} are the fuel pressure and the electron degenerate pressure, respectively), is expected to be as low as possible. In this case, the mass and the total internal energy are, respectively, given by

$$M_c = 4\pi H_c^3/3\rho_c, \tag{1.36}$$

$$\eta_c E_{Lc} = 3.3 \times 10^{12} \alpha_c \rho_c^{2/3} M_c, \tag{1.37}$$

where $H_c \equiv \rho_c R_c$ is the areal mass density of the main DT sphere, and η_c and E_{Lc} have the same definitions as for the igniting hot spot. The total drive energy is then given by $E_d = E_{Lc} + E_{Ls}$. Once the hot spot is successfully ignited, a burn wave is expected to propagate through the main fuel, and the burn fraction can be roughly estimated by $\Phi = H_c/(H_c + H_0)$ with $H_c = (3\rho_c^2 M_c/4\pi)^{1/3}$ and the constant $H_0 = 7\,\text{g/cm}^2$. The energy gain is finally evaluated to be

$$G = \Phi M_c \varepsilon_0/E_d, \tag{1.38}$$

where $\varepsilon_0 = 3.4 \times 10^{18}$ (erg/g) is a constant, and the small fraction of the thermonuclear energy released from the hot spot is ignored for simplicity.

Figure 1.25 shows the energy gain thus obtained as a function of E_d for different values of ρ_s as indicated by the labels, where $\alpha_c = 3$, $\rho_c = 200\,\text{g/cm}^3$, $T_s = 5\,\text{keV}$, $\eta_c = 10\%$, and $\eta_s = 7\%$ are assumed as an example. At the igniting points, where the gain curves begin to rise steeply, the total driver energies are attributed almost only to E_{Ls}. Then the minimum igniting energy scales as $T_s/\eta_s\rho_s^2$ (see Equation 1.35). In particular, it is crucial (Figure 1.25) to achieve the compressed density of the igniting plasma of the order of $\rho_s \sim 70$–$100\,\text{g/cm}^3$ for a high gain at $E_d = $ a few to several 100 kJ. Here it should be noted that this final impactor density is expected to be achieved at the end of all the compression processes of the adiabatic compression due to the ablation and the geometrical compression effect and the shock compression. In particular, an acceleration at a high Mach number seems to be necessary for the

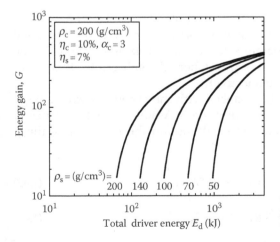

FIGURE 1.25
Gain curves expected for the impact ignition targets for different combinations of ρ_s.

geometrical compression (Basko and Meyer-ter-Vehn, 2002). Meanwhile, the density compression of the main fuel as high as $\rho_c = 200\,g/cm^3$ or equivalently 1000 times the solid density is still challenging but not too optimistic, because the density level of the order of 1000 times the solid density has been experimentally established (Azechi et al., 1991) though it has been obtained under pure spherical experiments with D_2-doped plastic shell targets.

1.7.2 Implosion Dynamics of the Impactor

The dynamics of such a spherical implosion (Murakami and Nishihara, 1987) is characterized by the mass $M(t)$, the radius $R(t)$, and the velocity $v_i(t)$. The "payload" consisting of a fuel shell coated by an ablator is driven by the ablation pressure P_a. The ablation layer is sustained by the quasistationary Chapman–Jouguet (CJ) deflagration wave, where the exhaust fluid velocity is equal to the local sound speed c_{CJ}. Then the dynamics is described by the normalized quantities $\tilde{M} = M/M_0$ and $\tilde{R} = R/R_0$ (the subscript 0 denotes the initial value), in the form

$$v_i = \chi c_{CJ} \ln \tilde{M}, \tag{1.39}$$

$$\tilde{M}(1 - \ln \tilde{M}) = 1 - \beta(1 - \tilde{R}^3)/3, \tag{1.40}$$

where $\chi = P_a/\rho_{CJ}c_{CJ}^2$ in Equation 1.39 is a dimensionless parameter characterizing the ablation layer, which weakly depends on time and specific driving schemes. For most practical cases of interest $\chi \approx 1.5$ has been found from numerical simulations to be a good approximation [Murakami and Nishihara, 1987]. The parameter β in Equation 1.40, defined by

$$\beta = \frac{1}{\chi}\frac{\rho_{CJ}}{\rho_0}\frac{R_0}{\Delta R_0}, \tag{1.41}$$

is the unique dimensionless parameter to describe the implosion within the framework under consideration. In Equation 1.41, $\rho_0\Delta R_0$ denotes the integrated initial areal mass density, and therefore, for the case of ablator-coated DT shell target, $\rho_0\Delta R_0 = (\rho\Delta R)_{f0} + (\rho\Delta R)_{a0} = (\rho\Delta R)_0$, where the subscripts "f" and "a" denote the fuel and the ablator, respectively.

The absorbed laser intensity I_a, balances with the exhaust plasma energy flux $I_a = 4\rho_{CJ}c_{CJ}^3$ at the specific heat ratio $\gamma = 5/3$. Assuming that ρ_{CJ} is nearly equal to the critical density, one can easily obtain the following scalings (Manheimer et al., 1982):

$$\rho_{CJ}(g/cm^3) = 3.7 \times 10^{-3}\lambda_L^{-2}, \tag{1.42}$$

$$c_{CJ}(cm/s) = 8.8 \times 10^7 (I_{a15}\lambda_L^2)^{1/3}, \tag{1.43}$$

$$P_a(\text{Mbar}) = 43(I_{a15}\lambda_L)^{2/3}, \tag{1.44}$$

where I_{a15} and λ_L (laser wave length) are in units of 10^{15} W/cm^2 and $1\,\mu$m, respectively. The DT fuel in the vicinity of the ablation surface is compressed by the ablation pressure $P_a(\text{Mbar}) = 2.2\alpha_s\rho_a^{5/3}$ (α_s is the so-called isentrope parameter of the in-flight impact shell), to the density,

$$\rho_a(\text{g/cm}^3) = 6.0(I_{a15}\lambda_L^{-1})^{2/5}\alpha_s^{-5/3}. \tag{1.45}$$

In the present scheme, a superhigh velocity beyond 10^8 cm/s is necessary to be achieved. One can then roughly estimate a possible parameter window on the laser condition assuming $0.15 < \tilde{M} < 0.30$, which leads to a peak hydrodynamic efficiency based on the rocket model in the form

$$0.08 \le I_{a15}\lambda_L^2 \le 0.3, \tag{1.46}$$

where $v_i = 1.1 \times 10^8$ cm/s is assumed. If we furthermore assume $\lambda_L = 0.25\,\mu$m, Equations 1.13 through 1.15 give the following parameter ranges:

$$1.3 \le I_{a15} \le 5.0, \quad 130 \le P_a(\text{Mbar}) < 320, \, 3.9 \le \rho_a(\text{g/cm}^3) \le 6.8, \tag{1.47}$$

where $\alpha_s = 6$ is assumed in obtaining the range of ρ_a. Moreover, together with $0.15 \le \tilde{M} \le 0.30$, Equation 1.40 prescribes the initial target structure in the form $1.2 \le \beta \le 1.9$, where $\tilde{R} = 1/2$ is specified. This then gives with the help of Equations 1.41 and 1.42,

$$470 \le \frac{R_0}{\lambda_L^2(\rho\Delta R)_0} \le 780, \tag{1.48}$$

where λ_L and ρ are in units of μm and g/cm^3, respectively.

1.7.3 Suppression of RT Instability with Double Ablation

It is a most crucial problem in ICF how to suppress the RT instability during the acceleration phase of the spherical shell. Recently, an interesting physical phenomenon has been found (Fujioka et al., 2004): When a small amount of high-Z material is doped in the ablator material, x-rays are generated in the laser-irradiated high-temperature region. The generated x-rays are transported to the inner and higher density region of the target and their energy is deposited there, causing radiation-driven ablation, if the doped material and its density are properly chosen. Previously, the double ablation was viewed as an undesired phenomenon because a second unstable surface is introduced in addition to the original ablation surface. However, as is well known, a radiation-driven ablation surface is much more stable than a generic

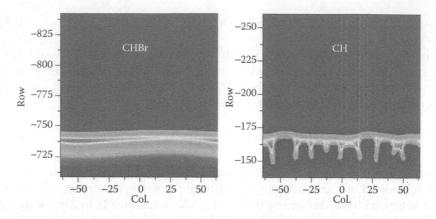

FIGURE 1.26
Two-dimensional hydrodynamic simulation for planar shell acceleration. The planar targets are irradiated from the bottom side at intensity of 4×10^{14} W/cm^2.

electron-driven ablation surface because of the large ablation velocity and the long density scale length. On the other hand, the ablation velocity at this electron ablation is extremely enhanced by two factors. (1) The mass flux from the upper stream is large because of the radiation-driven ablation. (2) The electron ablation density is much less than the upper stream density. Figure 1.26 is a comparison of 2D simulation with the use of a 25 μm thick planar target to show how the RT instability grows in time; the right and left targets are made of polystyrene (CH) and 3 atom % Br-doped polystyrene (CHBr), respectively. When these planar targets are observed as shown in Figure 1.4, they are remote from their initial positions by distances of about a few hundred micrometers. The laser is assumed to be normally incident on the target and the laser absorption is calculated with 1D ray tracing. The wavelength and amplitude of the perturbation imposed initially on the target surface are 80 and 0.8 μm, respectively. As is expected, the 2D simulation shows that the ablation surface of the CHBr target (left) is almost completely stabilized, whereas the deformation on the ablation of the CH target (right) is significant. Here it should be noted that the ablation velocity of the CHBr target is three times larger than that of the CH target.

In an experiment, perturbation growth in planar CHBr and CH targets has been diagnosed with face-on backlighting using 1.5 keV x-rays emitted from a tin plasma (Figure 1.27). The data at a relatively long perturbation wavelength of 50 μm showed that the perturbation growth on the CHBr is much lower than that of the CH target. The difference was even more pronounced at a shorter perturbation wavelength of 20–25 μm, where the growth rate of the generic CH target has its maximum. The perturbation growth in CHBr was almost completely stabilized as shown in Figure 1.27.

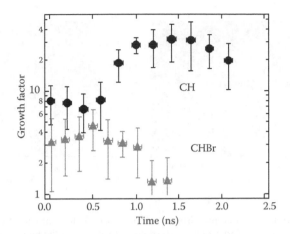

FIGURE 1.27

Comparison of the growth rates of the RT instability obtained for CH and CHBr targets. (From Fujioka, S., *Phys. Rev. Lett.*, 92, 195001, 2004. With permission.)

1.7.4 2D Integrated Hydrodynamics Simulation

We have conducted preliminary 2D hydrodynamic simulations (Nagatomo, 2003) to demonstrate the basic implosion performance of the impact ignition target. Due to limitation of CPU time, radiation transport is switched off. Moreover, for simplicity, the shell material is composed of only plastic (CH). In the radial and the azimuthal (180°) directions, 300 mesh points are assigned, respectively. The outer radius/shell thickness of the impact CH shell (left hemisphere) and the main CH shell (right hemisphere) are respectively chosen to be 750/30 μm and 660/21 μm. Simple Gaussian laser pulses at $\lambda_L = 0.35$ μm and $\tau_L = 2.5$ ns, which normally illuminate both the shells, are 1D ray-traced; there is no time delay between the two pulses. Applied laser intensities (energies) are 3×10^{15} W/cm² (100 kJ) on the impact shell and 9×10^{14} W/cm² (100 kJ) on the main shell. The cone guide is made of solid gold with a thickness of 10 μm with the full cone angle of 130°, and with a diameter of 150 μm for the open hole at its apex.

Figure 1.28a shows the isocontour map of the density (upper half plane in g/cm³) and the temperature (lower half plane in keV) shortly before the impact. At this stage, the velocity of the impact shell almost reaches its peak value ~6×10^7 cm/s, while the implosion velocity of the main shell is about 3×10^7 cm/s; those peak velocities are both achieved at around the Gaussian peak. Correspondingly, the ablation pressure of 110 Mbar and the in-flight density of 7 g/cm³ are observed for the impact shell.

Figure 1.28b shows the isocontour map of the density and the temperature as in Figure 1.28a but shortly after the impact at around the peak compression. The main CH core is structured such that the high-density spherical shell with $T = 1$–2 keV and $\rho = 200$–300 g/cm³ surrounds the central low-density core

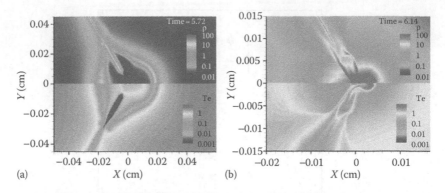

FIGURE 1.28
Isocontour map of the plasma quantities at a time shortly before (a) and after (b) the impact. The upper and lower half planes present the density (g/cm³) and the temperature (keV) distributions, respectively. (From Murakami, M., Nagatomo, H., Azechi, A., Ogando, F., and Eliezer, S., *Nucl. Fusion*, 46, 99, 2006. With permission.)

with $T = 3$–$4\,keV$ and $\rho = 50$–$100\,g/cm^3$. Meanwhile, the impact CH plasma is characterized by $T = 2$–$5\,keV$ and $\rho = 200$–$400\,g/cm^3$.

Because of the present crude design of the laser pulse and the target structure, the results here are still far from an optimized performance expected for the impact ignition target. Nevertheless, it should be stressed that the two separately imploded shells are synchronously collided at the center and a dense and hot plasma is generated on the impact collision. In spite of the deficiencies that dwell in this preliminary hydrodynamic simulation, the 2D simulation results demonstrated here indicate the potential of the impact ignition scheme and thus the necessity of further study on this scheme. It would be appropriate to note some additional key issues to the conventional ones that should be addressed in such a future study: (1) acceleration of the impact shell in the cone geometry to superhigh velocities, what is the maximum attainable velocity and density? To which extent can the RT instability be suppressed under such new ideas as the bromine-doped targets (Fujioka et al., 2004) or cocktail (multiwavelengths) laser illumination (Azechi et al., 2004), (2) conversion efficiency from the kinetic to the thermal energy as mentioned above, and (3) hydrodynamic instability and radiation transport problem raised under the superhigh velocities of the order of $10^8\,cm/s$ specifically at the interface between the impact shell and the cone wall.

1.7.5 Experimental Results on Superhigh Velocities and Neutron Yields

As the first step toward proof-of-principle of impact ignition, i.e., to achieve a high-velocity of the order of $10^8\,cm/s$, preliminary experiments have been conducted recently at HIPER facility (Murakami and Nagatomo, 2005). In the HIPER facility, 12 beams of the GEKKO-XII laser facility are combined into virtually one single beam in order that a large target area can be irradiated at a high intensity of the order of 10^{15} W/cm². The effective focusing optics of the combined beams is $f/3$. Originally, among the 12 beams, three frequency-doubled ($\lambda_L = 0.53\,\mu m$) beams were expected to be used as a foot pulse for precompression of the plane target; they did not work this time because of a system trouble. In the experiments, therefore, the remaining nine frequency-tripled ($\lambda_L = 0.35\,\mu m$) beams were split to eight bundled for the target irradiation and one for back lighting. Each of the nine beams was tuned to provide a rectangular-like super-Gaussian pulse with a pulse length of 2.5 ns and an intensity of about 5×10^{13} W/cm², and the peak total intensity was thus about 4×10^{14} W/cm². Two different types of planar targets with a variation of thickness of 12–20 μm were used: one made of polystyrene with a mass density of 1.06 g/cm³ and the other made of Br-doped plastic targets ($C_{50}H_{41.5}Br_{3.3}$) with a mass density of 1.35 g/cm³. With the use of the latter targets, highly stable accelerations were expected owing to a newly found physical effect that substantially suppresses the RT instability growth, in which a double ablation structure plays a crucial role (Fujioka et al., 2004). Indeed, a maximum velocity has been attained with a 14 μm thick Br-doped plastic target: Figure 1.29 shows the raw data obtained by an x-ray streak camera. In Figure 1.29, the maximum velocity of about 6.5×10^7 cm/s can be observed at the end of the laser pulse, which is the highest velocity ever achieved. One-dimensional hydrodynamic simulation also predicted a maximum velocity of about 7×10^7 cm/s, which indirectly proves that the acceleration was rather uniform and stable.

Besides the superhigh velocity experiments, we have recently done experiments with an integrated target, which is composed of two CD semispherical targets bonded by the gold cone as can be seen in Figure 1.30. The gold cone has an open hole of 50 μm in diameter at the tip. In the experiments, three beams for 1.0 kJ and nine beams for 3.2 kJ of Gekko XII laser system with 0.35 μm of wavelength and 1.3 ns (FWHM; full width at half maximum) of

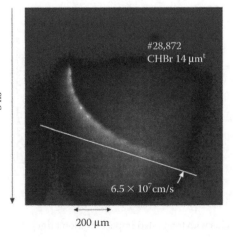

FIGURE 1.29

X-ray streak image showing an ablative acceleration of a planar Br-doped plastic target, in which the maximum velocity, 650 km/s, has been achieved. The applied laser intensity was 4×1014 W/cm² at $\lambda_L = 0.35\,\mu m$. (From Murakami, M., et al., *Plasma Phys. Control. Fusion*, 47, B815, 2005. With permission.)

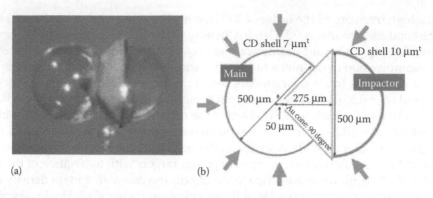

(a)   (b)

FIGURE 1.30
(a) Side view of the integrated target. (b) Schematic view of the target.

pulse length in Gaussian pulses were used to irradiate the impactor and the main fuel hemispheres, respectively. Figure 1.31 shows the timing relation between the implosion laser and the impactor laser (Figure 1.31a) and the resultant neutron yield as a function of the laser timing. For example, the laser timing of –1 ns means that the impactor laser beams are irradiated 1 ns prior to the implosion lasers. As can be seen in Figure 1.31, there is an optimum laser timing to maximize the neutron yield to achieve the neutron yield of 2×10^6. The neutrons are produced mainly in the impactor CD shell, which follows and supports the very original idea of the impact-FI scheme. Furthermore, Figure 1.31b shows that the laser timing is a crucial parameter in designing the impact ignition target.

Figure 1.32 shows the comparison of neutron yields between the impact-FI thus obtained (the highest record is indicated by the horizontal dotted line)

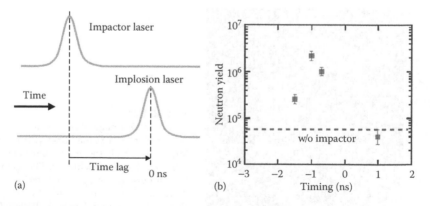

(a)   (b)

FIGURE 1.31
(a) Definition of pulse timing. (b) Neutron yields with integrated impact ignition target.

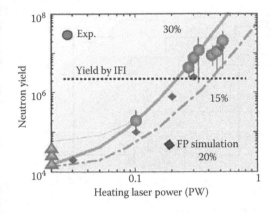

FIGURE 1.32
Comparison of neutron yields between impact and PW FI.

and the PW FI (solid circles). In the experiments of Figure 1.31, since only three beams were irradiated on the impactor shell, the 2D behavior of the acceleration is expected to be quite nonuniform. In experiments in the near future, however, more beams will be irradiated on the impactor shell, and even higher neutron yields can be expected.

1.8 IFE Power Plant Development

1.8.1 IFE Power Plant Systems

An IFE power plant consists of four major, separate, but interconnected subsystems (elements of a power plant) as shown in Figure 1.33, the functions of which are as follows:

1. The driver, usually either a laser or particle accelerator, converts electrical power into short pulses of light or particles and delivers them to the fuel pellet in the proper spatial and temporal form to cause implosion, ignition, and thermonuclear burn, i.e., fusion.

2. In the pellet factory, fuel pellets are manufactured, filled with DT fuel, sent to the reactor, and then injected into the reaction chamber.

3. In the reaction chamber, the injected fuel pellet (target) is tracked, i.e., its position, flight direction, and velocity are measured precisely. Driver beams are directed to the target to implode it and to produce thermonuclear energy with a repetition rate of a few times a second. The thermonuclear emissions are captured in a surrounding structure called a blanket, and their energy in converted into thermal energy (heat). Tritium is also produced in the blanket.

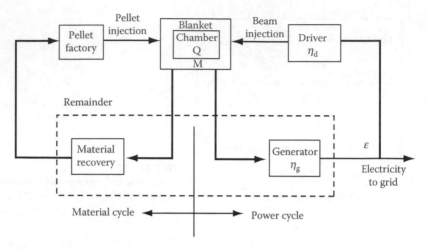

FIGURE 1.33
IFE power plant system. (From Nakai, S. and Mima, K., *Rep. Prog. Phys.*, 67, 321, 2004, figure 20.
With permission.)

4. In the remainder of the plant, two major processes for material and
   energy are performed. Tritium and some other target materials are
   extracted from the recirculating blanket fluid material and from the
   reaction chamber exhaust gases. Then these extracted materials are
   recycled to the target factory. The thermal energy in the blanket fluid
   is converted into electricity, a portion of which is conditioned and
   recirculated to power the driver.

Corresponding to the major processes, two cycles are formed in a power
plant as shown in Figure 1.33: the target material cycle and the power cycle.
In the material cycle, the entire target disintegrates due to implosion fusion
in the reaction chamber, and a portion of the DT fuel is burned. Target debris
is deposited in the chamber gas or in the blanket fluid. Tritium fuel is manufactured from Li in the blanket. Tritium and other target materials are
extracted, conditioned, and returned to the target factory for use in new
target fabrication.

In the power cycle, a series of energy conversion processes occurs, each
with a characteristic efficiency and multiplication as shown in Figure 1.33.
Electrical energy is converted to laser light or particle beam energy in the
driver with an efficiency η_d. The driver energy produces thermonuclear
energy by fuel pellet implosion with a target gain Q. The thermonuclear
energy is converted into thermal energy in the blanket fluid together with
the tritium breeding and the energy multiplication with a factor M through
the neutron reactions. Finally, the heat energy is converted into electricity
with an efficiency η_g. A portion, ε, of the gross electricity produced must be
recirculated to the power driver, completing the power cycle. In the power
cycle, the basic condition of the power balance is shown by

$$\eta_d QM\eta_g \varepsilon = 1. \tag{1.49}$$

The recirculation power fraction ε, should be less than 25% for economic reasons. If ε is too large, then the cost of electricity (COE) sold rises rapidly because much of the plant equipment is used simply to generate electricity for the driver itself. The blanket gain M, is about 1.05–1.25 depending on the design of the blanket material and structure. The efficiency of the turbine generator η_g, ranges from 30% to 40%. Therefore, the product $\eta_d Q$, should be

$$\eta_d Q \geq 10. \tag{1.50}$$

This product determines the minimum target gain necessary for any given driver efficiency. The estimated target gain is shown in Figure 1.6. The typical values predicted are $Q \approx 30$ for the indirect drive, $Q \approx 100$ for the direct drive, and $Q \approx 200$ for the FI. Corresponding to each gain value, the necessary condition for driver efficiency is given for an indirect drive with heavy ion beam (HIB) accelerator to be 30%–40%, for a direct driver with laser to be 10% and for FI, if it works well, 5%. The higher driver efficiency η_d, with a higher target gain Q, at a smaller driver energy, provides the condition for smaller recirculation power fraction ε, and a lower driver cost, which gives us a competitive COE.

An IFE power plant must be safe and must have minimum impact on the environment. Ensuring these features requires that all the system components of the plant and materials used must be examined carefully for their potential negative impacts. The economic, safety, and environment (ESE) aspects should be examined in the technical design of each specific type of power plant. It should be noted that these aspects can be evaluated reasonably quantitatively with the recent developments in physics and technology related to the elements of IFE power plants.

1.8.2 Driver Development

The specifications of the laser driver for a commercial power plant are (1) total energy (MJ/pulse), (2) intensity (10^{14}–10^{15} W/cm^2 on target), (3) pulse shape (tailored in 20–40 ns pulses), (4) wavelength (0.3–0.5 μm), (5) spatial uniformity of irradiation (<1%), (6) efficiency (>10% for direct implosions, >5% for FI), (7) repetitive operation (~10 Hz), (8) cost (capital and operating costs, including life and maintainability), and (9) reliability or availability (Hogan et al., 1995).

In the above specifications, items (1) through (5) are those required for the research phase to demonstrate ignition and fusion gain. Items (6) through (9) are those required for a driver for a laser fusion power plant. An advanced solid-state laser (Krupke, 1989) has demonstrated a breakthrough in diode laser pumping and feasible prospects of a power plant driver (Naito et al., 1992). Figure 1.34 shows the progress of flash lamp pumped

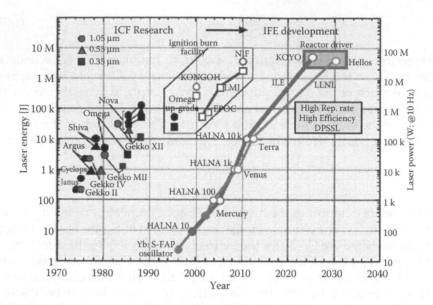

FIGURE 1.34
Glass laser system for implosion experiments and DPSSL development for power plants in the world. (From Nakai, S. and Mima, K., *Rep. Prog. Phys.*, 67, 321, 2004, figure 21. With permission.)

glass lasers on the left-hand side, which are basically single-shot lasers, and the expected development of diode-pumped solid-state lasers (DPPSLs) on the right-hand side.

The applicability of the solid-state lasers for reactor drivers depends strongly on the capability of achieving uniform irradiation on the surface of the fuel pellet for its stable implosion. Several important concepts and technologies for better irradiation have been developed. They are: (1) RPP (distributed phase plate) to obtain a smooth envelope distribution in a far-field pattern even with a beam of nonuniform intensity distribution at near field, (2) smoothing by SSD to obtain time-averaged uniformity of intensity even with a speckle structure of the instantaneous pattern, and (3) pulse tailoring in intensity (picket fence to control the shock heating of the fuel shell) in wavelength (two-color irradiation to control nonlocal heat transport) or in coherency (partially coherent light [PCL] at the foot to reduce the perturbation by the initial imprint). It should be noted that the Omega 60-beam achieved an on-target uniformity better than 1% rms with an improved power balance of 60 beams below 5% rms and implementation of several smoothing techniques (McCrory et al., 2002).

For better and more practical drivers of DPSSL for laser fusion power plants in commercial use, there are two major key issues. The first is high-power laser diodes for pumping and the other is solid-state laser material. Laser diodes have progressed and achieved the required specification for pumping. A cost reduction in the diode laser of 10–100 times less than the present level is required, and this is expected to be possible with the increase in demand for

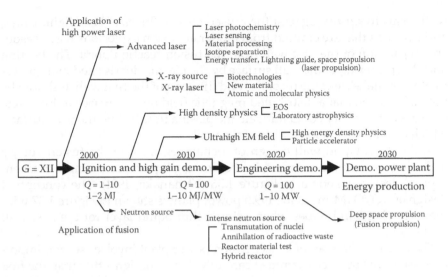

FIGURE 1.35
Application of high-power laser and spin-offs of IFE development. (From Nakai, S. and Mima, K., *Rep. Prog. Phys.*, 67, 321, 2004, figure 22. With permission.)

application of LD (Laser Diode) and DPSSL in industry and science, as shown in Figure 1.35, along with increased laser fusion power plant development.

1.8.3 Reaction Chamber

In the laser fusion power plant, small fusion targets of 5–6 mm diameter are successively injected and shot by laser beams. Following the implosion and fusion explosion of a fuel pellet, successive pulses of different energy species hit the first wall of the chamber. They are reflected or scattered laser light, x-ray from fusion plasma, neutron, and plasma debris. The energy spectrum and pulse shape of each energy species for typical high-gain implosions are estimated with burning simulation codes.

The unique feature of the IFE reaction chamber is the technical possibility of utilizing the wetted surface for the plasma-facing wall. This comes from the mitigated requirement for vacuum in the chamber for IFE, in contrast with MFE chamber, where high vacuum is important for better energy confinement. The use of a liquid wall gives us flexibility in designing various kinds of chamber concepts such as a thin liquid layer on porous solid (Booth, 1972), a woven fabric wall (Mima et al., 1992), or a thick liquid layer of 30–50 cm that is flowing down, forming a cavity for the fusion reaction (Yamanaka, 1981; Moir et al., 1994).

On the other hand, the use of a dry wall for an IFE reaction chamber needs special care because of the high peak intensity of the wall loading due to the pulsed release of the various species of energy from the implosion fusion. The dynamic responses of the dry wall are analyzed to examine the possibility of its use as the first wall of a reaction chamber (Kozaki et al., 2002). The dry wall seems to survive against the pulsed energy fluxes of laser light, x-rays, and neutrons that are emitted from moderate fusion explosions with reasonable separation of the first wall from the fusion reaction point. The heating and the sputtering of the wall surface by the plasma debris need more investigation to make clear the response and life span of the first wall. It should be noted, however, that a distributed magnetic field on the surface or low-pressure gas in the cavity could mitigate the plasma flow hitting the wall surface, which could lead to a longer life of the wall.

In the thick liquid wall design of a chamber, most of the fusion energy including neutrons is absorbed in the flowing liquid. An example of a conceptual design is shown in Figure 1.36 (Yamanaka, 1981). The conceptual design of KOYO (Mima et al., 1992) power plant is shown in Figure 1.37 with the technology and issues when using a thin liquid layer for the first wall protection.

The reaction chamber of a laser fusion power plant involves several important issues that must be examined carefully during design, other than the first wall protection described above. They are summarized as follows: (1) mechanical and thermal responses of the reaction chamber such as the dynamics of the surface of the first wall, which includes absorption of x-ray and plasma debris, evaporation and recondensation of the surface, shock formation

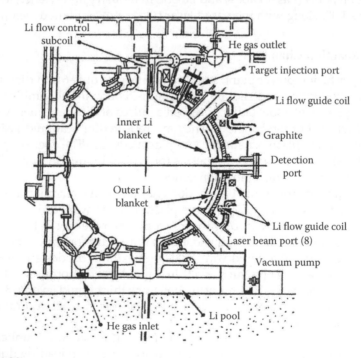

FIGURE 1.36
Thick liquid wall chamber "SENRI". Magnetic field guides the thick Li flow to follow the curved structure wall. (From Nakai, S. and Mima, K., *Rep. Prog. Phys.*, 67, 321, 2004, figure 24. With permission.)

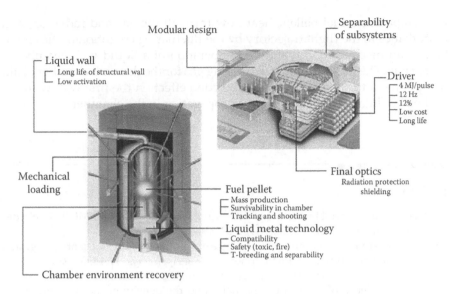

FIGURE 1.37
Key technologies and issues for IFE power plant KOYO. (From Nakai, S. and Mima, K., *Rep. Prog. Phys.*, 67, 321, 2004, figure 25. With permission.)

from the surface inward, and fatigue of the structural material under repeated stress, (2) radiation damage, neutron activation, and shielding, (3) restoration of the chamber environment for driver beam propagation and pellet injection, (4) damage protection of the final elements to guide the driver beams, and (5) tritium breeding in the blanket and extraction. All these technical issues span a wide field of physics and engineering and have many engineering problems in common with the reactor of magnetic fusion energy (MFE). Collaboration and cooperation with industry and MFE reactor development are essential and effective in processing with the IFE reactor.

1.8.4 Fuel Pellet

Precision fabrication of the structured fuel pellet and target for laser implosion and the related physics experiments have been well developed. Recent progress in micromachining, fine coating, and characterizing techniques for microstructures in industry has interacted well with the target technology for laser fusion, in both directions.

The next step toward the power plant is to demonstrate repetitive injection, tracking and shooting with a repetitively fired laser. This is a very challenging development for industry also. Mass production of fuel pellets at low cost is also an important issue for realistic IFE power plant development. The tritium handling technology is common with MFE, and we can share the development process except for the precision cryogenic fuel technology.

The effects of the residual gas in the chamber on the flying cryogenic pellet have been evaluated. They are the condensation effect of the chamber gas on low-temperature fuel pellets, heat flow through the gas and radiation, and disturbance to the flight trajectory by the fluctuating condition of the chamber. It can be concluded that 3–5 Hz operation of a liquid wall chamber is acceptable. The pressure of the protecting gas for the dry wall chamber could be optimized to compromise the protection effect on the first wall with the fuel pellet flight and the laser beam propagation in the chamber.

References

Atkinson, R.d'E. and Houtermans, F.G., Zur Frage der Aufbau-Möglichkeit der Elemente in Sternen, *Z. Phys.* 54, 1929, 656.

Atzeni, S., Thermonuclear burn performance of volume-ignited and centrally ignited bare deuterium-tritium microspheres, *Jpn. J. Appl. Phys.* 34, 1995, 1980.

Atzeni, S., Inertial fusion fast ignitor: Igniting pulse parameter window vs the penetration depth of the heating particles and the density of the precompressed fuel, *Phys. Plasmas* 6, 1999, 3316.

Azechi, H., Jitsuno, T., Kanabe, T., Katayama, M., Mima, K., Miyanaga, N., Nakai, M., Nakai, S., Nakaishi, H., Nakatsuka, M., Nishiguchi, A., Norrays, P.A., Setsuhara, Y., Takagi, M., Yamanaka, M., and Yamanaka, C., High-density compression experiments at ILE, Osaka, *Laser Part. Beams* 9, 1991, 193.

Azechi, H., Shiraga, H., Nakai, M., Shigemori, K., Fujioka, S., Sakaiya, T., Tamari, Y., Ohtani, K., Murakami, M., Sunahara, A., Nagatomo, H., Nishihara, K., Miyanaga, N., and Izawa, Y., Suppression of the Rayleigh? Taylor instability and its implication for the impact ignition, *Plasma Phys. Control. Fusion* 46, 2004, B245.

Basov, N.G. and Krokhin, O.N., Condition for heating up of a plasma by the radiation from an optical generator, *Sov. Phys.—JETP* 19, 1964, 123.

Basko, M.M. and Meyer-ter-Vehn, J., Asymptotic scaling laws for imploding thin fluid shells, *Phys. Rev. Lett.* 88, 2002, 244502.

Basov, N., Kriukov, P., Zakharov, S., Senatsky, Yu., Tchekalin, S., Experiments on the observation of neutron emission at a focus of high-power laser radiation on a lithium deuteride surface, *IEEE J. Quantum Electron.* 4, 1968, 864.

Basov, N.G., Gus'kov, S.Yu., and Feokistov, L.P., Thermonuclear gain of ICF targets with direct heating of ignitor, *J. Sov. Laser Res.* 13, 1992, 396.

Berk, H.I., Fusion, magnetic confinement, in *Encyclopedia of Applied Physics*, Vol. 6, VCH Publishers, New York, 1993, pp. 575–607.

Bethe, H.A., Resonance effects in nuclear processes, *Rev. Mod. Phys.* 9, 1937, 167.

Bethe, H.A., Energy production in stars, *Phys. Rev.* 55, 1939, 434.

Bethe, H.A. and Peierls, R., The neutrino, *Nature* 133, 1934, 689.

Betti, R., Goncharov, V.N., McCrory, R.L., and Verdon, C.P., Growth rates of the ablative Rayleigh-Taylor instability in inertial confinement fusion, *Phys. Plasmas* 5, 1998, 1446.

Bigot, B., Inertial fusion science in Europe, *J. Phys. IV France* 133, 2006, 3.

Bodner, S.E. et al., Naval Research Laboratory 1998 NRL/MR/6730-98-8113.

Bodner, S.E., Colombant, D.G., Schmitt, A.J., and Klapisch, M., High-gain direct-drive target design for laser fusion, *Phys. Plasmas* 7, 2000, 2298.

Booth, L.A., Central station power generation by laser driven fusion, Vol. 1, Report LA-4858-MS (Los Alamos, NM: Los Alamos National Laboratory, 1972).

Burbidge, E.M., Burbidge, G.R., Fowler, W.A., and Hoyle, F., Synthesis of the elements in stars, *Rev. Mod. Phys.* 29, 1957, 547.

Caruso, A. and Pais, V.A., The ignition of dense DT fuel by injected triggers, *Nucl. Fusion* 36, 1996, 745.

Dawson, J.M., On the production of plasma by giant pulse lasers, *Phys. Fluids* 7, 1964, 981.

Deutsch, C., Inertial confinement fusion driven by intense ion beams, *Ann. Phys.* 11, 1986, 1–111.

Deutsch, C., Furukawa, H., Mima, K., Murakami, M., and Nishihara, K., *Phys. Rev. Lett.*, 77, 1996, pp. 2483–2486.

Dolan, T.J., *Fusion Research*, Pergamon Press, New York, 1982.

Eliezer, S., Laser fusion for pedestrians, Israel Atomic Energy Commission Report 1A-1374, 1992.

Eliezer, S., *The Interaction of High-Power Lasers with Plasmas*, Institute of Physics Publishing, Bristol, 2002.

Eliezer, S. and Eliezer, Y., *The Fourth State of Matter*, 2nd edn., Institute of Physics Publishing, Bristol, 2001.

Eliezer, E., Henis, Z., and Martinez-Val, J.M., Effects of tritium seeding of advanced fusion fuels, *Nucl. Fusion* 37, 1997, 985.

Eliezer, S., Henis, Z., Martinez-Val, J.M., and Vorobeichik, I., Effects of different nuclear reactions on internal tritium breeding in deuterium fusion, *Nucl. Fusion* 40, 2000, 195.

Eliezer, S., Murakami, M., and Martinez-Val, J.M., Equation of state and optimum compression in inertial fusion energy, *Laser Part. Beams* 25, 2007, 585.

Floux, F., *Laser Interaction and Related Plasma Phenomena*, Schwarz, H. and Hora, H. (eds.), Plenum, New York, 1971, p. 409.

Fowler, T.K., *The Fusion Quest*, John Hopkins University Press, Baltimore, 1997.

Fujioka, S., Sunahara, A., Nishihara, K., Ohnishi, N., Johzaki, T., Shiraga, H., Shigemori, K., Nakai, M., Ikegawa, T., Murakami, M., Nagai, K., Norimatsu, T., Azechi, H., and Sakaiya, T., Suppression of the Rayleigh-Taylor instability due to self-radiation in a multiablation target, *Phys. Rev. Lett.* 92, 2004, 195001.

Gamow, G. and Teller, E., The rate of selective thermonuclear reactions, *Phys. Rev.* 53, 1938, 608.

Glasstone, S. and Lovberg, R.H., *Controlled Thermonuclear Reactions*, Van Nostrand, New York, 1960.

Gus'kov, S.Yu., Direct ignition of inertial fusion targets by a laser-plasma ion stream, *Quantum Electron.* N31 (10), 2001, 885.

Harms, A.A., Schoepf, K.F., Miley, G.H., and Kingdom, D.R., *Principles of Fusion Energy*, World Scientific, Singapore, 2000.

Harrison, E.R., Alternative approach to the problem of producing controlled thermonuclear power, *Phys. Rev. Lett.* 11, 1963, 535.

Haynam, C.A., Wegner, P.J., Auerbach, J.M., Bowers, M.W., Dixit, S.N., Erbert, G.V., Heestand, G.M., Henesian, M.A., Hermann, M.R., Jancaitis, K.S., Manes, K.R., Marshall, C.D., Mehta, N.C., Menapace, J., Moses, E., Murray, J.R., Nostrand, M.C., Orth, C.D., Patterson, R., Sacks, R.A., Shaw, M.J., Spaeth, M., Sutton, S.B., Williams, W.H., Widmayer, C.C., White, R.K., Yang, S.T., and Van Wonterghem, B.M., National ignition facility laser performance status, *Appl. Opt.* 46, 2007, 3276.

Hazeltine, R.D. and Meiss, J.D., *Plasma Confinement*, Perseus Books, Reading, MA, 1992.

Herman, R., *Fusion, the Search of Endless Energy*, Cambridge University Press, Cambridge, 1991.

Hogan, W.J., Coutant, J., Nakai, S., Rozanov, V.B., and Velarde, G., *Energy from Inertial Fusion*, IAEA, Vienna, 1995, ISBN 92-0-100794-9.

Hora, H., *Laser Plasma Physics, Forces and the Nonlinearity Principle*, SPIE Press, Bellingham, WA, 2000.

Izawa, Y., et al., *19th IAEA Fusion Energy Conference*, Lyon, France, 14–19 October 2002, IAEA-CN-94/IF/P-04.

Kadomtsev, B.B., *Tokamak Plasma, A Complex System*, Institute of Physics Publishing, Bristol, 1992.

Kastler, C.R., Paramagnetic resonance and the effect of resonance on the magnetic birefringence of radio waves, *Acad. Sci. Paris* 285, 1964, 489.

Key, M.K., Kable, M.D., Cowan, T.E., Estabrook, K.G., Hammel, B.A., Hatchett, S.P., Henry, E.A., Hinkel, D.E., Kilkenny, J.D., Koch, J.A., Kruer, W.L., Langdon, A.B., Lasinski, B.F., Lee, R.W., MacGowan, B.J., MacKinnon, A., Moody, J.D., Moran, M.J., Offenberger, A.A., Pennington, D.M., Perry, M.D., Phillips, T.J., Sangster, T.C., Singh, M.S., Stoyer, M.A., Tabak, M., Tietbohl, G.L., Tsukamoto, M., Wharton, K., and Wilks, S.C., Hot electron production and heating by hot electrons in fast ignitor research, *Phys. Plasmas* 5, 1998, 1966.

Kitagawa, Y., Sentoku, Y., Akamatsu, S., Mori, M., Tohyama, Y., Kodama, R., Tanaka, K.A., Fujita, H., Yoshida, H., Matsuo, S., Jitsuno, T., Kawasaki, T., Sakabe, S., Nishimura, H., Izawa, Y., Mima, K., and Yamanaka, T., Progress of fast ignitor studies and petawatt laser construction at Osaka University, *Phys. Plasmas* 9, 2002, 2202.

Kodama, R., Mima, K., Kitagawa, Y., Fujita, H., Norimatsu, T., Jitsuno, T., Yoshida, H., Kawamura, T., Heya, M., Izumi, N., Habara, H., Mori, M., Izawa, Y., and Yamanaka, T., Super-penetration of ultra-intense laser light in long scale-length plasmas relevant to fast ignitor, IAEA-CN-77/IFP/09, *18th IAEA FEC*, Sorrento, Italy, 2000.

Kodama, R., Norreys, P.A., Mima, K., Dangor, A.E., Evans, R.G., Fujita, H., Kitagawa, Y., Kulshelnick, K., Miyakoshi, T., Miyanaga, N., Norimatsu, T., Rose, S.J., Shozaki, T., Shigemori, K., Sunahara, A., Tampo, M., Tanaka, K.A., Toyama, Y., Yamanaka, T., and Zepf, M., Fast heating of super solid density matter as a step toward laser fusion ignition, *Nature* 412, 2001, 798.

Kodama, R., Shiraga, H., Shigemori, K., Toyama, Y., Fujioka, S., Azechi, H., Fujita, H., Habara, H., Hall, T., Izawa, Y., Jitsuno, T., Kitagawa, Y., Krushelnick, K.M., Lancaster, K.L., Mima, K., Nagai, K., Nakai, M., Nishimura, H., Norimatsu, T., Norreys, P.A., Sakabe, S., Tanaka, K.A., Youssef, A., Zepf, M., Yamanaka, T., Fast heating scalable to laser fusion ignition, *Nature* 418, 2002, 933.

Kozaki, Y., Furukawa, H., Yamamoto, K., Johzaki, T., Yamanaka, M., and Yamanaka, T., Issues and design windows of laser fusion reactor chambers, in *19th IAEA Fusion Energy Conference*, Lyon, France, 14–19 October 2002, IAEA-CN-94/FT/P-1–25.

Krupke, W.F., solid state laser driver for an ICF reactor, *Fusion Technol.* 15, 1989, 377.

Landau, L., Origin of stellar energy, *Nature* 141, 1938, 333.

Liberman, M.A., DeGroot, J.S., Toor, A., and Spielman, R.B., *Physics of High Density Z-Pinch Plasmas*, Springer, New York, 1999.

Lindl, J.D., *Inertial Confinement Fusion*, Springer, New York, 1998.

Mahady, A.I., Takabe, H., and Mima, K., Pulse heating and ignition for off-centre ignited targets, *Nucl. Fusion* 39, 1999, 467.

Maiman, T.H., Stimulated optical emission in Ruby, *Nature* 187, 1960, 493.

Maisonnier, C., Macroparticle accelerators and thermonuclear fusion, *Nuovo Cimento*, 42, 1966, 332.

Manheimer, W.M., Colombant, D.G., and Gardner, J.H., Steady-state planar ablative flow, *Phys. Fluids* 25, 1982, 1644.

McCrory, R.L. *et al.*, *19th IAEA Fusion Energy Conference*, Lyon, France, 14–19 October 2002, IAEA-CN-94/IF-1.

McKenty, P.W., Goncharov, V.N., Town, R.P.J., Skupsky, S., Betti, R., and McCrory, R.L., Analysis of a direct-drive ignition capsule designed for the National Ignition Facility, *Phys. Plasmas* 8, 2001, 2315.

Meyer-ter-Vhen, J., On the energy gain of fusion targets: The model of Kidder and Bodner improved, *Nucl. Fusion* 22, 1982, 561.

Mima, K., Kitagawa, Y., Takabe, H., Yamanaka, M., Nishihara, K., Naito, K., Hashimoto, T., Norimatsu, N., Yamanaka, T., Izawa, Y., Murakami, M., Nishiguchi, A., Yamanaka, C., Kosaki, Y., Yoshikawa, K., and Nakajima, H., Design of ICF reactor driven by laser diode pumped solid state laser, in *14th IAEA Fusion Energy Conference*, 1992, Wurzburg, Germany, 30 September–1 October, IAEA-CN-56/G-2-3.

Mima, K., Fujita, H., Izawa, Y., Jitsuno, T., Kitagawa, Y., Kodama, R., Miyanaga, N., Nagai, K., Nagatomo, H., Nakatsuka, M., Nishimura, H., Norimatsu, T., Shigemori, K., Takeda, T., Yoshida, H., Yamanaka, T., Tanaka, K.A., and Sakabe, S., Fast ignition experimental and theoretical researches toward FIREX, IAEA-CN-94/IF3, *19th FEC*, Lyon, France, 14–19 October, 2002.

Moir, R.W. *et al.*, HYHFE-II: A molten-salt inertial fusion energy power plant design-final report, *Fusion Technol.* 25, 1994, 5.

Murakami, M. and Nagatomo, H., A new twist for inertial fusion energy: Impact ignition, *Nucl. Instrum. Meth. Phys. Res. A* 544, 2005, 67.

Murakami, M. and Nishihara, K., Efficient shell implosion and target design, *Jpn. J. Appl. Phys.* 26, 1987, 1132.

Murakami, M., et al., Towards realization of hyper-velocities for impact fast ignition, *Plasma Phys. Control Fusion*, 47, B815, 2005.

Murakami, M., Nagatomo, H., Azechi, A., Ogando, F., and Eliezer, S., Innovative ignition scheme for ICF-impact fast ignition, *Nucl. Fusion* 46, 2006, 99.

Naito, K., Yamanaka, M., Nakatsuka, M., Kanabe, T., Mima, K., Yamanaka, C., and Nakai, S., Conceptual design studies of a laser diode pumped solid state laser system for the laser fusion reactor driver, *Jpn. J. Appl. Phys.* 31, 1992, 259.

Nakai, S. and Mima, K., Laser driven inertial fusion energy: Present and prospective, *Rep. Prog. Phys.* 67, 321–349, 2004, Figures 2 through 25.

Nakai, S. and Takabe, H., Principles of inertial confinement fusion—physics of implosion and the concept of inertial fusion energy, *Rep. Prog. Phys.* 59, 1996, 1071.

Nishimura, H., Azechi, H., Fujita, K., Heya, M., Izumi, N., Miyanaga, N., Nakai, M., Nakai, S., Nakatsuka, M., Nishihara, K., Norimatsu, T., Ochi, Y., Shigemori, K., Shiraga, H., Takabe, H., Yamanaka, T., Mima, K., and Yamanaka, C., Indirect-direct hybrid target experiments with Gekko XII laser, in *17th IAEA Fusion Energy Conference*, Yokohama, Japan, 19–24 October 1998, IAEA-CN-69/IF/04.

Norreys, P.A., Allott, R., Clarke, J., Collier, J., Neely, D., Rose, S.J., Zepf, M., Santala, M., Bell, A.R., Krushelnick, K., Dangor, A.E., Woolsey, N.G., Evans, R.G., Habara, H., Norimatsu, T., and Kodama, R., Experimental studies of the advanced fast ignitor scheme, *Phys. Plasmas* 7, 2000, 3721.

Nuckolls, J., Wood, L., Thiessen, A., and Zimmerman, G., Laser compression of matter to super-high densities: Thermonuclear applications, *Nature* 239, 1972, 139.

Post, R.F., Controlled fusion research—an application of the physics of high temperature plasma, *Rev. Mod. Phys.* 28, 1956, 338.

Puhkov, A. and Meyer-ter-Vehn, J., laser hole boring into overdense plasma and relativistic electron currents for fast ignition of ICF targets, *Phys. Rev. Lett.* 79, 1997, 2686.

Regan, S.P., Marozas, J.A., Kelly, J.H., Boehly, T.R., Donaldson, W.R., Jaanimagi, P.A., Keck, R.L., Kessler, T.J., Meyerhofer, D.D., Seka, W., Skupsky, S., and Smalyuk, V.A., Experimental investigation of smoothing by spectral dispersion, *J. Opt. Soc. Am. B* 17, 2000, 1483.

Rosen, M.D., The physics issues that determine inertial confinement fusion target gain and driver requirement: A tutorial, *Phys. Plasmas* 6, 1999, 1690.

Rosenbruth, M.N., *New Ideas in Tokamak Confinement*, Springer, Berlin, 1994.

Roth, M., Cowan, T. E., Key, M. H., Hatchett, S.P., Brown, C., Fountain, W., Johnson, J., Pennington, D.M., Snavely, R.A., Wilks, S.C., Yasuike, K., Ruhl, H., Pegoraro, F., Bulanov, S.V., Campbell, E.M., Perry, M.D., and Powell, H., Fast ignition by intense laser-accelerated proton beams, *Phys. Rev. Lett.* 86, 2001, 436.

Sentoku, Y., Liseikina, T.V., Esirkepov, T.Z., Mima, K., et al., High density collimated beams of relativistic ions produced by petawatt laser pulses in plasmas, *Phys. Rev. E.*, 62, 59, 2000, pp. 7271–7281.

Sentoku, Y., Mima, K., Kaw, P., and Nishikawa, K., Anomalous resistivity resulting from MeV-electron transport in overdense plasma, *Phys. Rev. Lett.* 90, 2003, 155001.

Shigemori, K., Azechi, H., Fujioka, S., Nakai, M., Nishihara, K., Nishikino, M., Sakaiya, T., Shiraga, H., and Yamanaka, T., Reduction of Rayleigh-Taylor growth rate by multi-color laser irradiation, *Bull. Am. Phys. Soc.* 46, 2001, 286.

Skupsky, S., Short, R.W., Kessler, T., Craxton, R.S., Letzring, S., and Soures, J.M., Improved laser-beam uniformity using the angular dispersion of frequency-modulated light, *J. Appl. Phys.* 66, 1989, 3456.

Stacey, W., *Fusion: An Introduction to the Physics and Techniques of Magnetic Confinement Fusion*, Wiley, New York, 1984.

Tabak, M., Hammer, J., Glinsky, M.E., Kruer, W.L., Wilks, S.C., Woodworth, J., Campbell, E.M., Perry, M.D., and Mason, R.J., Ignition and high gain with ultra-powerful lasers, *Phys. Plasmas* 1, 1994, 1626.

Takabe, H., Mima, K., Montierth, L., and Morse, R.L., Self-consistent growth rate of the Rayleigh-Taylor instability in an ablatively accelerating plasma, *Phys. Fluids* 28, 1985, 3676.

Tanaka, K.A., Kodama, R., Fujita, H., Heya, M., Izumi, N., Kato, Y., Kitagawa, Y., Mima, K., Miyanaga, N., Norimatsu, T., Pukhov, A., Sunahara, A., Takahashi, K., Allen, M., Habara, H., Iwatani, T., Matsusita, T., Miyakosi, T., Mori, M., Setoguchi, H., Sonomoto, T., Tanpo, M., Tohyama, S., Azuma, H., Kawasaki, T., Komeno, T., Maekawa, O., Matsuo, S., Shozaki, T., Suzuki, Ka., Yoshida, H., and Yamanaka, T., Studies of ultra-intense laser plasma interactions for fast ignition, *Phys. Plasmas* 7, 2000, 2014.

Teller, E., *Fusion*, Academic, New York, 1981.

Velarde, G., Ronen, Y., and Martinez-Val, J.M., *Nuclear Fusion by Inertial Confinement, a Comprehensive Treatise*, CRC Press, Boca Raton, FL, 1993.

Velarde, P., Martinez-Val, J.M., Eliezer, S., Piera, M., Guillen, J., Cobo, M.D., Ogando, F., Crisol, A., Gonzalez, L., Prieto, J., and Velarde, G., Hypervelocity jets from conical hollow charges, *AIP Conf. Proc.* 406 (1), 1997, 182.

Yamanaka, C., Report ILE-8127P Institute of Laser Engineering, Osaka, Japan, 1981.

Yamanaka, T., Internal report of ILE, Oskaka University, 1983.

Yamanaka, C., *Introduction to Laser Fusion*, Harwood, London, 1991.

Yamanaka, T. et al., *19th IAEA Fusion Energy Conference*, Lyon, France, 14–19 October 2002, IAEA-CN-94/OV/3-1.

2

Accelerators

James Koga

CONTENTS

For nearly 100 years, acceleration of charged particles using electromagnetic fields has been carried out. The goal of this field has been to continually achieve higher energies to examine, in finer detail, the basic constituents of matter. Great progress has been made; however, due to limitations on the maximum electric fields that can be obtained by conventional techniques larger and larger accelerator systems have become necessary to achieve higher energies. The cost of such systems is so huge that multinational collaboration and construction over many years is required. One method to achieve higher accelerating fields is using laser–electron and laser–plasma interactions. In recent years, rapid progress in this type of acceleration has been brought about by the development of ultrashort high-power lasers via the chirped pulse amplification method (Strickland and Mourou, 1985). This technique allows generation of high irradiance lasers of ultrashort duration.

The emphasis of this chapter will be on the use of numerical techniques to get the essence of the physical mechanisms by which charged particles are accelerated by lasers. Specifically, we will concentrate on the numerical integration of the equations of motion for a single electron interacting with a laser and the particle-in-cell (PIC) technique for laser–plasma interactions as tools to demonstrate the various properties. The reader can refer to the various excellent reviews of the PIC technique (Dawson, 1983; Birdsall and Langdon, 1985) for a description of this method and to a detailed numerical methods review for descriptions of numerical integration techniques (Press et al., 1992). Since a detailed description of the various laser–plasma acceleration mechanisms for all types of charged particles would require an entire book in itself, in this chapter we give an overview of the various mechanisms through which electrons can be accelerated by an ultraintense short pulse laser and address recent achievements in this field. The reader can refer to reviews of various theoretical and experimental aspects of laser–plasma acceleration and the references therein for more details (Esarey et al., 1996; Umstadter, 2001; Mourou et al., 2006; Joshi, 2007). For a brief outline of the relevant details, we recommend the work of Esarey (1999). For simplicity, most of the examples given are focused on one-dimensional (1D) aspects of the acceleration mechanisms. This chapter is divided into two sections: single electron motion in a laser wave and laser wake-field acceleration.

2.1 Single Electron Motion in a Laser Wave

We first describe the interaction of a single electron with a laser pulse. When a laser pulse interacts with a charged particle, it accelerates the particle based on the Lorentz equation (ignoring radiation reaction effects) (Jackson, 1999): $(d\vec{p}/dt) = q(\vec{E} + \vec{\beta} \times \vec{B})$, where $\vec{p} = \gamma m v$ is the momentum, m is the electron mass, q is the charge, $(\vec{E}, \vec{B})$ are the electromagnetic fields, and $\vec{\beta} = \vec{v}/c$ and $\gamma = 1/\sqrt{1-\beta^2}$ are the Lorentz factors with $\vec{v}$ being the velocity. The equation of motion can be analytically integrated for a finite size laser pulse of the form (Hartemann et al., 1995)

$$\vec{E} = \hat{z} E_0 g(\varphi) \sin(\varphi)$$

$$\vec{B} = -\hat{y} \frac{E_0}{c} g(\varphi) \sin(\varphi),$$

where
$\varphi(t) = \omega \left[t - \left(x(t)/c \right) \right]$
$g(\varphi) = \cos^2 \{ (\pi/2)[(\varphi/\omega\Delta\tau) - 1] \}$
ω is the laser frequency
$\Delta\tau$ is the pulse width

The integrated equations of motion are

$$\gamma \beta_z = a(\varphi)\sin(\varphi) + b(\varphi)\cos(\varphi) + q,$$

where
$a(\varphi) = p\kappa\sin\{\pi[(\varphi/\Delta\varphi) - 1]\}$
$b(\varphi) = p\cos\{\pi[(\varphi/\Delta\varphi) - 1]\} + (a_0/2)$

$$q = \kappa^2 p$$
$$\kappa = \pi/\Delta\varphi$$
$$\Delta\varphi = \omega\Delta\tau$$
$$p = (a_0/2)(1/1-\kappa^2)$$

The other quantities, $\gamma\beta_x$ and γ, can be obtained by using the equations

$$\gamma(1-\beta_x) = \text{constant} = \gamma_0(1-\beta_0) \quad \text{and} \quad \gamma(\varphi) = \gamma_0\left\{1 + [\gamma\beta_z]^2\left[\frac{1-\beta_0}{2}\right]\right\}.$$

The maximum energy is

$$\gamma_{\max} \approx \gamma_0\left\{1 + a_0^2\left[\frac{1+\beta_0}{2}\right]\right\}$$

and the maximum extent of motion transverse to the propagation direction (quiver) is

$$z_{\max} \approx a_0\gamma_0(1+\beta_0)\frac{\lambda}{2\pi},$$

where a_0 is the unitless laser amplitude defined by

$$a_0 = \frac{eE_0}{mc\omega} = \frac{eE_0\lambda}{2\pi mc^2} = 0.85 \times 10^{-9}\sqrt{I[\text{W}/\text{cm}^2]}\lambda[\mu\text{m}]$$

Figure 2.1 shows a plot of the analytical solutions for the interaction of an initially stationary electron with a finite duration laser pulse, where $x - ct$ is normalized by the laser wavelength λ_0. The laser pulse has a duration of $\Delta\tau = 30\,\text{fs}$ and wavelength $1\,\mu\text{m}$ with a peak amplitude $a_0 = 0.5$. The dark thick line represents the electron's momentum along the propagation direction $\gamma\beta_x$. The light line represents the laser pulse. It should be noted that the electron is accelerated during the interaction with the laser pulse, but after the pulse has passed the electron is again stationary with no net acceleration. This is a consequence of the fact that we are using a plane wave laser pulse without including the radiation damping. In an infinite plane geometry, even with a strong electromagnetic wave, an electron has no energy gain according to the Woodward–Lawson theorem (Woodward, 1947; Lawson, 1979; Esarey et al., 1995b).

Another way to view the action of a laser pulse on an electron is calculating the average force on the electron. This force on an electron can be derived following the procedure found in Chen (1977). For nonrelativistic electrons, $\bar{\beta} \ll 1$, the second term in the Lorentz equation involving the magnetic field can be ignored and we get

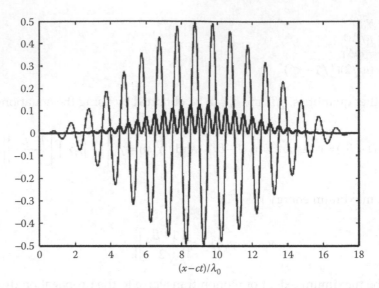

FIGURE 2.1
Laser pulse (light line) and electron momentum (dark thick line) where the x-axis is normalized by λ_0 and the y-axis is the normalized amplitude a_0 for the laser and $\gamma\beta_x$ for the momentum.

$$m_e \frac{d\vec{v}_1}{dt} = -e\vec{E}(\vec{x},t),$$

where for the electron $q = -e$. Assuming that the electric field of laser pulse is of the form $\vec{E}(\vec{x},t) = \vec{E}(\vec{x})\cos(\omega_0 t)$, where ω_0 is the laser frequency and integrating, we get

$$\vec{v}_1 = -\frac{e}{m_e\omega_0}\vec{E}(\vec{x})\sin(\omega_0 t).$$

Integrating this equation, one gets

$$\delta\vec{x} = \frac{e}{m_e\omega_0^2}\vec{E}(\vec{x})\cos(\omega_0 t).$$

The second-order terms are

$$m_e\frac{d\vec{v}_2}{dt} = -e\left(\vec{E}_1(\vec{x},t) + \frac{\vec{v}_1}{c}\times\vec{B}_1(\vec{x},t)\right),$$

where $\vec{E}_1(\vec{x},t)$ is obtained from a Taylor series expansion about the initial position $\vec{x}_0$: $\vec{E}(\vec{x},t) = \vec{E}_0(\vec{x}_0,t) + \delta\vec{x}\cdot\vec{\nabla}\vec{E}(\vec{x},t)|_{\vec{x}=\vec{x}_0} + \cdots$, so that $\vec{E}_1(\vec{x},t) = \delta\vec{x}\cdot\vec{\nabla}\vec{E}(\vec{x},t)|_{\vec{x}=\vec{x}_0}$. The magnetic field is obtained from Maxwell's equations $(\partial\vec{B}_1(\vec{x},t)/\partial t) = c\vec{\nabla}\times\vec{E}_0(\vec{x},t)$ from which we get $\vec{B}_1(\vec{x}) = (c/\omega_0)\vec{\nabla}\times\vec{E}_0(\vec{x})$. The resulting equations become

$$m_e \frac{d\vec{v}_2}{dt} = -e\left[(\delta\vec{x}\cdot\vec{\nabla})\,\vec{E}_1(\vec{x}) + \frac{\vec{v}_1(\vec{x})}{c}\times\vec{B}_1(\vec{x})\right].$$

Averaging this over time, one gets

$$m_e\left\langle\frac{d\vec{v}_2}{dt}\right\rangle = -\frac{e^2}{m_e\omega_0^2}\frac{1}{2}\left[(\vec{E}_1(\vec{x})\cdot\vec{\nabla})\,\vec{E}_1(\vec{x}) + \vec{E}_1(\vec{x})\times(\vec{\nabla}\times\vec{E}_1(\vec{x}))\right],$$

which simplifies to $m_e\langle d\vec{v}_2/dt\rangle = -(e^2/4m_e\omega_0^2)\vec{\nabla}E_1^2(\vec{x})$.

This is the so-called ponderomotive force. Note that this force is independent of the sign of the charge and is dependent on the gradient of the square of the electric field. The force is positive on the front side of the pulse and negative on the backside.

Here, we describe a few mechanisms by which net acceleration can occur. In the above calculation, the electron starts from rest in vacuum and interacts with the laser pulse. For a 1D laser pulse, if an electron starts from rest and encounters both the front and back of the pulse, the net force on the electron is zero. However, if the electron is initially bound in an atom, there is a possibility for the electron to suddenly appear within the laser pulse due to ionization such as tunneling ionization (Ammosov et al., 1986). Figure 2.2 shows traces of $\gamma\beta_x$ versus $(x - x_0 - ct)/\lambda_0$, where x_0 is the initial position of the laser pulse traveling in the $+x$ direction for the case of an initially free electron (dotted line) and bound electron (solid line) interacting with a Gaussian laser pulse where $g(\varphi) = \exp\{-[(\varphi/\omega\Delta\tau)]^2\}$ in the previous laser pulse. The Lorentz equation was integrated numerically using an adaptive Runge–Kutta

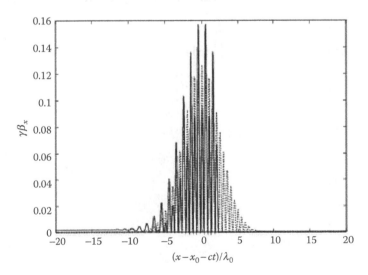

FIGURE 2.2
Comparison between an initially free electron (dotted line) and a bound electron (solid line) interacting with *a*.

scheme (Press et al., 1992). The bound electron becomes free via the tunneling ionization of Ar^{+7} (Ammosov et al., 1986). The laser pulse has a wavelength of 1 μm and a duration of $\Delta\tau = 16.7$ fs with $a_0 = 2.8$. It can be seen that in the ionization case there is a net acceleration of the electron, $\gamma\beta_x \approx 0.186$ (95 keV). Such ionization injection resulting in net acceleration of electrons has been observed in experiments where electrons up to 340 keV were observed with a laser pulse of wavelength $\lambda_0 = 1.054$ μm, irradiance 3×10^{18} W/cm² ($a_0 = 1.55$), and pulse width of 400 fs (Moore et al., 1999). Simulation studies have shown that by using highly charged ions with electrons that are not ionized until very high irradiances, GeV electron energies are possible (Hu and Starace, 2002, 2006).

Another way net acceleration can occur is by having a frequency variation (chirp) in the laser pulse (Khachatryan et al., 2004). As can be seen from the ponderomotive force, it is inversely proportional to the square of the laser frequency. Intuitively, we can see that if the frequency is properly varied along the pulse then there can be a net force on the electron. Figure 2.3 shows the interaction of an electron with Gaussian pulse with a linear chirp where

$$\vec{E} = \hat{z} E_0 g(\varsigma) \sin[\Omega(\varsigma)\varsigma]$$
$$\vec{B} = -\hat{y} \frac{E_0}{c} g(\varsigma) \sin[\Omega(\varsigma)\varsigma],$$

where
$$\varsigma = (x/c) - t$$
$$\Omega(\varsigma) = \omega[1 + (\Delta\Omega/\Delta\tau)\varsigma]$$

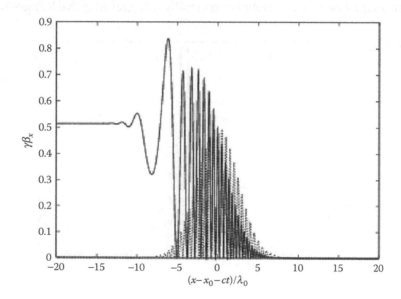

FIGURE 2.3
Comparison between an electron interacting with a laser pulse with no chirp (dotted line) and a linear chirp (solid line).

$$g(\varsigma) = \exp\{-[(\varsigma/\Delta\tau)]^2\}$$

ω is the central laser frequency

$\Delta\tau$ is the pulse width

$\Delta\Omega$ is the chirp

In the figure, the laser pulse has a central wavelength of $1\,\mu m$, a duration of $\Delta\tau = 16.7\,fs$ with $a_0 = 1$. Two cases are shown: without a chirp (dotted line) and with a chirp where $\Delta\Omega = 0.35$ (solid line). It can be seen that with the chirp there is a net acceleration of the electron, $\gamma\beta_x \approx 0.51$ (262 keV). In the calculation $\Delta\tau$ is the same for both cases. However, from a practical point of view adding a chirp to an existing pulse lengthens it. In addition, it has been shown that a circularly polarized chirped laser pulse can achieve much higher electron energies than a linearly polarized laser pulse (Gupta et al., 2007).

Up to this point we have been using a 1D plane wave laser pulse for simplicity. The production of 1D plane waves requires a current source of infinite extent (Feynman et al., 1964). However, in reality, focusing laser pulses are 3D in nature because they are generated from finite size current sources. In experiments using a laser pulse of wavelength $\lambda_0 = 1.056\,\mu m$, irradiance $10^{19}\,W/cm^2$ ($a_0 \approx 3$), and temporal duration of 300 fs, electrons in vacuum have been accelerated to MeV energy levels (Malka et al., 1997). It has been shown that in a tightly focused laser pulse an electron gains energy (Salamin et al., 2002; He et al., 2003). It has also been shown that an electron in vacuum can be accelerated by coherent dipole radiation, which satisfies Maxwell's equations, and that net acceleration occurs when there is significant curvature of the wave front (Troha et al., 1999). Where the wave front curvature is small, approaching that of a plane wave, there is no net acceleration (Troha et al., 1999). In the case of a focusing laser pulse, the laser fields become much more involved (e.g., Narozhny and Fofanov, 2000; Pang et al., 2002; Salamin et al., 2002). As an example, we choose (Salamin et al., 2002)

$$E_x = E\left\{S_0 + \varepsilon^2\left[\xi^2 S_2 + \frac{\rho^4 S_3}{4}\right] + \varepsilon^4\left[\frac{S_2}{8} - \frac{\rho^2 S_3}{4} - \frac{\rho^2\left(\rho^2 - 16\xi^2\right)S_4}{16} - \frac{\rho^4\left(\rho^2 + 2\xi^2\right)S_5}{8} + \frac{\rho^8 S_6}{32}\right]\right\}$$

$$E_y = E\xi\nu\left\{\varepsilon^2\left[S_2\right] + \varepsilon^4\left[\rho^2 S_4 - \frac{\rho^4 S_5}{4}\right]\right\}$$

$$E_z = E\xi\left\{\varepsilon[C_1] + \varepsilon^3\left[-\frac{C_2}{2} + \rho^2 C_3 - \frac{\rho^4 C_4}{4}\right] + \varepsilon^5\left[-\frac{3C_3}{8} - \frac{3\rho^2 C_4}{8} + \frac{17\rho^4 C_5}{16} - \frac{3\rho^6 C_6}{8} + \frac{\rho^8 C_7}{32}\right]\right\},$$

$$B_x = 0$$

$$B_y = E\left\{S_0 + \varepsilon^2\left[\frac{\rho^2 S_2}{2} - \frac{\rho^4 S_3}{4}\right] + \varepsilon^4\left[-\frac{S_2}{8} + \frac{\rho^2 S_3}{4} + \frac{5\rho^2 S_4}{16} - \frac{\rho^6 S_5}{4} + \frac{\rho^8 S_6}{32}\right]\right\}$$

$$B_z = E\nu\left\{\varepsilon[C_1] + \varepsilon^3\left[\frac{C_2}{2} + \frac{\rho^2 C_3}{2} + \frac{\rho^4 C_4}{4}\right] + \varepsilon^5\left[\frac{3C_3}{8} + \frac{3\rho^2 C_4}{8} + \frac{3\rho^4 C_5}{16} - \frac{\rho^6 C_6}{4} + \frac{\rho^8 C_7}{32}\right]\right\},$$

where

$$E = E_0(w/w_0)g(\eta)\exp[-(r^2/w^2)]$$
$$S_n = (w/w_0)^n \sin[\psi + n\psi_G]$$
$$C_n = (w/w_0)^n \cos[\psi + n\psi_G]$$
$$w = w_0\sqrt{1+(z/z_r)^2}$$
$$k = \omega/c$$
$$r^2 = x^2 + y^2$$
$$\rho = r/w_0$$
$$z_r = kw_0^2/2$$
$$\varepsilon = w_0/z_r$$
$$\xi = x/w_0$$
$$v = y/w_0$$
$$\zeta = z/z_r$$
$$\psi = \psi_0 + \psi_P - \psi_R + \psi_G$$
$$\psi_P = \omega t - kz = \eta$$
$$\psi_R = kr^2/2R$$
$$\psi_G = \tan^{-1}\zeta$$
$$R(z) = z + (z_r^2/z)$$
$$\psi_0 \text{ is a constant}$$
$$E_0 \text{ is the peak electric field}$$

This focusing field is accurate up to order ε^5. Figure 2.4 shows the interaction of an electron with this focusing pulse of infinite duration $g(\eta) = 1$, where the initial electron trajectory and laser parameters are the same as those in Salamin et al. (2002). The waist of the laser pulse envelope, w, is indicated by the dotted line where the laser is propagating in the z-direction, and where

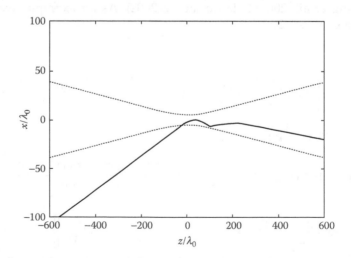

FIGURE 2.4
$z-x$ Coordinates of an electron (solid line) interacting with a laser pulse (dotted line).

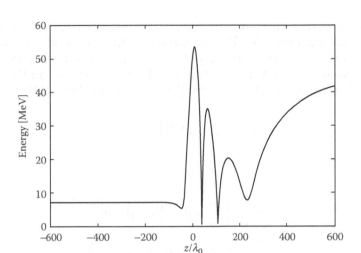

FIGURE 2.5
Trace of the electron energy as a function of z for the electron shown in Figure 2.4.

the wavelength is $\lambda = 1\,\mu m$ and the waist at the focus is $w_0 = 5\,\mu m$ ($\varepsilon = 0.06$). The normalized peak amplitude is $a_0 = 10$. The electron starts at $(x_0, y_0, z_0) = [-1, 0, \tan(\theta)]0.5\,cm$, where $\theta = 10$ is the angle between the x_0 and z_0 axis. The initial energy of the electron is 7.15 MeV ($\gamma = 15$), where the initial momentum is $[\sin(\theta), 0, \cos(\theta)]\gamma\beta$, where $\gamma\beta$ is the total momentum. In the figure, it can be seen that the electron is trapped in the laser pulse. For lower initial electron energies, the electron is reflected by the laser pulse, and for higher energies, it passes through the laser pulse. Figure 2.5 shows the evolution of electron energy as a function of z. It can be seen that the electron energy has increased to above 40 MeV after being trapped in the laser envelope. By taking into account the 3D nature of the laser pulse, we see that net acceleration of single electrons in vacuum can be attained.

Other electron acceleration methods in vacuum involve (1) multiple laser pulses (Hora, 1988), (2) a combination of external static magnetic fields and a laser pulse (Davydovskii, 1963; Kolomenskii and Lebedev, 1963; Roberts and Buchsbaum, 1964; Apollonov et al., 1988; Gupta and Ryu, 2005) or a laser pulse with a chirp that enhances the energy gain (Gupta and Suk, 2006a,b), (3) extraction of the laser pulse from the high electrons by the interaction of the laser pulse with a thin foil (Vshivkov et al., 1998a,b; Yu et al., 2000), (4) a combination of a tapered (varying wavelength) wiggler magnet and laser pulse (Singh and Tripathi, 2004) or chirped laser pulse for higher energy (Gupta and Suk, 2006a,b), and (5) two chirped laser pulses intersecting at an optimum angle in which electron energy gains up to 250 MeV have been shown to be theoretically achieved for a laser with a normalized amplitude as low as $a_0 = 3$ and with an initial electron energy of $\gamma_0 = 3$ (Gupta and Suk, 2007). Readers can refer to the cited references and the articles cited therein for more detailed explanations of these mechanisms.

In the previous sections, we used the Lorentz equation to describe the force acting on charged particles. However, at ultrahigh laser irradiances radiation reaction must be taken into account, that is, the radiation emitted by the accelerated charged particle is so large that this radiation changes the particle's motion. The equations of motion are (Landau and Lifshitz, 1994)

$$\frac{d\vec{p}}{dt} = -e\left\{\vec{E} + \vec{\beta}\times\vec{B}\right\} + \vec{f}_{RD}$$

where

$$\vec{f}_{RD} = -e\frac{2r_e}{3}\gamma\left(\frac{\partial}{c\partial t} + \vec{\beta}\cdot\vec{\nabla}\right)(\vec{E} + \vec{\beta}\times\vec{B})$$

$$+\frac{2r_e^2}{3}\left\{\vec{E}(\vec{\beta}\cdot\vec{E}) + (\vec{E} + \vec{\beta}\times\vec{B})\times\vec{B}\right\},$$

$$-\frac{2r_e^2}{3}\gamma^2\vec{\beta}\left\{(\vec{E} + \vec{\beta}\times\vec{B})^2 - (\vec{\beta}\cdot\vec{E})^2\right\}$$

and $r_e = (e^2/mc^2)$, which is the classical electron radius. Using these equations it was shown that radiation damping effects can lead to particle acceleration from the radiation pressure of a laser pulse with a very large normalized amplitude such that the energy increases on infinite timescales $(t \to \infty)$ as $\gamma \sim t^{3/5}$ (Bulanov et al., 2004).

2.2   Laser Wake-Field Acceleration

In 1979, Tajima and Dawson proposed a new method to accelerate charged particles using laser–plasma interaction (Tajima and Dawson, 1979). In this method, a short laser pulse interacts with a plasma generating a strong oscillating electrostatic field in its wake (wake-field) much in the same manner as a boat generates a wake when it travels through water. The novelty of this technique is that the electrostatic fields generated by such a process can be several orders of magnitude larger than that which can be generated by conventional radio frequency cavities in accelerators. This allows a substantial scaling down of the accelerator size.

2.2.1   Wake-Field Excitation

First, we describe how an ultrashort laser pulse induces large electrostatic fields in plasma. Figure 2.6 shows a 1D PIC simulation of a linearly polarized laser pulse propagating in plasma consisting of electrons and ions. The simulation solves Maxwell's equations for electromagnetic fields via fast Fourier

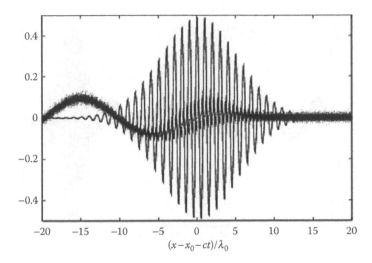

FIGURE 2.6
Pushing of plasma electrons by a laser pulse.

transforms with periodic boundary conditions and uses finite size macro-particles to represent the plasma electrons and ions that are advanced by the Lorentz force equation (see Dawson, 1983 for details). In the figure, the transverse electric field of a short Gaussian pulse laser (solid line) with an envelope of the form $g(x,t) = \exp\{-\frac{1}{2}[(x-x_0-ct/c\Delta\tau)]^2\}$ and pulse duration of $\Delta\tau = 16.7$ fs is propagating from the left to the right along the x-axis, which is normalized by $\lambda_0 = 1\,\mu m$ where x_0 is the initial position of the peak of the laser pulse. The laser pulse started in vacuum. The vertical axis represents the normalized amplitude, a_0, for the laser, which is 0.5 in this case. The individual dots in the figure represent plasma electrons, where the vertical axis represents the electron momentum in the propagation direction $\gamma\beta_x$. The background plasma density is $n_e = 5 \times 10^{18}\,\mathrm{cm}^{-3}$, where the ion-to-electron mass ratio is 1836 and the initial temperature of the plasma is 5 keV to avoid unphysical numerical heating (see Dawson, 1983). The motion can be compared to the vacuum case in Figure 2.1. In the vacuum case, the electron returns to zero momentum after interacting with the laser pulse. However, in the plasma case the electrons have net momentum after the laser pulse has passed, as seen in Figure 2.6 and the expanded view in Figure 2.7.

As opposed to the case in vacuum, the velocity perturbations induced in the electrons by the laser pulse result in the generation of an electrostatic field. The laser pulse induces charge density perturbations $\delta n/n_0$, where n_0 is the initial background plasma density behind the laser in the background plasma, as indicated by the dotted line where the density has been smoothed due to simulation fluctuations (Figure 2.8). As can be seen in Figure 2.8, the charge perturbations are in the form of a sinusoidal type of wave. These charge perturbations result in an electrostatic field or wake-field (bold line)

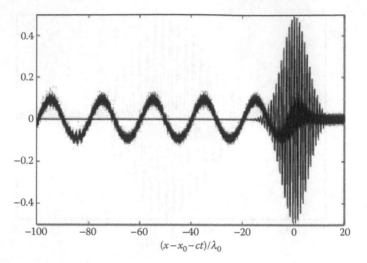

FIGURE 2.7
Expanded view of plasma electrons behind the laser pulse.

in the figure; it is plotted as E/E_0 where $E_0 = cm_e\omega_p/e$ is the nonrelativistic maximum attainable field in a cold plasma (Dawson, 1959). This electrostatic field can then be used to accelerate charged particles.

The generation of the wake-field directly from the laser pulse depends on the laser pulse length, L (Tajima and Dawson, 1979): $L \leq \pi\, (c/\omega_p)$, where ω_p is the plasma frequency, $\omega_p^2 = (4\pi n_e/m_e)$, c is the speed of light, and L is the laser pulse length. With the current density, the laser pulse length is $L = c\Delta\tau = 5.0\,\mu m$, which is shorter than $7.46\,\mu m$.

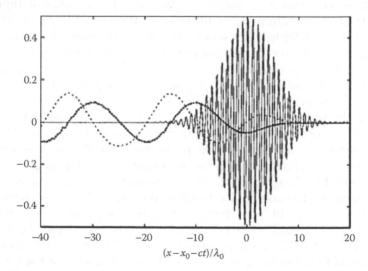

FIGURE 2.8
Density perturbations (dotted line) and wake-field (bold line) induced by the laser pulse (solid line).

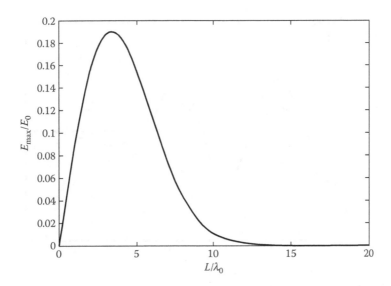

FIGURE 2.9
Variation of maximum electric field with laser pulse length, L, for a circularly polarized laser pulse.

For a Gaussian circularly polarized pulse with $a_0 \ll 1$ of the form $a^2 = a_0^2 \exp(-(\zeta^2/L^2))$ where $\zeta = x - ct$, the maximum wake-field amplitude, E_{max}, far enough behind the pulse ($\zeta \gg L$) is given by (Gorbunov and Kirsanov, 1987)

$$\frac{E_{max}}{E_0} = \left(\frac{\sqrt{\pi} a_0^2}{2} \right) k_p L \exp\left(-\frac{k_p^2 L^2}{4} \right),$$

where $k_p = \omega_p/c$. Using the current parameters, $E_{max}/E_0 = 0.1539$, which is close to the simulation results. The discrepancy can be attributed to the fact that in the simulations a linearly polarized laser pulse is used.

From the expression for the maximum electric field it can be seen that the amplitude of the wake-field generated behind the laser pulse depends on the pulse length. Figure 2.9 shows a plot of the maximum field as a function of laser pulse length for $a_0 = 0.5$ and density $5 \times 10^{18}\,\text{cm}^{-3}$. The optimum laser pulse length at which the wake-field is maximized is given by (Gorbunov and Kirsanov, 1987) $L = (\lambda_p/\pi)\sqrt{2}$, where $\lambda_p = 2\pi c/\omega_p$ and the maximum field is

$$\frac{E_{max}}{E_0} = a_0^2 \sqrt{\frac{\pi}{2e}} \cong 0.76 a_0^2.$$

Figure 2.10 shows simulation results of the variation of the wake-field with the laser pulse length, where the parameters are the same as those in the

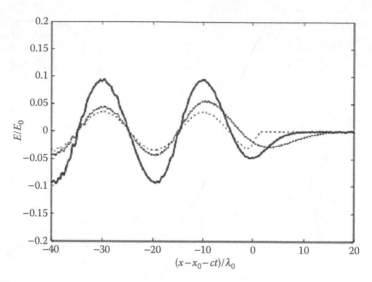

FIGURE 2.10

Wake-field amplitude dependence on laser pulse length, L, where $L = 1$ (dotted line), 5 (bold line), and 9 (solid line) μm.

previous simulation. Three cases are shown with $L = 1$, 5, and 9 μm represented by the dotted line, bold solid line, and solid line, respectively. It can be seen that the amplitude of the wake-field is smaller than the maximum value when the laser pulse is longer or shorter than the optimum pulse length, which is 6.7 μm.

The optimum laser pulse length and the maximum wake-field amplitude also vary depending on the shape and polarization of the laser pulse. For circularly polarized square pulses, the maximum field becomes (Bulanov et al., 1989; Sprangle et al., 1990; Ting et al., 1990; Berezhiani and Murusidze, 1990) $E_{max}/E_0 = a_0^2(1+a_0^2)^{-1/2}$, where for linear polarization a_0^2 is replaced by $a_0^2/2$, which for $a_0^2 \ll 1$ becomes $E_{max}/E_0 = a_0^2$ and the optimum pulse length is $L \cong \lambda_{Np}/2$, where λ_{Np} is the nonlinear plasma wavelength given by

$$\lambda_{Np} = \lambda_p \begin{cases} 1, & \text{for } E_{max}/E_0 \ll 1 \\ (2/\pi)E_{max}/E_0, & \text{for } E_{max}/E_0 \gg 1 \end{cases}.$$

For circularly polarized laser pulses of the form $a^2 = a_0^2 \sin^2(\pi\zeta/L)\sin^2(k\zeta)$ with $-L < \zeta < 0$ or otherwise 0 for $a_0^2 \ll 1$, the optimum pulse length is $L \cong 0.75\lambda_{Np}$ and the maximum wake-field is $E_{max}/E_0 = 0.82a_0^2$ (Sprangle et al., 1990; Ting et al., 1990). For asymmetric-shaped pulses, larger wake-fields and less distortion than for symmetric pulses have also been shown (Berezhiani and Murusidze, 1990).

When the pulse length is long compared with the plasma wavelength, $L > \lambda_p$, the wake-field generated behind the laser pulse is small. However, due to

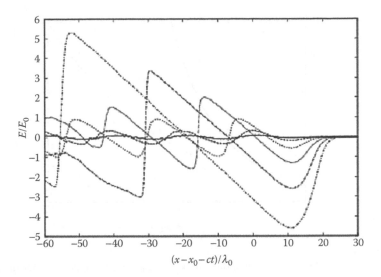

FIGURE 2.11
Variation of the wake-field amplitude as a function of laser amplitude.

forward Raman scattering in the 1D limit (Mori et al., 1994, 1996), or an envelope self-modulation instability in two dimensions (Esarey et al., 1994; Andreev et al., 1994, 1995), the long laser pulse breaks up into a train of short pulses allowing the generation of a large wake-field. The reader is directed to the references for a detailed description of the mechanisms by which this occurs.

As seen in the case of square linearly polarized laser pulses when the laser amplitude increases, the maximum wake-field also increases as $E_{max}/E_0 = a_0^2/2(1 + a_0^2/2)^{-1/2}$.

This can be seen in Figure 2.11, where laser pulses with amplitudes of $a_0 = 0.5, 1, 2, 4, 8,$ and 16 are shown, and the corresponding lines increase in amplitude and length with increasing a_0. The corresponding E_{max}/E_0 are 0.11, 0.41, 1.15, 2.67, 5.57, and 11.27, respectively, using the equation for square pulses. These values are somewhat close to the simulation values, where the differences are most likely due to the different pulse shapes.

Note that as the amplitude increases the trailing wake-field becomes less sinusoidal and longer. For $a_0 = 8$ and 16, one gets $\lambda_{Np}/\lambda_0 = 70.46$ and 142.5, respectively, which is close to the simulation result in Figure 2.11, where half of the wake-field wave for $a_0 = 16$ is shown. This lengthening with a_0 is due to relativistic effects. Recall that $\gamma_{max} \approx \gamma_0\{1+a_0^2[(1+\beta_0)/2]\}$ for a plane wave in vacuum. For large a_0 the electrons are accelerated to high energies so that they become more massive ($\gamma_{max} \gg 1$). At these large amplitudes, we enter the relativistic engineering regime (see Mourou et al., 2006 for a detailed explanation).

An increase in the wake-field amplitude can also be brought about by the use of multiple laser pulses. One method involves the use of a train of laser pulses

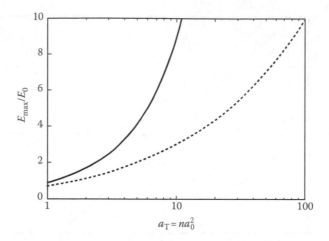

FIGURE 2.12
Maximum wake-field as a function of the number of pulses for $n = 1$ (dotted line) and 3 (solid line), where $a_T = na_0^2$.

(Berezhiani and Murusidze, 1992; Nakajima, 1992; Umstadter et al., 1994, 1995). It has been shown analytically that a train of appropriately positioned square laser pulses can generate a wake-field of the form behind the nth pulse (Umstader et al., 1994, 1995): $E_n/E_0 = x_n^{1/2} - x_n^{-1/2}$, where $x_n = \prod_{i=1}^{n} \gamma_{\perp i}^2, \gamma_{\perp i}^2 = 1 + a_i^2$, and a_i is the normalized amplitude of the ith pulse. The optimized pulse lengths, nonlinear plasma wavelengths, and the position of the nth + 1 pulse are (Umstader et al., 1994, 1995) $L_n = (2/k_p)x_n^{1/2} E_2(\rho_n)$, $\lambda_{Nn} = (4/k_p)x_n^{1/2} E_2(\hat{\rho}_n)$, and $(2l + 1)\lambda_{Nn}/2$, respectively, where E_2 is the complete elliptic integral of the second kind, $\rho_n^2 = 1 - \gamma_{\perp n}^2 x_n^{-1/2}$, $\hat{\rho}_n^2 = 1 - x_n^{-1/2}$, and l is an integer. It can be seen that the maximum field increases rapidly with the pulse number as shown in Figure 2.12.

It has also been shown theoretically and via simulations that the laser-generated wake-field amplitude can be further enhanced by the collision of two counterpropagating laser pulses (Shvets et al., 1999; Nagashima et al., 2001). One is a laser pulse with pulse length less than the plasma wavelength and the other is a long laser pulse. Both pulses have relatively low irradiance levels ~10^{16} W/cm². The energy from the long pulse goes into the generation of the wake-field generated by the short pulse. If the long laser pulse has a frequency difference with the short laser pulse of $\Delta\omega \approx \pm 2\omega_p$, the wake-field enhancement is observed up to a factor of 40 times the standard wake-field case (Nagashima et al., 2001).

2.2.2 Wake-Field Acceleration

In Section 2.2.1, we have described how a short pulse laser or multiple pulses can excite a wake-field in plasma. In this section, we describe how electrons can achieve net acceleration. In a standard linear wake-field there is no net acceleration of electrons in the plasma. After interacting with the laser pulse the electrons simply oscillate in time. Net acceleration occurs when the electrons are injected into the wake-field in some manner with velocities comparable to the phase velocity of the wave. This is analogous to surfers who must attain some minimum velocity to be "injected" into an ocean wave.

The phase velocity of the excited wake wave, v_{ph}, is equal to the group velocity of the laser pulse v_g (Tajima and Dawson, 1979). In the case of a plane electromagnetic wave in an unmagnetized plasma with infinitesimal amplitude, it is $v_g = c\sqrt{1 - \omega_p^2/\omega_0^2}$. For a finite amplitude wave, the expression is more complicated and has been investigated by Decker and Mori (1994). For a relativistically strong circularly polarized electromagnetic wave, it is (Akhiezer and Polovin, 1956) $v_g = c\sqrt{1 - \omega_p^2/(\omega_0^2\sqrt{1 + a_0^2})}$. It can be seen that for higher plasma density $\omega_p^2/\omega_0^2 \lesssim 1$ the necessary velocity for injection of electrons into the wake wave is lower than for low plasma density, $\omega_p^2/\omega_0^2 \ll 1$. In addition, the group velocity can be reduced by 3D effects due to finite spot size effects (Esarey et al., 1995a).

Once an electron is trapped in the potential well generated by the wake-field, there are limits on the maximum energy that the electron can attain. One limiting factor is the maximum electric field that can be maintained by the plasma. For laser amplitudes where the wake-field is linear, $E_x = E_0 \sin[\omega_p(x/v_{ph} - t)]$ and for a laser pulse propagating in the x-direction, the maximum field can be obtained using Poisson's equation $\vec{\nabla} \cdot \vec{E} = 4\pi(n_0 - n_e)$ and the assumption that all the plasma electrons oscillate in the wave with a wave number $k_p = \omega_p/c$ (Dawson, 1959; Tajima and Dawson, 1979): $(\omega_p/c)E_0 = 4\pi e n_0$ where n_0 is the uniform background plasma density. This is the nonrelativistic maximum field that can be obtained in a plasma. For nonlinear plasma waves (relativistic) the limiting field derived using the 1D relativistic cold fluid equations is higher and given by (Akhiezer and Polovin, 1956) $E_{WB} = \sqrt{2(\gamma_{ph} - 1)}E_0$, where $\gamma_{ph} = (1 - v_{ph}^2/c^2)^{-1/2}$ is the relativistic factor of the wake wave with phase velocity $v_{ph} < c$. When $a_0 \gtrsim 1$, then one is in the nonlinear relativistic wake-field regime. Such a fast wave does not (easily) break because the electron momentum increases while its velocity is still at c (see Mourou et al., 2006 and references therein). Plasma finite temperature effects can reduce this field (Katsouleas and Mori, 1988).

Once the electron is trapped and accelerated by the wake-field it will increase in velocity approaching the speed of light and eventually overtake the wake wave, which is moving with $v_{ph} < c$. The electron will reach regions in the wake-field where the electron is decelerated by the wake-field, limiting the maximum energy that it can attain. The distance over which this occurs is called the dephasing length L_d. This distance is defined as the distance over which an electron slips half of the plasma period. This distance can be estimated by a highly relativistic electron $v \approx c$ where the dephasing time is given by $\omega_p(c/v_p - 1)t_d = \pi$ so that $L_d = ct_d \approx \gamma_p^2 \lambda_p$ for $\gamma_p \gg 1$ and the resulting maximum energy gain is given by (Tajima and Dawson, 1979) $W_{max} \approx eE_{max}L_d \approx 2\pi\gamma_{ph}^2 (E_{max}/E_0) m_e c^2$, where E_{max} is the maximum electric field of the wake wave. A detailed analysis has shown that the maximum energy of an electron trapped in the wake wave is given by (Esarey and Pilloff, 1995)

$$\gamma_{max} \cong 2\gamma_p^2 \begin{cases} 2(E_{max}/E_0), & \text{for } 1/4\gamma_p^2 \ll (E_{max}/E_0)^2 \ll 2 \\ (E_{max}/E_0)^2, & \text{for } 2 \ll (E_{max}/E_0)^2 \end{cases},$$

which implies that higher energy electrons can be generated in nonlinear plasma waves. A rough estimate of the dephasing distance also shows variation based on the wake-field amplitude (Esarey and Pilloff, 1995):

$$L_d = \gamma_p^2 \lambda_{Np} \begin{cases} 2/\pi, & \text{for } (E_{max}/E_0) \ll 1 \\ 1/2, & \text{for } (E_{max}/E_0) \gg 1 \end{cases}$$

where λ_{Np} was given previously. In addition, it has been shown that some electrons, which are injected into the first plasma period behind the laser pulse, still retain high energies compared with electrons injected into plasma periods further behind the laser pulse even after propagating over the dephasing distance, and that these electrons can overtake the laser pulse (Bulanov et al., 2005).

As a laser pulse propagates in plasma, it loses energy due to absorption by the plasma. The laser pulse depletion length, l_{de}, is estimated by (Bulanov et al., 1993) $l_{de} \approx L(\omega_0/\omega_p)^2$, where L is the laser pulse length. If this length is less than the dephasing distance, this will reduce the maximum attainable energy for the electron, since the laser pulse is absorbed by the plasma over this distance. The depletion length based on the amplitude of a circularly polarized square laser pulse of length $L \cong \lambda_{Np}/2$ is given by (Ting et al., 1990; Bulanov et al., 1992; Teychenne et al., 1994)

$$l_{de} \approx (\omega_0/\omega_p)^2 \lambda_p \begin{cases} a_0^{-2} & \text{for } a_0^2 \ll 1 \\ a_0/3\pi & \text{for } a_0^2 \gg 1. \end{cases}$$

Up to this point we have been considering 1D laser pulses. In 3D the laser pulses are focused and defocused. The typical length over which Gaussian-shaped laser pulses focus and defocus is given by the Rayleigh length in vacuum (see for example Kogelnik and Li, 1966): $z_r = kw_0^2/2$. If the waist size, w_0, is very small, then the Rayleigh length could be smaller than the dephasing distance L_d, and thus limit the maximum energy that the electron could obtain due to the shortening of the effective acceleration distance. A way around this involves modifying the background plasma in which the laser pulse propagates so that optical guiding can occur. This can be achieved for long laser pulses in which $L > \lambda_p$ by relativistic self-focusing (Max et al., 1974) or a preformed plasma channel (Sprangle et al., 1992) when the power of the laser pulse, P, satisfies the condition (Esarey et al., 1994) $P > P_c (1 - \Delta n/\Delta n_c)$, where $P_c \approx 17(\lambda_p/\lambda_0)^2$ GW is the critical power for relativistic self-focusing and Δn is the plasma channel depth where $\Delta n_c = 1/\pi r_e r_0^2$ is the critical channel depth and r_e is the classical electron radius 2.8179×10^{-13} cm with the assumption of a parabolic channel of the form $n_i = n_0(1 + \Delta n r^2/r_0^2)$. It has been shown both theoretically (Naumova et al., 2001) and experimentally (Chen et al., 2007) that even with relativistic self-focusing, a laser pulse, through a hosing instability, can bend away from the pulse direction limiting the distance over which electron acceleration can occur. In addition, for short laser pulses $L < \lambda_p$ relativistic self-focusing has been shown to be ineffective (Sprangle et al., 1990). Therefore, optical guiding requires $\Delta n \geq \Delta n_c$ (Sprangle et al., 1992).

Injection of electrons into the wake wave can occur by several mechanisms. One simple method is to inject electrons externally with velocities comparable to the phase velocity of the wake-field, which has been demonstrated experimentally (Nakajima et al., 1995). However, current conventional accelerator electron beams are much longer than the wake-field length, so they are injected into a variety of injection points receiving varying degrees of acceleration.

Another method is self-injection of background plasma electrons into the accelerating phase of the plasma wave. The group velocity of a relativistically strong electromagnetic wave is given by (Akhiezer and Polovin, 1956) $v_g = c\sqrt{1 - \omega_p^2/(\omega_0^2\sqrt{1+a_0^2})}$, which depends on the laser wave amplitude a_0. When plasma electrons are accelerated by the laser pulse to velocities above this value, they can then be self-injected into the accelerating phase of the wake-field. As a result the amplitude of the laser pulse, which generates such a self-injection due to a matching between the electron velocity and wake wave velocity, is given by (Zhidkov et al., 2004): $a_0 > (2^{1/4}\omega_0/\omega_p)^{2/3}$. This corresponds to the wave-breaking condition in which the electron displacement inside the wave becomes equal to or greater than the wake wave wavelength (see Mourou et al., 2006 and references therein). In 3D the wake-field transverse structure comes into play so that transverse wake wave breaking and electron injection into the acceleration phase can occur at wave amplitudes much lower than in 1D (Bulanov et al., 1997). The 3D wake-field structure (relativistic flying mirror) has been shown to be useful for frequency upshifting of counterpropagating lasers both theoretically (Bulanov et al., 2003) and experimentally (Kando et al., 2007). Wave-break injection has been observed in experiments where broad electron energy distributions have been observed (Modena et al., 1995) and where energies up to 200 MeV have been measured (Malka et al., 2002). Recently, electron bunches with quasi-monoenergetic distributions with low transverse emittance have been observed experimentally (Faure et al., 2004; Geddes et al., 2004; Mangles et al., 2004; Miura et al., 2005; Yamazaki et al., 2005). Simulations have also shown these quasi-monoenergetic electrons (Pukhov and Meyer-Ter-Vehn, 2002; Zhidkov et al., 2004). In the experiments the electron energy ranged between 10 and 170 MeV, the laser irradiance varied from 10^{18} to 10^{19} W/cm^2, the pulse lengths were between 30 and 55 fs with plasma density ranges between 10^{18} and 10^{20} cm^{-3}. Within these experimental conditions, background electrons get into the acceleration phase of the wake wave as a result of self-injection via wave breaking (Mourou et al., 2006). By using a gas-filled capillary discharge waveguide to optically guide a 40 TW laser pulse 1 GeV electrons with a small energy spread have also been obtained (Leemans et al., 2006). The disadvantage of these methods for generating quasi-monoenergetic electron beams is the uncontrollability. There is a shot-to-shot variation in the charge and energy of the electrons with some shots producing no accelerated electrons.

One way of controlling the injection of the electrons into the accelerating phase of the wake-field is by using multiple laser pulses (Umstadter et al., 1996; Esarey et al., 1997, 1999; Schroeder et al., 1999; Zhang et al., 2003; Kotaki et al., 2004; Fubiani et al., 2004; Dodd et al., 2004; Faure et al., 2006). A primary laser pulse can be used to generate the wake-field, while secondary laser pulses can be used to ponderomotively kick some of the background electrons into the appropriate accelerating phase of the wake wave. The methods differ in the number (one or two) and the angle of propagation with respect to the primary pulse of the secondary pulses (perpendicular, counterpropagating, or both for multiple secondary pulses). As an example, we provide a case with one primary pulse counterpropagating with a single secondary pulse (Kotaki et al., 2004; Fubiani et al., 2004; Faure et al., 2006).

Figure 2.13 shows a 1D PIC simulation where the injection of some background electrons has occurred. In the figure the transverse electric field of a short Gaussian pulse laser (solid line) with a pulse duration of $\Delta\tau = 16.7$ fs is propagating from the left to the right along the x-axis, which is normalized by $\lambda_0 = 1\,\mu m$, where x_0 is the initial position of the peak of the primary laser

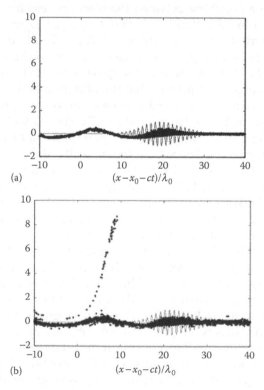

(a) $(x - x_0 - ct)/\lambda_0$

(b) $(x - x_0 - ct)/\lambda_0$

FIGURE 2.13
(a) Before and (b) after the collision with the second laser pulse.

pulse. The laser pulse started in vacuum. The vertical axis represents the normalized amplitude for the laser, which is $a_0 = 1.0$ for the primary laser pulse and $a_1 = 0.4$ for the counterpropagating secondary laser pulse. The individual dots in the figure represent plasma electrons where the vertical axis represents the electron momentum in the propagation direction $\gamma \beta_x$. The background plasma density is $n_e = 5 \times 10^{18}$ cm^{-3}. Figure 2.13a shows the laser and background plasma electrons before the collision with the secondary laser pulse, and Figure 2.13b shows them after the collision. One can see that some background electrons have been injected into the wake and have been accelerated. Such a type of controlled injection has been demonstrated in experiments where monoenergetic electron beams with energies tunable between 15 and 250 MeV, durations of 10 fs, and small divergence were obtained (Faure et al., 2006).

Inhomogeneous plasmas are another method by which electrons can be injected into the accelerating phase of a wake wave (Bulanov et al., 1998; Suk et al., 2001; Shen et al., 2007). In one case a smoothly inhomogeneous plasma, where the wavelength of a relativistically strong Langmuir wave depends on time, was used so that when the wavelength becomes of the order of the quiver amplitude of the electrons the wake begins to break, allowing injection of electrons (Bulanov et al., 1998).

Another method involves using a sharp density gradient to inject electrons (Suk et al., 2001). For a sharp density gradient, nanowires in background plasma have been proposed for use as injectors (Shen et al., 2007).

2.3 Conclusions

In this chapter, we have attempted to give an overview of the acceleration methods for electrons interacting with very high irradiance laser pulses. Emphasis was placed on numerical examples to give an idea of the essence of the acceleration mechanisms. It has been shown through various techniques that relatively large electron energies can be achieved. This field is currently undergoing a renaissance similar to the early days when the laser was first developed in the 1960s. This is due to the rapid development of ultrahigh irradiance laser pulses and, in part, due to the rapid development of large-scale massively parallel computers, which allow simulation models to guide and analyze experiments. It is hoped that this chapter may contribute to developing interest in this exciting field among students.

Acknowledgments

The author acknowledges useful discussions with and comments from T. Tajima, S. Bulanov, T. Esirkepov, H. Daido, and M. Yamagiwa. The author also acknowledges support from the Ministry of Education, Science, Sports and Culture of Japan, Grant-in-Aid for Specially Promoted Research No. 15002013.

References

Akhiezer, A. I. and R. V. Polovin. 1956. Theory of wave motion of an electron plasma. *Zh. Eksp. Teor. Fiz.* 30:915–928. *Sov. Phys. JETP* 3:696–705.

Ammosov, M. V., N. B. Delone, and V. P. Krainov. 1986. Tunnel ionization of complex atoms and atomic ions by an alternating electromagnetic field. *Sov. Phys. JETP* 64:1191–1194.

Andreev, N. E., L. M. Gorbunov, V. I. Kirsanov, A. A. Pogosova, and R. R. Ramazashvili. 1994. The theory of laser self-resonant wake field excitation. *Phys. Scr.* 49:101–109.

Andreev, N. E., V. I. Kirsanov, and L. M. Gorbunov. 1995. Stimulated processes and self modulation of a short intense laser pulse in the laser wake-field accelerator. *Phys. Plasmas* 2:2573–2582.

Apollonov, V. V., A. I. Artem'ev, Yu. L. Kalachev, A. M. Prokhorov, and M. V. Fedorov. 1988. Electron acceleration in a strong laser field and a static transverse magnetic field. *JETP Lett.* 47:91–94.

Berezhiani, V. I. and I. G. Murusidze. 1990. Relativistic wake-field generation by an intense laser pulse in a plasma. *Phys. Lett. A* 148:338–340.

Berezhiani, V. I. and I. G. Murusidze. 1992. Interaction of highly relativistic short laser pulses with plasmas and nonlinear wake-field generation. *Phys. Scr. Lett.* 45:87–90.

Birdsall, C. K. and A. B. Langdon. 1985. *Plasma Physics via Computer Simulation*. New York: McGraw-Hill.

Bulanov, S.V., V. I. Kirsanov, and A. S. Sakharov. 1989. Excitation of ultrarelativistic plasma waves by pulse of electromagnetic radiation. *JETP Lett.* 50:198–201.

Bulanov, S. V., I. N. Inovenkov, V. I. Kirsanov, N. M. Naumova, and A. S. Sakharov. 1992. Nonlinear depletion of ultrashort and relativistically strong laser pulses in an underdense plasma. *Phys. Fluids B* 4:1935–1942.

Bulanov, S. V., V. I. Kirsanov, N. M. Naumova, A. S. Sakharov, H. A. Shah, and I. N. Inovenkov. 1993. Stationary shock-front of a relativistically strong electromagnetic radiation in an underdense plasma. *Phys. Scr.* 47:209–213.

Bulanov, S. V., F. Pegoraro, A. M. Pukhov, and A. S. Sakharov. 1997. Transverse-wake wave breaking. *Phys. Rev. Lett.* 78:4205–4208.

Bulanov, S. V., N. Naumova, F. Pegoraro, and J. Sakai. 1998. Particle injection into the wave acceleration phase due to nonlinear wake wave breaking. *Phys. Rev. E* 58: R5257–R5260.

Bulanov, S. V., T. Zh. Esirkepov, and T. Tajima. 2003. Light intensification towards the Schwinger limit. *Phys. Rev. Lett.* 91:085001-1–085001-4.

Bulanov, S. V., T. Zh. Esirkepov, J. Koga, and T. Tajima. 2004. Interaction of electromagnetic waves with plasma in the radiation-dominated regime. *Plasma Phys. Rep.* 30:196–213.

Bulanov, S. V., M. Yamagiwa, T. Zh. Esirkepov, J. K. Koga, M. Kando, Y. Ueshima, K. Saito, and D. Wakabayashi. 2005. Spectral and dynamical features of the electron bunch accelerated by a short-pulse high intensity laser in an underdense plasma. *Phys. Plasmas* 12:073103-1–073103-11.

Chen, F. F. 1977. *Introduction to Plasma Physics*. New York: Plenum Press.

Chen, L. M., H. Kotaki, K. Nakajima, J. Koga, S. V. Bulanov, T. Tajima, Y. Q. Gu, H. S. Peng, X. X. Wang, T. S. Wen, H. J. Liu, C. Y. Jiao, C. G. Zhang, X. J. Huang, Y. Guo, K. N. Zhou, J. F. Hua, W. M. An, C. X. Tang, and Y. Z. Lin. 2007. Self-guiding of 100 TW femtosecond laser pulses in centimeter-scale underdense plasma. *Phys. Plasmas* 14:040703-1–40703-4.

Davydovskii, V. Ya. 1963. Possibility of resonance acceleration of charged particles by electromagnetic waves in a constant magnetic field. *Sov. Phys. JETP* 16:629–630.

Dawson, J. M. 1959. Nonlinear electron oscillations in a cold plasma. *Phys. Rev.* 133:383–387.

Dawson, J. M. 1983. Particle simulation of plasmas. *Rev. Mod. Phys.* 55:403–447.

Decker, C. D. and W. B. Mori. 1994. Group velocity of large amplitude electromagnetic waves in a plasma. *Phys. Rev. Lett.* 72:490–493.

Dodd, E. S., J. K. Kim, and D. Umstadter. 2004. Simulation of ultrashort electron pulse generation from optical injection into wake-field plasma waves. *Phys. Rev. E* 70: 056410-1–056410-13.

Esarey, E. 1999. Plasma accelerators. In *Handbook of Accelerator Physics and Engineering*, A. W. Chao and M. Tigner (eds.). Singapore: World Scientific Publishing, pp. 543–547.

Esarey, E. and M. Pilloff. 1995. Trapping and acceleration in nonlinear plasma waves. *Phys. Plasmas* 2:1432–1435.

Esarey, E., J. Krall, and P. Sprangle. 1994. Envelope analysis of intense laser pulse self-modulation in plasmas. *Phys. Rev. Lett.* 72:2887–2890.

Esarey, E., P. Sprangle, and J. Krall. 1995a. Laser acceleration of electrons in vacuum. *Phys. Rev. E* 52:5443–5453.

Esarey, E., P. Sprangle, M. Pilloff, and J. Krall. 1995b. Theory and group velocity of ultrashort, tightly focused laser pulses. *J. Opt. Soc. Am. B: Opt. Phys.* 12: 1695–1703.

Esarey, E., P. Sprangle, J. Krall, and A. Ting. 1996. Overview of plasma-based accelerator concepts. *IEEE Trans. Plasma Sci.* 24:252–288.

Esarey, E., R. F. Hubbard, W. P. Leemans, A. Ting, and P. Sprangle. 1997. Electron injection into plasma wakefields by colliding laser pulses. *Phys. Rev. Lett.* 79:2682–2685.

Esarey, E., C. B. Schroeder, W. P. Leemans, and B. Hafizi. 1999. Laser-induced electron trapping in plasma-based accelerators. *Phys. Plasmas* 6:2262–2268.

Faure, J., Y. Glinec, A. Pukhov, S. Kiselev, S. Gordienko, E. Lefebvre, J.-P. Rousseau, F. Burgy, and V. Malka. 2004. A laser-plasma accelerator producing monoenergetic electron beams. *Nature (London)* 431:541–544.

Faure, J., C. Rechatin, A. Norlin, A. Lifschitz, Y. Glinec, and V. Malka. 2006. Controlled injection and acceleration of electrons in plasma wakefields by colliding laser pulses. *Nature* 444:737–739.

Feynman, R. P., R. B. Leighton, and M. Sands. 1964. *The Feynman Lectures on Physics*, Vol. II-20. Reading, MA: Addison-Wesley.

Fubiani, G., E. Esarey, C. B. Schroeder, and W. P. Leemans. 2004. Beat wave injection of electrons into plasma waves using two interfering laser pulses. *Phys. Rev. E* 70:016402-1–016402-12.

Geddes, C. G. R., Cs. Toth, J. van Tilborg, E. Esarey, C. B. Schroeder, D. Bruhwiler, C. Nieter, J. Cary, and W. P. Leemans. 2004. High-quality electron beams from a laser wakefield accelerator using plasma-channel guiding. *Nature (London)* 431:538–541.

Gorbunov, L. M. and V. I. Kirsanov. 1987. Excitation of plasma waves by an electromagnetic wave packet. *Zh. Eksp. Teor. Fiz.* 93:509–518. *Sov. Phys. JETP* 66:290–294.

Gupta, D. N. and C. Ryu. 2005. Electron acceleration by a circularly polarized laser pulse in the presence of an obliquely incident magnetic field in vacuum. *Phys. Plasmas* 12:053103-1–053103-5.

Gupta, D. N. and H. Suk. 2006a. Combined role of frequency variation and magnetic field on laser electron acceleration. *Phys. Plasmas* 13:013105-1–013105-6.

Gupta, D. N. and H. Suk. 2006b. Frequency chirping for resonance-enhanced electron energy during laser acceleration. *Phys. Plasmas* 13:044507-1–044507-2.

Gupta, D. N. and H. Suk. 2007. Electron acceleration to high energy by using two chirped lasers. *Laser Part. Beams* 25:31–36.

Gupta, D. N., M. S. Hur, and H. Suk. 2007. Electron acceleration by a chirped Gaussian laser pulse in vacuum (comment) [*Phys. Plasmas* 13, 123108 (2006)]. *Phys. Plasmas* 14:044701-1–044791-2.

Hartemann, F. V., S. N. Fochs, G.P. Le Sage, N. C. Luhmann, J. G. Woodworth, M. D. Perry, Y. J. Chen, and A. K. Kerman. 1995. Nonlinear ponderomotive scattering of relativistic electrons by an intense laser field at focus. *Phys. Rev. E* 51:4833–4843.

He, F., W. Yu, P. Lu, H. Xu, L. Qian, B. Shen, X. Yuan, R. Li, and Z. Xu. 2003. Ponderomotive acceleration of electrons by a tightly focused intense laser beam. *Phys. Rev. E* 68:046407-1–046407-5.

Hora, H. 1988. Particle acceleration by superposition of frequency-controlled laser pulses. *Nature* 333:337–338.

Hu, S. X. and A. F. Starace. 2002. GeV electrons from ultraintense laser interaction with highly charged ions. *Phys. Rev. Lett.* 88:245003-1–245003-4.

Hu, S. X. and A. F. Starace. 2006. Laser acceleration of electrons to giga-electron-volt energies using highly charged ions. *Phys. Rev. E* 73:066502-1–066502-14.

Jackson, J.D. 1999. *Classical Electrodynamics*. New York: Wiley.

Joshi, C. 2007. The development of laser- and beam-driven plasma accelerators as an experimental field. *Phys. Plasmas* 14:055501-1–055501-14.

Kando, M., Y. Fukuda, A. S. Pirozhkov, J. Ma, I. Daito, L.-M. Chen, T. Zh. Esirkepov, K. Ogura, T. Homma, Y. Hayashi, H. Kotaki, A. Sagisaka, M. Mori, J. K. Koga, H. Daido, S. V. Bulanov, T. Kimura, Y. Kato, and T. Tajima. 2007. Demonstration of laser-frequency upshift by electron-density modulations in a plasma wakefield. *Phys. Rev. Lett.* 99:135001-1–135001-4.

Katsouleas, T. and W. B. Mori. 1988. Wave-breaking amplitude of relativistic oscillations in a thermal plasma. *Phys. Rev. Lett.* 61:90–93.

Khachatryan, A. G., F. A. van Goor, and K. -J. Boller. 2004. Interaction of free charged particles with a chirped electromagnetic pulse. *Phys. Rev. E* 70: 067601-1–067601-4.

Kogelnik, H. and T. Li. 1966. Laser beams and resonators. *Appl. Opt.* 5:1550–1567.

Kolomenskii, A. A. and A. N. Lebedev. 1963. Resonance effects associated with particle motion in a plane electromagnetic wave. *Sov. Phys. JETP* 17:179–184.

Kotaki, H., S. Masuda, M. Kando, J. K. Koga, and K. Nakajima. 2004. Head-on injection of a high quality electron beam by the interaction of two laser pulses. *Phys. Plasmas* 6:3296–3302.

Landau, L. D. and E. M. Lifshitz. 1994. *The Classical Theory of Fields*. New York: Pergamon.

Lawson, J. D. 1979. Laser and accelerators. *IEEE Trans. Nucl. Sci.* NS-26:4217–4219.

Leemans, W. P., B. Nagler, A. J. Gonsalves, Cs. Toth, K. Nakamura, C. G. R. Geddes, E. Esarey, C. B. Schroeder, and S. M. Hooker. GeV electron beams from a centimetre-scale accelerator. *Nat. Phys.* 2:696–699.

Malka, G., E. Lefebvre, and J. L. Miquel. 1997. Experimental observation of electrons accelerated in vacuum to relativistic energies by a high-intensity laser. *Phys. Rev. Lett.* 78:3314–3317.

Malka, V., S. Fritzler, E. Lefebvre, M. -M. Aleonard, F. Burgy, J. -P. Chambaret, J. -F. Chemin, K. Krushelnick, G. Malka, S. P. D. Mangles, Z. Najmudin, M. Pittman, J. -P. Rousseau, J. -N. Scheurer, B. Walton, and A. E. Dangor. 2002. Electron acceleration by a wake field forced by an intense ultrashort laser pulse. *Science* 298:1596–1600.

Mangles, S. P. D., C. D. Murphy, Z. Najmudin, A. G. R. Thomas, J. L. Collier, A. E. Dangor, E. J. Divall, P. S. Foster, J. G. Gallacher, C. J. Hooker, D. A. Jaroszynski, A. J. Langley, W. B. Mori, P. A. Norreys, F. S. Tsung, R. Viskup, B. R. Walton, and K. Krushelnick. 2004. Monoenergetic beams of relativistic electrons from intense laser-plasma interactions. *Nature (London)* 431:535–538.

Max, C. E., J. Arons, and A. B. Langdon. 1974. Self-modulation and self-focusing of electromagnetic waves in plasmas. *Phys. Rev. Lett.* 33:209–212.

Miura, E., K. Koyama, S. Kato, N. Saito, M. Adachi, Y. Kawada, T. Nakamura, and M. Tanimoto. 2005. Demonstration of quasi-monoenergetic electron-beam generation in laser-driven plasma acceleration. *Appl. Phys. Lett.* 86:251501-1–251501-3.

Modena, A., Z. Najimudin, A. E. Dangor, C. E. Clayton, K. A. Marsh, C. Joshi, V. Malka, C. B. Darrow, C. Danson, D. Neely, and F. N. Walsh. 1995. Electron acceleration from the breaking of relativistic plasma waves. *Nature* 337:606–608.

Moore, C. I., A. Ting, S. J. McNaught, J. Qiu, H. R. Burris, and P. Sprangle. 1999. A laser-accelerator injector based on laser ionization and ponderomotive acceleration of electrons. *Phys. Rev. Lett.* 82:1688–1691.

Mori, W. B., C. D. Decker, D. E. Hinkel, and T. Katsouleas. 1994. Raman forward scattering of short-pulse high-intensity lasers. *Phys. Rev. Lett.* 72:1482–1485.

Mori, W. B., C. D. Decker, T. Katsouleas, and D. E. Hinkel. 1996. Spatial temporal theory of Raman forward scattering. *Phys. Plasmas* 3:1360–1372.

Mourou, G. A., T. Tajima, and S. V. Bulanov. 2006. Optics in the relativistic regime. *Rev. Mod. Phys.* 78:309–371.

Nagashima, K., J. Koga, and M. Kando. 2001. Numerical study of laser wake field generated by two colliding laser beams. *Phys. Rev. E* 64:066403-1–066403-4.

Nakajima, K. 1992. Plasma-wave resonator for particle-beam acceleration. *Phys. Rev. A* 45:1149–1156.

Nakajima, K., D. Fisher, T. Kawakubo, H. Nakanishi, A. Ogata, Y. Kato, Y. Kitagawa, R. Kodama, K. Mima, H. Shiraga, K. Suzuki, K. Yamakawa, T. Zhang, Y. Sakawa, T. Shoji, Y. Nishida, N. Yugami, M. Downer, and T. Tajima. 1995. Observation of ultrahigh gradient electron acceleration by a self-modulated intense short laser pulse. *Phys. Rev. Lett.* 74:4428–4431.

Narozhny, N. B. and M. S. Fofanov. 2000. Scattering of relativistic electrons by a focused laser pulse. *JETP* 90:753–768.

Naumova, N. M., J. Koga, K. Nakajima, T. Tajima, T. Zh. Esirkepov, S. V. Bulanov, and F. Pegoraro. 2001. Polarization, hosing and long time evolution of relativistic laser pulses. *Phys. Plasmas* 8:4149–4155.

Pang, J., Y. K. Ho, X. Q. Yuan, N. Cao, Q. Kong, P. X. Wang, L. Shao, E. H. Esarey, and A. M. Sessler. 2002. Subluminous phase velocity of a focused laser beam and vacuum laser acceleration. *Phys. Rev. E* 66:066501-1–066501-4.

Press, W. H., S. A. Teukolsky, W. T. Vetterling, and B. P. Flannery. 1992. *Numerical Recipes in FORTRAN*. Cambridge: Cambridge University Press.

Pukhov, A. and J. Meyer-Ter-Vehn. 2002. Laser wake field acceleration: the highly non-linear broken-wave regime. *Appl. Phys. B: Lasers Opt.* 74:355–361.

Roberts, C. S. and S. J. Buchsbaum. 1964. Motion of a charged particle in a constant magnetic field and a transverse electromagnetic wave propagating along the field. *Phys. Rev.* 135:A381–A389.

Salamin, Y. I., G. R. Mocken, and C. H. Keitel. 2002. Electron scattering and acceleration by a tightly focused laser beam. *Phys. Rev. STAB* 5:101301-1–101301-14.

Schroeder, C. B., P. B. Lee, J. S. Wurtele, E. Esarey, and W. P. Leemans. 1999. Generation of ultrashort electron bunches by colliding laser pulses. *Phys. Rev. E* 59: 6037–6047.

Shen, B., Y. Li, K. Nemeth, H. Shang, Y. Chae, R. Soliday, R. Crowell, E. Frank, W. Gropp, and J. Cary. 2007. Electron injection by a nanowire in the bubble regime. *Phys. Plasmas* 14:053115-1–053115-5.

Shvets, G., N. J. Fisch, A. Pukhov, and J. Meyer-ter-Vehn. 1999. Generation of periodic accelerating structures in plasma by colliding laser pulses. *Phys. Rev. E* 60: 2218–2223.

Singh, K. P. and V. K. Tipathi. 2004. Laser induced electron acceleration in a tapered magnetic wiggler. *Phys. Plasmas* 11:743–746.

Sprangle, P., E. Esarey, and A. Ting. 1990. Nonlinear theory of intense laser-plasma interactions. *Phys. Rev. Lett.* 64:2011–2014; Nonlinear interaction of intense laser pulses in plasmas. 1990. *Phys. Rev. A* 41:4463–4469.

Sprangle, P., E. Esarey, J. Krall, and G. Joyce. 1992. Propagation and guiding of intense laser pulses in plasmas. *Phys. Rev. Lett.* 69:2200–2203; Nonlinear interaction of intense laser pulses in plasmas. *Phys. Rev. A* 41:4463–4469.

Strickland, D. and G. Mourou. 1985. Compression of amplified chirped optical pulses. *Opt. Commun.* 56:219–221; doi:10.1016/0030-4018(85)90120-8.

Suk, H., N. Barov, J. B. Rosenzweig, and E. Esarey. 2001. Plasma electron trapping and acceleration in a plasma wake field using a density transition. *Phys. Rev. Lett.* 86:1011–1014.

Tajima, T. and J. M. Dawson. 1979. Laser electron accelerator. *Phys. Rev. Lett.* 43: 267–270. Available at: http://link.aps.org/abstract/PRL/v43/p267

Ting, A., E. Esarey, and P. Sprangle. 1990. Nonlinear wake-field generation and relativistic focusing of intense laser pulses in plasmas. *Phys. Fluids B* 2:1390–1394.

Troha, A. L., J. R. Van Meter, E. C. Landahl, R. M. Alvis, Z. A. Unterberg, K. Li, N. C. Luhmann, A. K. Kerman, and F. V. Hartemann. 1999. Vacuum electron acceleration by coherent dipole radiation. *Phys. Rev. E* 60:926–934.

Teychenne, D., G. Bonnaud, and J. Bobin. 1994. Electrostatic and kinetic energies in the wake wave of a short laser pulse. *Phys. Plasmas* 1:1771–1773.

Umstadter, D. 2001. Review of physics and applications of relativistic plasmas driven by ultra-intense lasers. *Phys. Plasmas* 8:1774–1785.

Umstadter, D., E. Esarey, and J. Kim. 1994. Nonlinear plasma waves resonantly driven by optimized laser pulse trains. *Phys. Rev. Lett.* 72:1224–1227.

Umstadter, D., J. Kim, E. Esarey, E. Dodd, and T. Neubert. 1995. Resonantly laser-driven plasma waves for electron acceleration. *Phys. Rev. E* 51:3484–3497.

Umstadter, D., J. K. Kim, and E. Dodd. 1996. Laser injection of ultrashort electron pulses into wakefield plasma waves. *Phys. Rev. Lett.* 76:2073–2076.

Vshivkov, A. V., N. M. Naumova, F. Pegoraro, and S. V. Bulanov. 1998a. Nonlinear interaction of ultra-intense laser pulses with a thin foil. *Nucl. Instrum. Methods Phys. Res. A* 410:493–498.

Vshivkov, A. V., N. Naumova, F. Pegoraro, and S. V. Bulanov. 1998b. Nonlinear electrodynamics of the interaction of ultra-intense laser pulses with a thin foil. *Phys. Plasmas* 5:2727–2741.

Woodward, P. M. 1947. A method for calculating the field over a plane aperture required to produce a given polar diagram. *J. IEEE* 93:1554–1558.

Yamazaki, A., H. Kotaki, I. Daito, M. Kando, S. V. Bulanov, T. Zh. Esirkepov, S. Kondo, S. Kanazawa, T. Homma, K. Nakajima, Y. Oishi, T. Nayuki, T. Fujii, and K. Nemoto. 2005. Quasi-monoenergetic electron beam generation during laser pulse interaction with very low density plasmas. *Phys. Plasmas* 12:093101-1–093101-5.

Yu, W., V. Bychenkov, Y. Sentoku, M. Y. Yu, Z. M. Sheng, and K. Mima. 2000. Electron acceleration by a short relativistic laser pulse at the front of solid targets. *Phys. Rev. Lett.* 85:570–573.

Zhang, P., N. Saleh, S. Chen, Z. Sheng, and D. Umstadter. 2003. An optical trap for relativistic plasma. *Phys. Plasmas* 10:2093–2099.

Zhidkov, A., J. Koga, K. Kinoshita, and M. Uesaka. 2004. Effect of self-injection on ultraintense laser wake-field acceleration. *Phys. Rev. E* 69:035401-1–035401-4.

3

X-Ray Sources

Hiroaki Nishimura

CONTENTS

3.1 Introduction

Intense electromagnetic radiations emanate from laser-produced plasma (LPP). Soon after the invention of lasers, the possibility of LPP as a new radiation source was investigated. In particular, new laser technologies such as oscillation with a Q-switch, mode locking, chirped pulse amplification to laser systems, and shorter duration and higher power laser pulses became available. As a consequence, one can generate intense radiation pulses extending from terahertz wave (~100 μm in wavelength), visible light, extreme ultraviolet light (EUV, ~10 nm), and x-rays (0.1–1 nm) to γ-rays (<<0.1 nm). Emanation of dominant radiations is controlled by optimizing laser-irradiation conditions and target materials. For example, conversion efficiency, defined by the ratio of the x-ray radiation emitted in whole space to laser energy onto the target, can be as high as 80% when a target consisting of high-Z materials such as gold is irradiated with frequency-tripled Nd:glass laser light at 10^{13} W/cm². Such a high conversion efficiency enables us to drive a fusion pellet to attain high fusion gain (Lindl, 1998). Intense radiation pulse is very useful to observe the dynamics of rapidly moving hot-dense materials such as laser-driven fusion pellets, live organisms, transient phenomena of shock-compressed crystalline matter, and objects of nondestructive inspections. LPP radiation is a compact pulse source, thus it has been extended to a wide variety of industrial and scientific applications.

In this chapter, an overview of the principles of x-ray generation in LPP is presented in Section 3.2, followed by various applications of LPP x-rays. We shall focus mostly on x-ray generation and application. Those of EUV are summarized in Chapter 5, and LPP pumped x-ray and γ-ray lasers are discussed in Chapter 6.

3.2   Principles of X-Ray Generation in LPP

3.2.1   Density and Temperature Profiles of LPP

When an intense laser pulse irradiates the surface of material, hot-dense plasma is created. Interesting plasma parameters for x-ray generation are (1) electron temperatures (T_e), mostly depending on laser intensity, from a few tens eV to several MeV, and (2) scale length, defined as $T_e / |\Delta T_e|$, from sub-μm to several tens μm. The discussion of x-ray generation in LPP is closely related to the spatial structures involved in the interaction of laser with plasma and energy transport. Typical density and temperature profiles of LPP are shown in Figure 3.1. The most common absorption mechanism is the collisional process (i.e., inverse bremsstrahlung). The absorption coefficient for this process is given by (Zel'dovich and Raizer, 1966)

$$\kappa_{v,ff} = \frac{4}{3}\left(\frac{2\pi}{3m_e kT_e}\right)^{1/2}\frac{n_e e^6}{hcm_e v^3}\sum_i n_i Z_i^2$$

$$= 3.43\times10^6 \frac{n_{e,cc}}{T_{e,eV}^{1/2}}\sum_i n_{i,cc}Z_i^2\frac{1}{v^3}\left[1-\exp\left(-\frac{hv}{kT_e}\right)\right]\ [\text{cm}^{-1}], \quad (3.1)$$

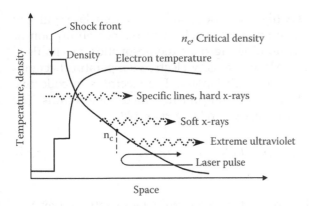

FIGURE 3.1
Schematic illustration of density and temperature profiles of LPP.

where

m_e and e are the electron mass and charge, respectively

k is the Stefan–Boltzmann constant

h is the Planck constant

c is the speed of light

v is the frequency of laser light

$n_{e,cc}$ and $n_{i,cc}$ are the electron and ion densities (in units of cm^{-3}), respectively

Z_i is the ionic charge number

The last term $[1 - \exp(-hv/kT_e)]$ represents correction for spontaneous emission, but it can be unity for laser light since hv for laser is close to a few eV while plasma temperature is in the order of a few tens eV or above. The absorption coefficient increases in proportion to the square of density so that absorption occurs mainly at a region near the critical density n_c (the electron density for which the local plasma frequency is equal to the laser frequency: $n_c \sim 10^{21}/\lambda^2_{L\,\mu m}$ [cm^{-3}] where $\lambda^2_{L\,\mu m}$ is the laser wavelength in units of micrometers). Transport of the absorbed energy to denser regions is made mostly by electron thermal conduction. The interaction configurations and characteristic parameters such as temperatures, densities, gradients, and scale lengths are dependent on timescales varying from sub-picoseconds to several nanoseconds. X-rays are emitted from the absorption, interaction, and transport regions. At densities nearly or slightly above the critical density, a large fraction of absorbed laser energy is reemitted as x-rays with energy between a few tens eV and a few keV. For high-Z plasmas, opacity is so high that low-energy radiations such as EUV radiation are distributed in a near blackbody continuum due to self-absorption. Even a group of fine lines fits to the blackbody continuum. For low-Z plasmas, energy regime from 1 to several keV is a rich line spectra from He-like and H-like ionic states so that line intensities, ratios, line shift, line shape, and x-ray spectroscopic imaging are used to diagnose plasma parameters such as temperature, density, and flow velocities.

When laser intensity is very high, say above 10^{15} W/cm^2 for a laser pulse of 1 μm in wavelength, collective processes such as resonance absorption or parametric instabilities start dominating over the absorption mechanism. This is because, with increase in laser intensity, electron temperature becomes higher and thus plasma becomes collisionless. The electrons in collective processes create an energetic electron component, called hot electrons. The temperatures of the hot electrons are given by (Kruer, 1984)

$$T_{hot} = 8.5(T_{bulk(keV)})^{1/4}(I_{15}\lambda_{L\,\mu m})^{0.39} \;[\text{keV}], \qquad (3.2)$$

where
   T_{hot} and $T_{bulk(keV)}$ are the hot and bulk electron temperatures (keV), respectively
   I_{15} is the laser intensity (10^{15} W/cm^2)

This formula was validated with experiments for intensities from 10^{14} to 10^{17} W/cm^2. Thus, T_{hot} over several tens keV is possible at high intensities. Furthermore, recent advances in laser technology have enabled us to attain laser irradiance above 10^{21} W/cm^2 using terawatt to petawatt lasers (a kJ energy in ps duration). The electron quivering velocity in the laser field $v_{osc} = e|E|/m_e 2\pi v$ (E is the electric field amplitude for laser light) can be very close to the speed of light so that the motion of the electron becomes relativistic for laser intensities above 10^{18} W/cm^2. The electron energy accelerated by a ponderomotive potential is given by (Wilks and Kruer, 1997)

$$T_{hot} = \left(\sqrt{1+0.36I_{18}\lambda_{L\,\mu m}^2}\right)511 \;[\text{keV}], \qquad (3.3)$$

where I_{18} is the laser intensity in units of 10^{18} W/cm^2. Thus, the hot-electron temperature can be an order of MeV. These hot electrons generated at high or ultrahigh intensities penetrate deeply in cold-dense materials and can cause inner-shell ionization. As a result, specific lines such as Kα lines are generated due to inner-shell transition of orbital electrons.

3.2.2   X-Ray Generation Processes

The transport of radiation from the emission region to the outside of the LPP source is an important consideration in spectroscopic analysis and practical applications. If the highly opaque region locates between the energy deposition region and the high emissivity region, one cannot efficiently extract radiation from the plasma. The radiation transfer equation can be written as (Zel'dovich and Raizer, 1966)

$$\frac{1}{c}\left(\frac{\partial I_v}{\partial t}+c\Omega\cdot\nabla I_v\right) = J_v\left(1+\frac{c^2}{2hv^3}I_v\right)-\kappa_v I_v, \qquad (3.4)$$

where
I_v is spectral intensity for the radiation frequency of v
Ω is the vector of radiation propagation
J_v is the emissivity
κ_v is the absorption coefficient

If photons propagate more rapidly than plasma motion, Equation 3.4 is simply rewritten for a one-dimensional propagation case as

$$\nabla I_v(x) = J_v(x) - \kappa_v I_v(x). \tag{3.5}$$

In this way, one can quantify radiation intensity by first obtaining the emissivity and the absorption coefficient (also called opacity), and then solving the radiation transfer function. The observable intensity for a slab of plasma of length L with uniform density and temperature, and the solution for the radiation intensity are simply given as

$$I_v = S_v \left[1 - \exp(-\tau_v) \right], \tag{3.6}$$

where
S_v is the source function
τ_v is the optical depth

They are defined as

$$S_v = J_v / \kappa_v \quad \text{and} \quad \tau_v = \kappa_v L. \tag{3.7}$$

For x-ray generation in LPP, three emission mechanisms and corresponding inverse processes are important. For emission processes, bremsstrahlung (or free–free transition), radiative recombination (free-bound transition), and radiative de-excitation (bound–bound transition) are important. In the same way inverse bremsstrahlung (*ff* transition), photoionization (*bf* transition), and radiative excitation (*bb* transition) are important for absorption processes. The emissivity J_v and opacity κ_v are provided by summing all emissivities $J_{v,ff}$, $J_{v,fb}$, and $J_{v,bb}$, and opacities $\kappa_{v,ff}$, $\kappa_{v,bf}$, and $\kappa_{v,bb}$.

3.2.2.1 Free–Free Transition

When a free electron is decelerated by an ion (Coulomb collision), the system emits continuum radiation since both the initial state and the final state of the electron are in the free levels. Assuming electrons are in Maxwell distribution with temperature T_e, emissivity $J_{v,ff}$ is given as

$$J_{v,\text{ff}} = \frac{32\pi}{3}\left(\frac{2\pi}{3kT_e}\right)^{1/2}\frac{n_e e^6}{m_e c^3}\sum_i n_i Z_i^2 \exp\left(-\frac{hv}{kT_e}\right)$$

$$= 5.05\times 10^{-41}\frac{n_{e,\text{cc}}}{T_{e,\text{eV}}^{1/2}}\sum_i n_{i,\text{cc}} Z_i^2 \exp\left(-\frac{hv}{kT_e}\right)\ [\text{erg/cm}^3/\text{s/Hz/sr}]\,, \qquad (3.8)$$

where $T_{e,\text{cc}}$, $n_{e,\text{cc}}$, and $n_{i,\text{cc}}$ are, respectively, the electron temperature in units of eV, electron density, and ion density in cm^{-3}. The sum is taken over all ionization stages. The opacity for the free–free process has already been given in Equation 3.1.

3.2.2.2 Free–Bound Transition

When a free electron is trapped in an orbit of an ion, excessive energy is released in the form of radiation. This process is called radiative recombination. Inverse process is photoionization. Because energy levels of the bound state are fixed, the sum of absorption cross sections shows a saw-tooth structure with a sharp onset at the ionization potential and slow decay as v^{-3}. The formula for the emissivity from a state in ion stage $i+1$ recombining to a state i is

$$J_v = 1.01\times 10^{-42}\,n_{i+1,\text{cc}}\,n_{e,\text{cc}}\,\frac{g_i}{g_{i+1}}\,g_{\text{fb}}\,\frac{I_{p,\text{eV}}^{5/2}}{T_{e,\text{eV}}^{3/2} Z}\exp\left(-\frac{h(v-v_{I_p})}{kT_e}\right)\quad \text{for } v\geq v_{I_p}$$

$$J_v = 0 \quad \text{for } v < v_{I_p}\,, \qquad\qquad\qquad\qquad\qquad\qquad\qquad\qquad\qquad (3.9)$$

where

I_p and v_{I_p} are the ionization potential and corresponding frequency of the state i in units of eV and Hz, respectively

g_i, g_{i+1} are statistical weights of the state i and $i+1$, respectively

g_{fb} is a Gaunt factor for free-bound transition (Karzas and Latter, 1961)

The opacity for the same transition is

$$\kappa_v = 4.13\times 10^{26}\,\frac{n_{i,\text{cc}} I_{p,\text{eV}}^{5/2}}{Z v^3}\,g_{\text{bf}}$$

$$\times\left[1 - 1.66\times 10^{-22}\frac{n_{i+1,\text{cc}}}{n_{i,\text{cc}}}\frac{g_i}{g_{i+1}}\frac{n_{e,\text{cc}}}{T_{e,\text{eV}}^{3/2}}\exp\left(-\frac{h(v-v_{I_p})}{kT_e}\right)\right]\quad \text{for } v\geq v_{I_p}$$

$$\kappa_v = 0 \quad \text{for } v < v_{I_p}\,. \qquad\qquad\qquad\qquad\qquad\qquad\qquad\qquad\qquad (3.10)$$

3.2.2.3 Bound–Bound Transition

Electron transition from the upper orbital level to the lower orbital levels yields a line emission. The emissivity for a bound–bound station between the upper level u and the lower level l is given by

$$J_v = n_{u,cc} A_{ul} \frac{h\nu_{ul}}{4\pi} \varphi(v), \tag{3.11}$$

where
  n_u is the population density of the upper state u (cm⁻³)
  A_{ul} is the spontaneous radiative rate (Hz)
  n_{ul} is the frequency of radiation for the transition u to l
  $\phi(v)$ is the line profile function

The opacity of the same transition is given by

$$\kappa_v = n_{l,cc} \left(1 - \frac{n_{u,cc}}{n_{l,cc}} \frac{g_l}{g_u} \right) f_{lu} \varphi(v), \tag{3.12}$$

where
  $n_{l,cc}$ is the population density of the upper state u (cm⁻³)
  g_u and g_l are the statistical weights for respective states
  f_{lu} is the absorption oscillator strength of the transition from the lower
    state l to the upper state u

3.3 Various Applications

3.3.1 Driver Source for Inertial Confinement Fusion Pellet

Systematic studies of inertial confinement fusion have been made aiming at the goal of achieving laboratory demonstration of ignition and burn of fusion pellets driven with an extremely uniform heat source. X-rays generated with high-Z materials are practically used to drive fusion pellets. Nonuniformities inherently involved in a drive laser beam are smoothed out via x-ray conversion and reemission processes in an x-ray confining cavity. Therefore, this scheme is called an indirect drive.

3.3.1.1 X-Ray Conversion with High-Z Plasmas

Interaction of laser light with high-Z materials provides an attractive subject of radiation hydrodynamics (a material state in which radiation emission and absorption are lightly coupled with hydrodynamic motions) and the atomic physics of hot-dense matter. Considerable work has been carried out on this issue to improve understanding of various aspects of x-ray conversion and transport. It has become generally understood that short wavelength laser irradiation on high-Z materials at moderate laser intensities leads to high x-ray conversion. An approach to investigating this subject generally involves comparison with numerical simulations. Through this benchmarking, the physical models are refined to provide quantitatively reliable estimations. One difficulty of this approach is that planar target experiments are affected by lateral energy transport and plasma expansion, particularly when the laser spot size is small in comparison with typical plasma-scale length. This lateral effect has been eliminated in uniform irradiation of a spherical target with multiple laser beams (Goldstone et al., 1987).

In some studies, absolutely calibrated broadband x-ray detectors consisting of vacuum diode and L- or K-edge absorption filters (typically $\tilde{\varepsilon}/\delta\varepsilon = 3$ at $\varepsilon = 0.8\,keV$) were used at a photon energy of 0.1–2 keV. Uncertainty the absolute value of the detailed spectral features is lost in the data analysis due to the broadband response. A transmission grating spectrograph (TGS) coupled with an absolutely calibrated film or a back-illumination charge-coupled device detector has an advantage in this respect, providing better resolutions (typically $\tilde{\varepsilon}/\delta\varepsilon = 10$ at $\varepsilon = 0.8\,keV$) (Eidmann et al., 1986).

Detailed studies on measurements of x-ray emission and transport in gold plasma were conducted using intense blue laser (Nishimura et al., 1991a). The absolute x-ray emission spectra were measured under three different conditions. The laser used in the experiment was the frequency-tripled (351 nm in wavelength) light from Gekko XII Nd:glass laser at the Institute of Laser Engineering, Osaka University. Twelve beams are arranged in a two-bundled illumination configuration. Two axial beams of the 12 are on the axis of symmetry, and 5 other beams on each side are arranged conically with a cone-half angle of 50°. The focal *f*-number of each beam was 3. The laser pulse waveform was approximately a Gaussian shape. The full width at the half maximum (FWHM) of the pulse was 0.75 ns. The laser energy per beam was varied from 30 to 500 J.

The main diagnostic instruments were TGS coupled with an x-ray film (KODAK Type 101-07) for time-integrated measurements of an x-ray streak camera for time-resolved measurements. The gratings were made of gold of 0.4 µm thickness with 1000 lines/mm pitch mounted on a pinhole disk of 25 or 50 µm. The spectral resolution was typically 0.2 nm. The observed spectral range was 50–1240 eV. Absolute calibration of the recording film was performed using an LPP x-ray source and a self-calibration bolometer to avoid errors arising from reciprocity failure (Nishimura et al., 1991c). The energy absorption of laser light was measured with 40 channel plasma calorimeters distributed over the inside surface of the target chamber.

3.3.1.1.1   *Incident-Angle Dependence of Laser Light Absorption and X-Ray Conversion*

Each of the clustered beams at one side was used to irradiate a planar gold foil at 20°–70° of incident angle. The laser energy was fixed to 300 J but the

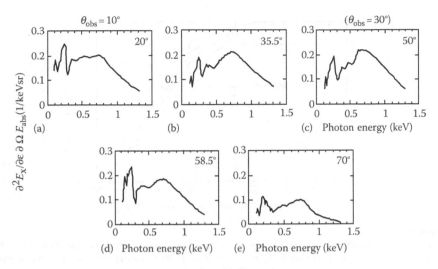

FIGURE 3.2

X-ray emission spectra from gold planar targets irradiated at different incident angles. The observation angle with respect to the target normal was fixed at 10° except for the 50° incidence. The laser intensity was 1.2×10^{15} W/cm² common to all cases. (From Nishimura, H., et al., *Phys. Rev. A*, 43(6), 3073, 1991a. With permission.)

focal position of the laser beam was displaced to keep the irradiance constant at 1.2×10^{15} W/cm² even for oblique incidence.

Figure 3.2 shows the observed spectra for each incidence angle. Note that the spectral fluence is the value measured at the observation angle and not converted to that at the target normal. The relative shape of the spectra does not vary greatly but the total amount of the x-ray emission decreased with the incidence angle, particularly at the angle greater than 50°. This reduction can also be seen in Figure 3.3, which shows the absorption of laser light and x-ray conversion rate (or efficiency) given as a function of incident angle. In this case, the total x-ray energy was derived by assuming the Lambertian distribution (i.e., $\cos \theta$ angular distribution, where θ is the angle measured from the target normal). The absorption and the conversion rates are nearly constant up to 50°, above which they decrease with an increase in the incidence angle.

The obliquely incident electromagnetic wave is reflected at low-density regions, and a major part of the incident energy is, hence, deposited in the lower-density region. Thus, the number of particles contributing to x-ray emission decreases. One can estimate absorption fraction f_A for the inverse bremsstrahlung at the incident angle of θ as (Kruer, 1984)

$$f_A = 1 - \exp\left(-a_0 \cos^2 \theta\right), \tag{3.13}$$

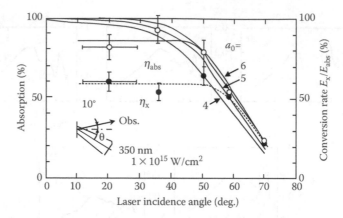

FIGURE 3.3
Laser light absorption and x-ray conversion rates versus laser incidence angle. The thin solid lines represent the calculated absorption rate for three different absorption parameters a_0. (From Nishimura, H., et al., *Phys. Rev. A*, 43(6), 3073, 1991a. With permission.)

where a_0 is the absorption coefficient for the normal incidence as

$$a_0 = 8v_{ei}^* L / 3c. \tag{3.14}$$

Here v_{ei}^* is the electron–ion collision frequency evaluated at the critical point. As is seen from solid curves in Figure 3.3 the best fit is obtained for $a_0 = 5$, corresponding to the scale length of 6 μm for T_e of 2 keV and an average Z of 60.

3.3.1.1.2   X-Ray Emission from Spherical Plasma

Gold-coated solid spheres were irradiated with 12 laser beams. The thickness of the gold coat was 1 μm and the diameter of the sphere was 250 μm. The focal point was displaced four radii beyond the target center. Effective laser intensity was estimated from an x-ray pinhole image of the spectral range from 1 to 4 keV. The total laser energy was varied from 370 to 2300 J to yield irradiance from 3.5×10^{14} to 1.6×10^{15} W/cm². Time-integrated spectra for three different cases are shown in Figure 3.4. For comparison the spectra calculated with the one-dimensional radiation hydrodynamic code ILESTA (Takabe et al., 1988) are also plotted. The spectrum is mainly composed of two emission bands corresponding to the O-shell transition lines (radiative transitions to O-shell orbits from their upper levels) at around 0.25 keV and to the N-shell lines, located at around 0.8 keV. With an increase in laser intensity, ionization proceeds and the population density of ions in the $n = 4$ levels becomes less. In this way the transition from the upper levels to $n = 4$ becomes dominant.

The open circles in Figure 3.5 show the x-ray conversion rates obtained with the spherical target. For comparison, results from identical experiments

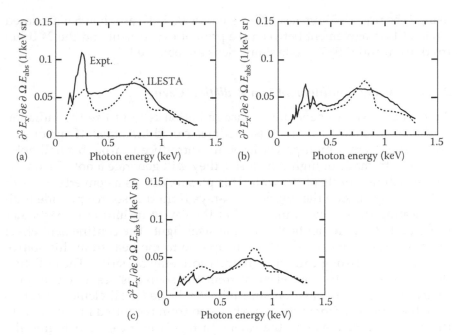

FIGURE 3.4

X-ray emission spectra from a gold-coated sphere at intensities of (a) 3.5×10^{14}, (b) 9.9×10^{14}, and (c) 1.6×10^{15} W/cm². Dashed lines are the spectra calculated by the ILESTA code. (From Nishimura, H., et al., *Phys. Rev. A*, 43(6), 3073, 1991a. With permission.)

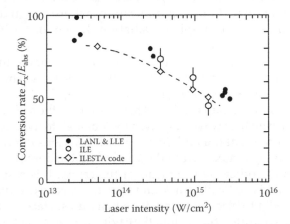

FIGURE 3.5

X-ray conversion rate derived from a spherical target irradiation. Good correspondence is seen between the present experiment and the ILESTA simulations as well as the result obtained from similar experiments. (From Nishimura, H., et al., *Phys. Rev. A*, 43(6), 3073, 1991a. With permission.)

using 24 beams of 351 nm laser (Goldstone et al., 1987) are shown with closed circles. Close agreement between the present experiment and the 24 beam irradiation, and ILESTA code simulations are obtained.

3.3.1.2   X-Ray Confinement in a Radiation Cavity

An effective way to generate intense radiation is to utilize the radiation confinement effect in a cavity (sometimes called *hohlraum*): If one or several beams from a high-power laser are introduced through small holes into a cavity made of high-Z material, they will generate a hot LPP on the laser-irradiated parts of the wall. This plasma acts as a converter of laser light to x-rays inside the cavity. The x-rays from this source provide indirect heating of the total inner wall of the cavity including in particular those parts that are not heated by the laser light. The confinement effect arises because the cavity wall heats up due to the heat from this source and becomes a strong emitter of thermal soft x-ray radiation. The radiation flux received by the wall element from the source is reemitted into the cavity. Because of reemission, an indirectly heated wall element receives radiation not only from the source but also from reemitted radiation from the other wall elements in the cavity. In this way it can radiate into the cavity a flux of thermal x-rays exceeding the received source flux. If the reemission coefficient of the hot wall material, sometimes called *albedo*, reaches a value of order unity, a large circulating flux of reemitted radiation forms in the cavity, forming a wall-emitting Planck radiation, which is connected with a correspondingly high temperature through the Stefan–Boltzmann law. The spatial distribution of the source energy connected with the existence of a large flux of reemitted photons will help to establish a spatially uniform temperature distribution in the cavity. This feature is of great importance for applications where a high degree of uniformity is required, for example in radiation-driven (or indirectly driven) inertial confinement fusion.

The key element for describing the radiation confinement in the cavity is the reemission coefficient of the x-ray heated wall. It is defined by the radiation-driven ablative heat wave propagating into the depth of the wall material (called Marshak wave). Pakula and Sigel have extended the self-similar solution of a radiation-driven heat wave first given by Marshak to include expansion of the heated material (Pakula and Sigel, 1985, 1986). Consider a wall element of a cavity receiving a flux S_s from the source and incident flux S_i of thermal radiation from the other wall elements of the cavity. It is assumed that the hot-wall material is optically thick and that the received radiation is completely converted into blackbody radiation in a thin surface cavity layer. This layer radiates a reemitted flux S_r into the cavity whereas a net heat flux S_{hw} of radiation diffuses into the wall. The quasistationary energy balance of this thin surface layer is given by

$$S_s + S_i = S_r + S_{hw}. \tag{3.15}$$

We allow localized holes in the cavity with a fractional hole area $n - 1$ but assume uniform energy distribution conditions in the cavity, that is, the source flux is the same for all wall elements and the reemitted radiation from a wall element falls uniformly on all other wall elements including holes. Under this model assumption, S_i and S_r are related by (Sigel et al., 1988)

$$S_i = (1 - n^{-1})S_r. \tag{3.16}$$

This is a well-known property of an Ulbricht sphere. Together with the energy balance relation described above, Equation 3.16 becomes

$$S_{hw} = S_s - n^{-1}S_r. \tag{3.17}$$

For a given heat flux S_{hw}, the temperature at the boundary between the ablation heat wave and the vacuum is given by the self-similar solutions for the ablative heat wave in gold (Pakul et al., 1985, 1986; Sigel et al., 1988). Note that for blackbody radiation $S_r = \sigma T^4$ (σ is the Stefan–Boltzmann constant), S_r and T are obtained as a function of the given quantities n^{-1}, S_s, and t (time included in Pakula's self-similar solutions). With the help of S_r, one finds the reemission coefficient r as

$$r \equiv \left(\frac{S_r}{S_s + S_i}\right) = \frac{S_r}{S_s + (1 - n^{-1})S_r}. \tag{3.18}$$

Alternatively one may introduce a quality factor $N = S_r/S_{hw}$. This is related to r by $N = r/(1 - r)$. In a closed cavity, where $S_{hw} = S_s$, N is equal to the ratio of the reemitted flux to the received source flux of the wall. Physically, it corresponds to the number of reemission of the source energy, inside the cavity. We may note that S_r and S_s are related through

$$S_r = \frac{S_s}{N^{-1} + n^{-1}}, \tag{3.19}$$

that is, for a given source flux S_s the reemitted flux S_r depends on the losses to the wall (represented by N^{-1}) and through the holes (by n^{-1}). Figure 3.6 shows reemission coefficients for a closed cavity ($n^{-1} = 0$) of gold as a function of the source flux at different times (Nishimura et al., 1991b). The reemission coefficients for open cavities are weakly dependent on the fractional open area and are smaller than the values shown in Figure 3.6. The boundary

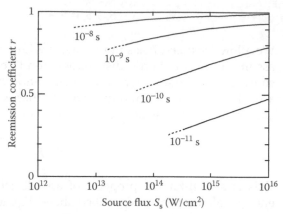

FIGURE 3.6

The reemission coefficients of the wall of a closed cavity of a gold, derived from Equation 3.19 and the wall temperature derived from the self-similar solution as a function of the source flux at different times. (From Nishimura, H., et al., *Phys. Rev. A*, 44(12), 8323, 1991b. With permission.)

between the solid and dashed lines shows a criterion from which on the lower source flux side the similarity solutions are gradually invalid.

In the experiment, gold-made cavities were irradiated with blue laser light from Gekko XII. Ten beams were used, forming two bundles of five beams of opposite sides. The bundle has a common horizontal axis with respect to which each beam makes an angle of 50°. The *f*-number of each beam was 3. The pulse waveform is a quasisquare of 0.87 ns at FWHM. The laser energy was about 360 J per beam. The targets used in this experiment were single-shell cavities of spherical shape with laser inlet holes and observation windows. The thickness of the wall was 10 μm. Cavity diameters were varied from 1 to 3 mm to yield average laser intensity S_L from 7×10^{12} to 2×10^{14} W/cm². The main diagnostic instruments were TGSs coupled with absolutely calibrated film or x-ray streak camera. The TGS film measured the time-integrated fluence of x-rays emitted from the inner wall of the cavity through the diagnostic hole. Assuming that the inner wall radiates according to a Lambertian distribution, one obtains the time-integrated value of the reemitted flux S_r of the wall elements. TGS-streak camera was absolutely calibrated against the data from TGS film. In this way, the time-resolved reemitted flux and then the brightness temperature of the wall element were finally obtained. The reemitted x-ray flux S_r at the time of maximum temperature is plotted in Figure 3.7. The highest temperature of 240 eV was attained. The experimental points lie above the line $S_r = S_L$. This means that the observed wall element emits more power in the form of thermal x-rays than corresponding to its share of the available laser power. This is possible only when the observed wall elements receive additional heating from the other wall elements in the cavity. For a quantitative comparison with the similarity solutions, the source flux was obtained as $S_s = \eta_x S_L$ where η_x is the overall conversion efficiency from the incident laser light to x-rays. Assuming the conversion efficiencies suggested by measurement of open geometry for laser intensities corresponding to those at

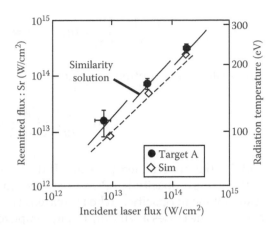

FIGURE 3.7
Reemitted flux S_r and corresponding brightness temperatures given as a function of incident average laser flux S_L. The predictions of the similarity model are represented by solid lines for each cavity size. The dashed line represents the case of $S_r = S_L$, above which x-ray confinement occurs. (From Nishimura, H., et al., *Phys. Rev. A*, 44(12), 8323, 1991b. With permission.)

the first bounce of the heating laser on the inner wall of the cavity, one obtains $\eta_x = 0.8$ for 3 mm, 0.7 for 2 mm, and 0.5 for 1 mm cavities. The predicted values of S_r are plotted with solid lines. As can be seen from Figure 3.7, there is good overall agreement between experiment and theory over a very wide parameter range.

This confinement effect is particularly important to attain high brightness temperatures and spatially uniform irradiation of materials set inside the cavity with thermal x-rays for applications such as inertial confinement fusion, equation of state, hydrodynamic stabilities, and physics of condensed matters.

3.3.1.3 X-Ray Enhancement with High-Z Mixture

Use of mixtures composed of high-Z elements has been suggested to obtain (1) higher conversion of laser light to x-rays, (2) higher reemission for thermal x-rays, which is essential for radiation confinement in a cavity, and (3) manipulation of the spectral shape for various applications by spectrum blending (Nishimura et al., 1992; Orzechowski et al., 1996). It is known theoretically and experimentally that emission spectra of laser-produced high-Z plasmas, such as gold, consist of two major emission bands originating from the transitions from upper electronic excited states to the N-shell (principal quantum number $n = 4$) and O-shell ($n = 5$) states. The spectra have a valley of weak emissions between these emission bands. In optically thick materials, the emission characteristics, the absorption, and the transport of radiation are closely related to the material opacity. Because the opacity in the weak emission regions is much smaller than the strong emission regions, the mean opacity of x-ray radiation is dominated by the opacity at the photon energy corresponding to the weak emission valley. The Rosseland mean opacity κ_R is given by (Zel'dovich and Raizer, 1966)

$$\frac{1}{\kappa_R} = \int_0^\infty \frac{1}{\kappa_\nu} G(u) du, \quad u = h\nu/kT,$$ (3.20)

where $G(u)$ is a weighting function

$$G(u) = \frac{15}{4\pi^4} \frac{u^4 e^{-u}}{(1-e^{-u})^2},$$ (3.21)

and T is the material temperature. If the low-opacity region is compensated with the high-opacity band from another material of different Z value, the resultant mean opacity can be increased. This is similar to a case where, on a cold day, one closes a window but in completely so that warm air escapes out of a room through a slightly opened gap. The x-ray conversion efficiency of LPPs may be given in a simple form $\eta_x = 0.5\alpha(1 + R)$. Here, α is the x-ray conversion efficiency at the energy deposition region and R is the reemission coefficient, which is given by (Sigel et al., 1990)

$$R = 1 - 7.4 \times 10^{-2} S_{s,14}^{13/16} - t_{ns}^{-1/2} C_R^{1/2},$$ (3.22)

where
  $S_{s,14}$ is the net radiation flux in units of $10^{14}\,W/cm^2$
  t_{ns} is the time in units of ns
  C_R is the ratio of mean opacities ($=\kappa_{BD}/\kappa_R$), where κ_{BD} is the Bernstein
    and Dyson maximum opacity (Sigel et al., 1990)

If the mean opacity κ_R approaches the maximum opacity, C_R decreases and hence R increases.

  The Rosseland mean opacity for a mixture of two elements was calculated on the basis of the screened-hydrogenic average-atom model (More, 1982) assuming local thermodynamic equilibrium conditions. The atomic states of two elements were determined in a self-consistent manner so that the ionosphere radius of each matter determines a unique chemical potential μ for the Fermi–Dirac distribution function. In the calculation, we have used the line smearing approximation with a Gaussian weighting function, giving the effective line width $\Delta\varepsilon$ as an adjustable parameter. A scaling law for $\Delta\varepsilon$ of the form $25 + 0.0283_0^{1.835}$ (Z_0, the atomic charge) was determined so that the Rosseland mean opacity for each of the constituent elements agreed with the SESAME opacity (SESAME, 1983).

  A mixture composed of gold and an element in the lanthanide series was chosen because measured emission spectra from these materials match well with each other. Results of the calculation for the Au–Sm mixture are shown in Figure 3.8 assuming $T = 0.2\,keV$ and a mass density $\rho = 0.1\,g/cm^3$ as a representative case. Figure 3.8a shows the frequency-dependent opacity for the mixture of Au/Sm = 0.999/0.001 (representing pure Au) and Figure 3.8b

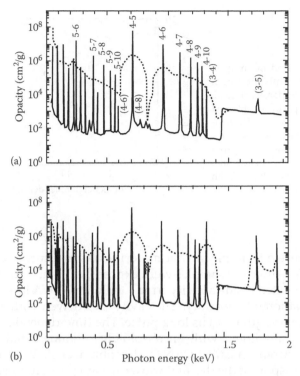

FIGURE 3.8
Calculated opacities for the Au–Sm mixtures at two different mixing ratios: (a) Au/Sm = 0.999/0.001 and (b) Au/Sm = 0.75/0.25. Dotted lines represent the corrected opacities after the line-smearing process. The numbers inset are the principal quantum numbers of the electronic energy states relevant to each absorption transition. The numbers in parentheses are for the samarium. (From Nishimura, H., et al. *Appl. Phys. Lett.*, 62(22), 1344, 1992. With permission.)

for Au/Sm = 0.75/0.25. The mixing ratio is defined as the ratio of atomic number densities. The low-opacity region around 0.3–0.5 keV seen in the case of high gold content (Figure 3.8a) can be enhanced with the absorption of radiation due to the N-shell transitions of samarium. Figure 3.9 shows the Rosseland mean opacities for the Au–Sm mixtures as a function of the fractional number density of gold for the same temperature and density applied in the case of Figure 3.8. We find that opacity values obtained for the mixtures at moderate mixing ratios are larger than those of two individual elements. Assuming $C_R^{Au}/C_R^{Au-Sm} = 2.9$, $C_R^{Au} = 7$, $S_{s,14} = 1$, and $t_{ns} = 1$, we obtain $R^{Au-Sm} = 0.88$, which should be compared with $R^{Au} = 0.80$. Although the increase in the reemission coefficient appears small, it is large enough to increase the quality factor of confinement N for a radiation confining cavity from $N = 4$ to 7.3.

To determine the Rosseland mean opacity of a mixture material, the propagation time of a radiation heat wave (i.e., Marshak wave) through a

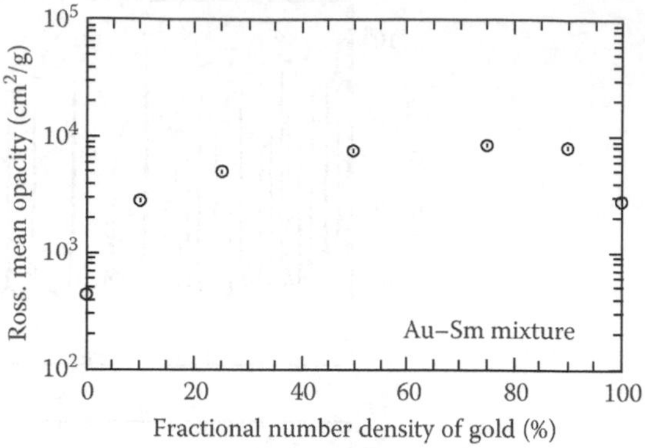

FIGURE 3.9
Calculated Rosseland mean opacity for the Au–Sm mixture versus fractional number density of gold. Higher mean opacity is obtained from the Au–Sm mixture. The temperature and density are the same as those in Figure 3.8. (From Nishimura, H., et al. *Appl. Phys. Lett.*, 62(22), 1344, 1992. With permission.)

well-characterized sample of that material was measured (Orzechowski et al., 1996). The mixture samples were exposed to Planckian radiation distribution generated inside a cavity irradiated with NOVA laser at Lawrence Livermore National Laboratory. The cavity was driven with about 27 kJ of 351 nm in wavelength in a 1 ns long pulse. The time-dependent cavity temperature was monitored with an absolutely calibrated multiple-channel soft x-ray spectrometer called DANTE (Kornblum et al., 1986). In the experiment, the transport of the thermal wave through pure Au foils and Au–Gd mixtures was investigated. The mixture samples were formed on a substrate by depositing very thin layers of the two elements alternatively. Finally, the substrate was removed to provide a freestanding sample. The areal density of each layer must be optically thin to the drive radiation duration. Two different samples corresponding to different atomic fractions of Au and Gd were fabricated in such a way that one sample comprises 146–200 layer pairs of Au and Gd. The overall thicknesses were 2.22–3.15 μm and the areal density of this sample was 2.97–3.15 mg/cm². The thermal radiation corresponding to the cavity drive was monitored with a TGS coupled with an x-ray streak camera to measure the burn-through time of the sample foils. In order to determine the ratio of the Rosseland mean opacity of the mixture to that of Au, the results of self-similar solutions were used. Figure 3.10 shows the ratio of the Rosseland mean opacities of the Au–Gd foil to that of Au for the two different concentrations of Gd. The errors associated with the measurement correspond to uncertainties in the streak camera sweep speed and in errors in determining the precise thickness of the foils. The solid curve in Figure 3.10 shows the calculated Rosseland mean opacity of the Au–Gd mixture as a function of Gd concentration and assuming a temperature of 250 eV and a density of 1.0 g/cm². These calculations indicate that the maximum improvement in opacity corresponds to a 50/50 mixture of Au and Gd. The overall improvement in the opacity (over that of pure Au at the same temperature and

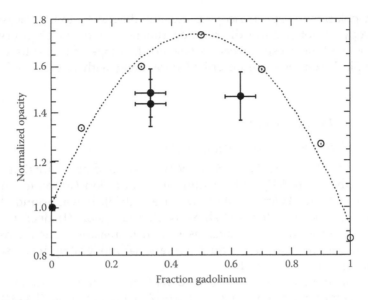

FIGURE 3.10
Rosseland mean opacity of composite normalized to that of pure Au at a temperature of 250 eV and a density of $1\,g/cm^2$. The open points correspond to the results of model calculations, and the line is a quadratic fit to those points. The solid points show the normalized opacity of the measurements. (From Orzechowski, T.J., et al., *Phys. Rev. Lett.*, 77(17), 3545, 1996. With permission.)

density) is a factor of 1.7. This curve is normalized to the Rosseland mean opacity of Au at $1.0\,g/cm^3$ and a temperature of 250 eV ($\kappa_{R,Au} = 1500\,cm^2/g$).

3.3.2 Applications to High-Energy Density Physics

Laser-produced x-ray source has been studied as a potential application in the research of high-energy density physics of hot-dense matters created with high-power lasers such as fusion research and laboratory astrophysics. The implosion velocity of the fusion pellet is typically from a few times 10^7 to $10^8\,cm/s$. In order to take an x-ray image with a spatial resolution of $10\,\mu m$, a temporal resolution better than a few 10 ps is necessary. This is accomplished either by a flash radiography technique using very short bursts of intense x-ray emission to freeze the motion of the pellet components or by the use of an x-ray imager with a very high temporal resolution. In addition, the density radius product of a fusion pellet is typically of the order of 0.01 to $0.3\,g/cm^2$ so that x-ray photon energy greater than a few keV is needed. Because of these require-ments, numerous studies on efficient x-ray source generation have been carried out. X-ray spectroscopy of LPP is one of the essential diagnostic methods. For example, x-ray emission from a laser-driven core provides information about the electron density and temperature of the compressed plasma.

X-ray polarization spectroscopy is a useful diagnostic tool to measure the anisotropy of hot-electron distribution function. A thorough overview of previous work on these subjects is beyond the scope of this subsection, so only typical examples are described here along with references for more details.

3.3.2.1 X-Ray Radiography

3.3.2.1.1 Planar Source for K- and L-Shell Lines

LPP x-ray sources from 1.4 to 8.6 keV were studied as a function of laser wavelengths (1060, 530, and 350 nm), pulse duration (100 ps to 2 ns), and intensity (10^{14}–10^{16} W/cm²) (Mathews et al., 1983). It was found that the conversion efficiency for K-shell x-ray line emission (1) decreases with increasing x-ray energy, (2) decreases with increasing laser intensity, (3) decreases rapidly with pulse duration, and (4) moderately increases with decreasing laser wavelength.

The measurements were performed with an absolutely calibrated x-ray crystal spectrograph. For Al-K and Au-M band radiation (1.4–3.0 keV) a KAP (Potassium Acid Phthalate) crystal was used. Ti spectra were measured with a PET (Pentaery thritol) crystal while those of Ni and Zn were provided with an LiF crystal. Kodak no-screen film was used to record the spectra. Assuming the plasmas are optically thin, observation geometry assumes little difference in measured yield as a function of observation angle. Figure 3.11 shows the Ti K-line conversion efficiency for various

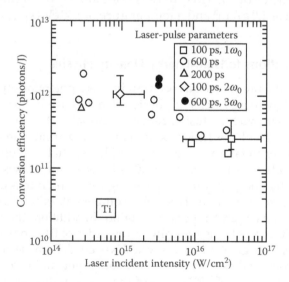

FIGURE 3.11
Ti K-line conversion efficiency as a function of laser intensity of the Ti foil. (From Mathews, D.L., et al. *J. Appl. Phys.*, 54(8), 4260, 1983. With permission.)

conditions. It seems that there is a threshold intensity below which there is insufficient plasma temperature generated as well as insufficient time available to produce a species of plasma ions that radiate K-line x-rays. This threshold may be critically sensitive to laser pulse duration. For shorter laser wavelength, a better conversion is attained because the shorter wavelength laser light deposits energy in denser regions, which in turn shortens the time of ionization. The intensity threshold is higher for the higher photon energy lines. Figure 3.12 shows comparison of Mathews' result (Mathews et al., 1983) with that of Yaakobi (Yaakobi et al., 1980). Overall improvements in the conversion efficiencies are attained for shorter wavelength lasers.

An experimental database on the x-ray conversion for much shorter pulse duration is provided for the $M \rightarrow L$ spectra of Y (2.00–2.35 keV), Pd (2.95–3.46 keV), and Cs (4.43–5.15 keV), and the $L \rightarrow K$ spectra of Cl (2.79 keV), Ti (4.749 keV), Mn (6.18 keV), and Ni (7.804 keV) (Pillion and Heiley, 1986).

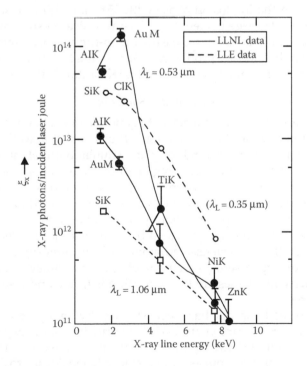

FIGURE 3.12
Comparison of Mathews' result (Mathews, et al., 1983) with that of Yaakobi (Yaakobi, et al., 1980). (From Mathews, D.L., et al., *Appl. Phys.* 54(8), 4260, 1983. With permission.)

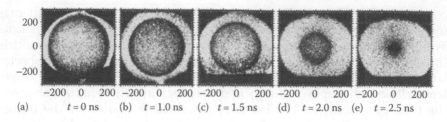

FIGURE 3.13
Sequential images of a Ge-doped capsule recorded at $t = 50.0$, 1.0, 1.5, 2.0, and 2.5 ns by the use of 4.7 keV x-ray from Ti plasma generated with two green laser pulses of 2 ns at 5×10^{14} W/cm^2. (From Back, C.A., et al., *Phys. Plasmas*, 10(5), 2047, 2003. With permission.)

To perform a quantitative study of hydrodynamic instabilities occurring during fusion capsule implosion, x-ray backlighting techniques are used to image laser-driven capsules. This is particularly useful when the capsule is driven with thermal x-rays generated with laser (i.e., in the case of indirectly driven fusion) since electron temperature of the ablation region is much lower than the direct-drive case. For example, x-ray backlight images of a full capsule through the implosion phase with 55 ps and 15 μm resolution were obtained with a laser-generated 4.7 keV x-ray pulse. Results are shown in Figure 3.13. They are utilized to measure the in-flight aspect ratios for doped ablators. The radial density profile was derived as a function of time by Abel inversion of the x-ray transmission profiles (Kalantar et al., 1997).

3.3.2.1.2 Volumetric Source for K- and L-Shell Lines

As seen above, solid targets enable us to generate multi-keV x-ray sources at various photon energies but exhibit low efficiencies. It is well known from x-ray conversion experiments and computer simulations that multi-keV x-rays originate from the hot underdense region of the ablated material of the solid target. In this case, the electron density n_e, important for x-ray generation, is lower than the critical density for the laser–plasma interaction. When $n_e < n_c$, laser absorption occurs predominantly by inverse bremsstrahlung, and a large volume of plasma can be heated by a supersonically bleaching wave (Back et al., 2001). Underdense radiators such as preformed plasma, gas, doped foam, or doped aerogel are more efficient multi-keV x-ray converters since hydrodynamic losses are lower, heating is faster, and laser absorption volume is larger.

Laser-preformed plasma has been studied by many researchers (e.g., Kodama et al., 1987; Girard et al., 2005). The principle of the experiment is to create an extended underdense volume from an initially solid material irradiated with a weak laser pulse. Conversion efficiencies up to 3.6% in 2π sr for energies above 4 keV with a preformed plasma have been attained, to be compared to the case without a prepulse where the CE is 1.5%. The best conversion efficiency up to 4.9% has been reached in this scheme with a main laser pulse of 351 nm wavelength at the intensity of 2.2×10^{15} W/cm² (Girard et al., 2005).

Another approach to create a low-density plasma is developed using low-density targets to maximize emission in the 4–7 keV photon range (Back et al., 2003). The radiators are produced by laser-heating gas, which is confined in an enclosure made of low-Z elements. They are underdense because they are formed from a gas whose laser-ionized electron density in the plasma is slightly less than the critical density of the laser, $n_c = 1.1 \times 10^{21} / \lambda^2_{L\mu m}$ cm⁻³. This enables an efficient supersonic heating of the gas to high temperatures. This technique allows efficient ionization of high-Z atoms, with less energy expended in kinetic energy and parasitic sub-keV emission losses.

Differences in plasma parameters between plasmas generated with a solid matter and low-density gas are the basis for achieving higher multi-keV conversion efficiency in underdense targets. In Figure 3.14, the temperature and density gradients for the underdense radiator, initially at 0.01 g/cm³ (~4 atm) and a Xe disk target initially at 3.06 g/cm³ (~solid density), are compared at 1 ns after the onset of laser irradiation (Back et al., 2003). No prepulse was imposed on both targets. A laser pulse of 350 nm at 1×10^{15} W/cm² is incident from the right-hand side onto a target. Multi-keV emission is typically created in high-temperature regions where the heated material becomes highly ionized. In the solid disk, the electron efficiently carries absorbed laser

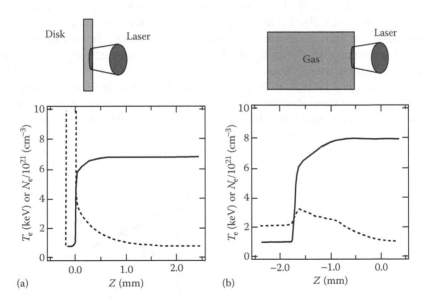

FIGURE 3.14
Calculated profiles of the electron density T_e (solid line) and the electron density n_e (dotted line) for (a) a cryogenic (i.e., solid) Xe disk and (b) Xe gas column irradiated for 1 ns with a laser pulse at 10^{15} W/cm². The gas column produces a large volume of high-temperature plasma leading to efficient multi-keV x-ray emission. (From Kalantar, D.H., et al., Rev. sci. Instrum., 68(1), 814, 1997. With permission.)

energy to much denser and cooler regions of the disk. The density is steep at the ablation front and only a small volume of the plasma contributes to multi-keV production. In the gas target, however, the laser absorption occurs at densities less than the critical density. The laser propagates into the gas and the volume for multi-keV production is two to three times larger.

In addition to such gas-fill targets, low-density materials such as aerogel are suggested to cover spectral gaps appearing between K-shell line emissions from gases such as Ar K-shell (3.1 keV), Kr L-shell (1.8 keV), Xe L-shell (4.5 keV), and Kr K-shell (13.3 keV). The measurements of Ti K-shell emitters in a low-density aerogel plasma is presented (Fournier et al., 2004). A density of 3.1 ± 0.1 mg/cm^3 in aerogel targets, which gives an ionized density relative to the laser critical density of $0.1n_c$ (in this case, $n_c = 9 \times 10^{21}$ cm^{-3} for frequency-tripled laser light used in the experiment). The targets show supersonic heating, as is the case for the gas-fill targets; x-ray output 1%–2% of the incident laser energy in the 4.67–5.0 keV is attained. Figure 3.15 summarizes the conversion efficiencies of 2–13 keV x-rays by the use of conventional planar targets and low-density targets (Yaakibi, 1980; Back et al., 2001, 2003; Fournier et al., 2004, 2006). It is clearly seen that the low-density targets or laser-preformed plasma are suitable to attain high conversion efficiency.

3.3.2.1.3 *Kα Line Radiation*

With an increase in $I_L \lambda_{L\mu m}^2$ above around 10^{15} W/cm^2 μm^2 (I_L is the laser intensity in units of W/cm^2), the collective processes in laser–plasma interaction such as resonance absorption or parametric instabilities start dominating over the absorption mechanism so that plasma electrons have a second component consisting of energetic electrons, that is, hot electrons. The created plasma becomes less collisional and, because of the long penetration depth of hot electrons, plasma is heated nonlocally. Bulk electron temperatures tend to be lower than that for the lower intensity, and x-ray emission mainly arises from inner-shell ionization and subsequent radiative de-excitation. As a result, the observed x-ray lines consist of the primary Kα line from singly ionized ions and the energy-shifted components from partially ionized ions

FIGURE 3.15
Summary of conversion efficiency for multi-keV K- and L-shell x-rays from plasma generated with frequency-tripled Nd:glass laser.

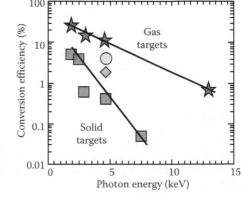

(Nishimura et al., 2003). These multi-keV x-rays are of primary interest in the new fields of ultrafast x-ray diffractometry and biomedical radiography such as x-ray imaging for high-energy density physics. Depending on their energy and the target material (Z), the electrons will typically penetrate several microns into the solid, generating bremsstrahlung and Kα line radiation as they slow down via collisions with cold atoms. The characteristic Kα or Lα radiations have received most attention because of their potential as a monoenergetic, pulse x-ray source.

An analytical model of sub-ps Kα x-ray generation from laser-irradiated foils is presented for questions about the slowdown process of hot electrons in the solid material and associated Kα line yield efficiency. Quantitative discussions are given for the photon emission yield in both forward and backward directions as a function of hot-electron temperature and target thickness. The optimum combination of hot-electron temperature and the target thickness to attain the highest conversion is studied (Salzmann et al., 2002). Figure 3.16 shows a comparison of the model prediction with the experiment. ε_f and ε_k are respectively Kα photons generated per electron and the conversion efficiency of Kα photons emitted per unit solid angle. As is seen, the experimental conversion efficiency is, in general, an order of magnitude smaller than the theoretical prediction.

The characteristics of 22–40 keV Kα x-ray sources are measured using 100 TW and petawatt high-intensity lasers (Park et al., 2006). The measurements show that the size of the Kα source from a simple foil target is larger than 60 μm. This size is too large for most radiography applications under a point projection scheme. It is found that the total Kα yield is independent of target thicknesses, verifying that refluxing of hot electrons through target foils

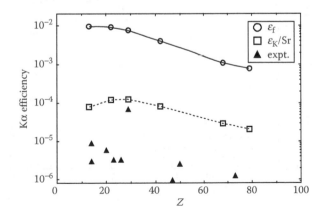

FIGURE 3.16
Kα generation efficiencies under optimal conditions for various target materials Z. Open circles represent photons per electron, open squares represent conversion efficiency of Kα photons emitted per unit solid angle, and closed triangles experimental measurements. (From Salzmann, D., et al., *Phys. Rev. E*, 65, 036402, 2002. With permission.)

plays a major role in photon generation. In fact, smaller radiating volumes emit brighter Kα radiation. Figure 3.17 shows a comparison of the experiments with simulations and the model predictions mentioned here. With an increase in laser intensity, the conversion efficiencies tend to increase, but a difference about an order of magnitude remains. Nevertheless, one-dimensional radiography experiments using small-edge-on foils resolved 10 μm features with high contrast.

A spectroscopic study of a Kα line is important to investigate the transport physics of hot electrons propagating in dense material. By using multilayered targets including Kα, trace material are often used to diagnose the penetration depth or distribution function of hot electrons (Hall et al., 1998; Wharton et al., 1998; Batani et al., 2000; Pisani et al., 2000). It is noteworthy that the Kα line and its ionization shift component are also useful to derive information about hot-electron transport and bulk electron temperature using intensity ratios of the shifted Kα lines in comparison with calculations by an atomic kinetic code specified for the shifted Kα lines (Kawamura et al., 2002).

3.3.2.1.4   Polarized X-Ray Emission

In ultrashort, high-intensity LPP, efficient energy transports in the dense plasma is one of the major issues. In this plasma, there are two components

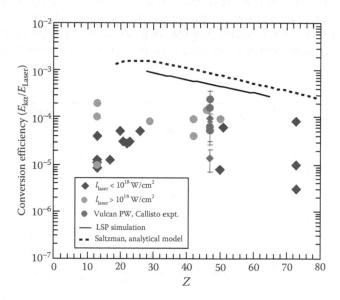

FIGURE 3.17
Comparison of Kα line conversion efficiency as function of target Z among experiments, simulations, and the model predictions (Park, 2006). The diamond plotting symbols represent data with $I_L < 10^{18}$ W/cm² and the circles $I_L > 10^{18}$ W/cm². The solid line represents the prediction from the simulation modeling and the dotted line represents the prediction from the analytical model. (From Park, H.-S., et al., *Phys. Plasmas*, 13, 056309, 2006. With permission.)

of electrons. One is the hot electrons generated by a collective process. They have highly anisotropic initial distribution. The other is the cold bulk electrons; these electrons flow as the return current for the hot electrons and are heated via an ohmic process in a relatively low-density region. Thus, velocity distribution of the cold electrons is substantially isotropic. Numerous experiments have been done to determine the hot-electron spectrum and the energy transport to the bulk electrons by observing the Kα line and its energy-shifted components from ultrahigh intensity laser-irradiated targets as discussed above.

X-ray line polarization spectroscopy is a powerful diagnostic tool to measure directly the anisotropy of the hot-election distribution function (VDF) (Kieffer et al., 1993; Inubushi et al., 2004). When plasma has anisotropy of electric–magnetic fields, polarized x-rays corresponding to the magnetic quantum number are emitted. According to the context of the polarization spectroscopy in an electron beam ion trap (Beiersdorfer and Slater, 2001), the polarization degree P is generally defined by

$$P = \frac{I_\pi - I_\sigma}{I_\pi + I_\sigma},$$
(3.23)

where

I_π is the intensity of line x-ray whose electric field is parallel to the direction of the electron beam

I_σ is that perpendicular to the beam

Hence, anisotropy of hot-electron VDF can be derived from the polarization degree P. Although the polarization degree is 60% at the highest for the beam-like distribution of hot electrons, ultrahigh intensity LPP is useful to generate the polarized x-ray source.

3.3.3 Applications to Oncology Research

The feasibility of phase-contrast imaging with an ultrafast laser-based hard x-ray source has been demonstrated (Toth et al., 2005). Hard x-rays are generated with a femtosecond laser pulse (10 TW, 60 fs, 10 Hz) focused onto a solid target in a very small spot. Such an x-ray source is compact and cheaper than a synchrotron orbit radiation (SOR), and thus has higher power and better x-ray spectrum control than a micro-focal x-ray tube. The Kα line at 17 keV from a solid Mo target and the in-line imaging geometry are utilized to realize phase-contrast images of test objects and biological samples. Figure 3.18 is an x-ray radiograph image of a bee, showing significant enhancement of interfaces due to an x-ray phase shift. This feature was hardly observable in absorption images. It is shown that an ultrafast laser-based x-ray source provides complementary information about the imaged objects and tools for laboratory or clinic-based biomedical imaging.

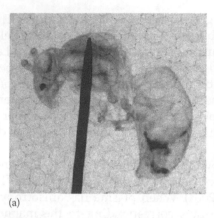 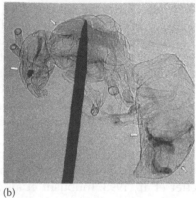

(a)   (b)

FIGURE 3.18
Projection images of a bee obtained with the Mo solid target: (a) absorption image and (b) phase-contrast image with 17 keV photon. The white arrows indicate features enhanced by phase-contrast imaging. (From Park, H.-S., et al., *Phys. Plasmas*, 13, 056309, 2006. With permission.)

3.3.4   Application of LPP MeV Radiation to Inspection

High-energy radiations are used for a wide variety of nondestructive inspection. Radioisotopes or x-ray tubes are commonly used if objects do not move. If they move, for example automobile engines, electron beam accelerators are used to generate a high-energy x-ray pulse. Intense laser (10^{21} W/cm^2) driven hard x-ray sources offer a new alternative to these conventional sources. It was demonstrated with a petawatt laser at the NOVA laser facility at Lawrence Livermore National Laboratory that short ~1 ps laser pulses focused on high-Z targets can produce hard, intense x-ray spectra that can be used for nondestructive radiograph (Perry et al., 1999). These laser-driven sources offer considerable simplicity in design and cost advantage for multiple axis views and have the potential for much higher spatial and temporal resolution than is achievable with accelerator sources. Figure 3.19 shows radiographs through

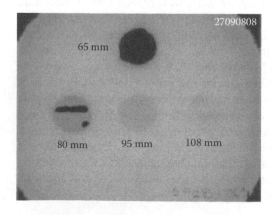

FIGURE 3.19
Radiograph taken with an x-ray film set at 45° to the PW laser axis through varying thicknesses of a lead attenuator. (From Perry, M.D., *Rev. Sci. Instrum.*, 70(1), 265, 1999. With permission.)

various thicknesses of a lead attenuator. These were taken with an x-ray film pack placed 81 cm from the target and at 45° with respect to the laser axis. The depth inset in the figure corresponds to the thickness of the lead attenuator set in front of the test object. For depths of 65, 80, and 108 mm, two additional features were machined into the object at 14 mm less depth of lead. This radiograph would correspond to a dose of >1 MeV photons.

3.4   Features of LPP X-Ray Sources

It might be worthwhile summarizing the features of LPP x-ray sources before concluding this chapter. One will remember SOR (Synchrotron Orbital Radiation) as an indispensable radiation source to perform basic research on material science, pharmacy, medicine, biology, and so on. SOR provides a successive pulse train of coherent radiation in continuum spectrum ranging from visible to very hard x-rays (several tens keV). Nominally, a monochromatic spectrum is extracted from a SOR beam using tunable monochromaters. However, nominally SOR is a large-scale machine and one needs some procedures to access the beam line. In contrast, the LPP x-ray source is a compact x-ray source, which one can rather easily access.

Merits of the LPP x-ray source might be

1. Duration of x-ray pulse is in the order of picosecond to nanosecond, which is much shorter than a conventional light source such as an x-ray tube. This feature enables us to take a snapshot image of rapidly moving materials with high spatial and temporal resolutions.

2. Size of the source is in the order of several to several hundred micrometers. A point x-ray source is favorable to handle the light in the form of diverging, transporting, and converging.

3. Source intensity is extremely bright in comparison with a nominal SOR, discharge plasma source, and x-ray tubes.

4. Conversion efficiency is relatively high. It can be a few to several tens percent, except for radiations arising from the inner-shell transitions.

5. Emission spectrum is rather easily varied with a choice of target materials. In the case of x-ray tube, only electrically conductive materials can be used as the anode.

6. X-rays are generated with a relatively compact machine of high repetition rate.

In spite of these merits, other issues prevent the LPP x-ray source from being used more widely in various research fields.

1. Debris, consisting of ions, neutral atoms, and fine particles from the target, contaminate surrounding things such as the target supply, collection mirrors, and diagnostic instruments so that the lifetime of the LPP source system is extremely shortened unless special treatment is given. To mitigate debris, filters, electromagnetic screens, gas-curtains, or rovers are set between the source plasma and specimens. Adoption of a target consisting of the minimum mass necessary only for x-ray generation is effective to substantially reduce debris.

2. LPP x-ray source is not a continuum spectral source. This limits application to some extent.

Despite these demerits, a wide variety of applications of LPP x-ray sources are under development, demonstrating its high potential.

References

Back, C. A., Grun, J., Decker, C., Suter, L. J., Davis, J., Landen, O. L., Wallace, R., Hsing, W. W., Laming, J. M., Feldman, U., Miller, M. C., and Wuest, C., 2001. Efficient multi-keV underdense laser-produced plasma radiators, *Phys. Rev. Lett.* 87(2), 275003.

Back, C. A., Davis, J., Grun, J., Suter, L. J., Landen, O. L., Hsing, W. W., and Miller, M. C., 2003. Multi-keV x-ray conversion efficiency in laser-produced plasma, *Phys. Plasmas* 10(5), 2047.

Batani, D., Davies, J. R., Bernardinello, A., Pisani, F., Koenig, M., Hall, T. A., Ellwi, S., Norreys, P., Rose, S., Djaoui, A., and Neel, D., 2000. Explanations for the observed increase in fast electron penetration in laser shock compressed materials , *Phys. Rev. E* 61, 5725.

Beiersdorfer, P. and Slater, M., 2001. Measurement of the electron cyclotron energy component of the electron beam of an electron beam ion trap , *Phys. Rev. E* 64, 066408.

Eidmann, K., Kishimoto, T., Hermann, P., Mizui, J., Pakula, R., Sigel, R., and Witkowski, S., 1986. Absolute soft x-ray measurements with a transmission grating spectrometer, *Laser Part. Beams* 4, 521.

Fournier, B. K., Constantin, C., Poco, J., Miller, M. C., Back, C. A., Suter, J. L., Satcher J., Davis, J., and Grun, J., 2004. Efficient multi-keV x-ray sources from Ti-doped aerogel targets, *Phys. Rev. Lett.* 92(16), 165005.

Fournier, K. B., Constantin, C., Back, C. A., Suter, L., Chung, H. -K., Miller, M. C., Froula, D. H., Gregori, G., Glenzer, S. H., Dewald, E. L., Landen, O. L., 2006. Electron-density scaling of conversion efficiency of laser energy into L-shell X-rays, *J. Quant. Spectrosc. Radiat. Transfer* 99, 186.

Girard, F., Jadaud, J. P., Naudy, M., Villette, B., Babonneau, D., Primout, M., Miller, M. C., Kauffman, R. L., Suter, L. J., Grun, J., and Davis, J., 2005. Multi-keV x-ray conversion efficiencies of laser-preexploded titanium foils, *Phys. Plasmas* 12, 092705.

Goldstone, P. D., Goldman, R. S., Mead, W. C., Cobble, J. A., Stradling, G., Day, R. H., Hauer, A., Richardson, M. C., Majoribanks, R. S., Jaanimagi, P. A., Keck, R. L., Marshall, F. J., Seka, W., Barnouin, O., Yaakobi, B., and Letzring, S. A., 1987. Dynamics of high-Z plasmas produced by a short-wavelength laser, *Phys. Rev. Lett.* 59, 56.

Hall, T. A., Ellwi, S., Batani, D., Bernardinello, A., Masella, V., Koenig, M., Benuzzi, A., Krishnan, J. F., Pisani, F., Djaoui, A., Norreys, P., Neely, D., Rose, S., Key, M., and Fews, P., 1998. Fast electron deposition in laser shock compressed plastic targets, *Phys. Rev. Lett.* 81, 1003.

Inubushi, Y., Nishimura, H., Ochiai, M., Fujioka, S., Izawa, Y., Kawamura, T., Shimizu, S., Hashida, H., and Sakabe, S., 2004. X-ray polarization spectroscopy for measurement of anisotropy of hot electrons generated with ultraintense laser pulse, *Rev. Sci. Instrum.* 75, 3699.

Kalantar, D. H., Haan, S. W., Hammel, B. A., Keane, C. J., Landen, O. L., and Munro, D. H., 1997. X-ray backlit imaging measurement of in-flight pusher density for an indirect drive capsule implosion, *Rev. Sci. Instrum.* 68(1), 814.

Karzas W. J. and Latter, R., 1961. Electron radiative transitions in a Coulomb field, *Astrophys. J. Suppl. Ser.*, 55, 167.

Kawamura, T., Nishimura, H., Koike, F., Ochi, Y., Matsui, R., Miao, W. Y., Okihara, S., Sakabe, S., Uschmann, I., Foerster, E., and Mima, K., 2002. Population kinetics on Kα lines of partially ionized Cl atoms, *Phys. Rev. E* 66, 016402.

Kieffer, J. C., Matte, J. P., Chaker, M., Beaudoin, Y., Chien, C. Y., Coe, S., Mourou, G., DUbau, J., and Inal, M. K., 1993. X-ray-line polarization spectroscopy in laser-produced plasmas, *Phys. Rev. E* 48, 4648.

Kodama, R., Mochizuki, T., Tanaka, K. A., and Yamanaka, C., 1987. Enhancement of keV ray emission in laser-produced plasmas by a weak prepulse laser, *Appl. Phys. Lett.* 50(12), 720.

Kornblum, H. N., Kauffman, R. L., and Smith, J. A., 1986. Measurement of 0.1–3-keV x rays from laser plasmas, *Rev. Sci. Instrum.* 57, 2179.

Kruer, W. L., 1984. *The Physics of Laser Plasma Interactions*, The Advanced Book Program, Westview Press, Cambridge, MA.

Lindl, J. D. 1998. *Inertial Confinement Fusion*, AIP Press, Springer, New York.

Mathews, D. L., Campbell, E. M., Ceglio, N. M., Hermes, G., Kauffman, R., Koppel, L., Lee, R., Manes, K., Rupert, V., Slivinsky, V. W., Turner, R., and Ze, F., 1983. Characterization of laser-produced plasma x-ray sources for use in x-ray radiography), *J. Appl. Phys.* 54(8), 4260.

More, R. M., 1982. Electronic energy-levels in dense plasmas, *J. Quant. Spectrosci. Radiat. Transfer* 27, 345.

Nishimura, H., Takabe, H., Kondo, K., Endo, T., Shiraga, H., Shigemori, K., Nishikawa, T., Kato, Y., and Nakai, S., 1991a. X-ray emission and transport in gold plasma generated by 351-nm laser irradiation, *Phys. Rev. A* 43(6), 3073.

Nishimura, H., Kato, Y., Takabe, H., Endo, T., Kondo, K., Shiraga, H., Sakabe, S., Jitsuno, T., Takagi, M., Yamanaka, C., Nakai, S., Sigel, R., Tsakiris, G. D., Massen, J., Murakami, M., Lavarenne, F., Fedosejevs, R., Meyer-ter-Vehn, J., and Witkowski, S., 1991b. X-ray confinement in a gold cavity heated by 351-nm laser light, *Phys. Rev. A* 44(12), 8323–8333.

Nishimura, H., Eidmann, K., Sugimoto, K., Schwanda, W., Toyoda, T., Taniguchi, K., Kato, Y., and Nakai, S., 1991c. Influence of exposure time on the sensitivity of Kodak 101 X-ray film, *J. X-Ray Sci. Technol.* 3, 14–18.

Nishimura, H., Endo, T., Shiraga, H., Kato, Y., and Nakai, S., 1992. X-ray emission from high-Z mixture plasmas generated with intense blue laser light, *Appl. Phys. Lett.* 62(22), 1344.

Nishimura, H., Kawamura, T., Matsuia, R., Ochi, Y., Okihara, S., Sakabe, S., Koike, F., Johzaki, T., Nagatomo, H., Mima, K., Uschmann, I., Foerster, E., 2003. Kα spectroscopy to study energy transport in ultrahigh-intensity laser produced plasmas, *J. Quant. Spectrosc. Radiat. Transfer* 81, 327.

Orzechowski, T. J., Rosen, M. D., Kornblum, H. N., Porter, J. L., Suter, L. J., Thiessen, A. R., and Wallace, R. J., 1996. The Rosseland mean opacity of a mixture of gold and gadolinium at high temperatures, *Phys. Rev. Lett.* 77(17), 3545.

Pakula, R. and Sigel, R., 1985. Self-similar expansion of dense matter due to heat transfer by nonlinear conduction, *Phys. Fluids*, 28, 232.

Pakula, R. and Sigel, R., 1986. Self-similar expansion of dense matter due to heat transfer by nonlinear conduction, *Phys. Fluids*; 29, 1340 (errata).

Park, H. -S., Chambers, D. M., Chung, H. -K., Clarke, R. J., Eagleton, R., Giraldez, E., Goldsack, T., Heathcote, R., Izumi, N., Key, M. H., King, J. A., Koch, J. A., Landen, O. L., Nikroo, A., Patel, P. K., Price, D. F., Remington, B. A., Robey, H. F., Snavely, R. A., Steinman, D. A., Stephens, R. B., Stoeck, C., Storm, M., Tabak, M., Theobald, W., Town, R. P. J., Wickersham, J. E., and Zhang, B. B., 2006. High-energy Kα radiography using high-intensity, short-pulse lasers, *Phys. Plasmas* 13, 056309.

Perry, M. D., Sefcik, J. A., Cowan, T., Hatchett, S., Hunt, A., Moran, M., Pennington, D., Snavely, R., and Wilks, S. C., 1999. Hard x-ray production from high intensity laser solid interactions, *Rev. Sci. Instrum.* 70(1), 265.

Pillion, D. W. and Heiley, C. J., 1986. Brightness and duration of x-ray line sources irradiated with intense 0.53-μm laser at 60 and 120 ps pulse width, *Phys. Rev. A* 34(6), 4886.

Pisani, F., Bernardinello, A., Batani, D., Antonicci, A., Martinolli, E., Koenig, M., Gremillet, L., Amiranoff, A., Baton, S., Davies, J., Hall, T., Scott, D., Norreys, P., Djaoui, A., Rousssseaux, C., Fews, P., Bandulett, H., and Pepin, H., 2000. Experimental evidence of electric inhibition in fast electron penetration and of electric-field-limited fast electron transport in dense matter, *Phys. Rev. E* 62, R5927.

Salzmann, D., Reich, Ch., Uschmann, I., Förster, E., and Gibbon, P., 2002. Theory of Ka generation by femtosecond laser-produced hot electrons in thin foils, *Phys. Rev. E* 65, 036402.

SESAME, 1983. Report on the Los Alamos equation-of-state library, T4-Group, Los Alamos National Laboratory Report No. LALP-83-4, Los Alamos, NM.

Sigel, R., Pakula, R., Sakabe, S., and Tsakiris, G. D., 1988. X-ray generation in a cavity heated by 1.3- or 0.44-μm laser light. III. Comparison of the experimental results with theoretical predictions for x-ray confinement, *Phys. Rev. A* 38, 5779.

Sigel, R., Eidmann, K., Lavarenne, F., and Schmalz, R. F., 1990. Conversion of laser light into soft x rays. Part I: Dimensional analysis, *Phys. Fluids B* 2, 199.

Takabe, H., Yamanaka, M., Mima, K., Yamanaka, C., Azechi, H., Miyanaga, N., Nakatsuka, M., Jitsuno, T., Norimatsu, T., Takagi, M., Nishimura, H., Nakai, M., Yabe, T., Sasaki, T., Yoshida, K., Nishihara, K., Kato, Y., Izawa, Y., Yamanaka, T., and Nakai, S., 1988. Scalings of implosion experiments for high neutron yield, *Phys. Fluids* 31, 2884.

Toth, R., Kieffer, J. C., Fourmaux, S., Ozaki, T., and Krol, A., 2005. In-line phase-contrast imaging with a laser-based hard x-ray source, *Rev. Sci. Instrum.* 76, 083701.

Wharton, K. B., Hatchett, S. P., Wilks, S. C., Key, M. H., Moody, J. D., Yanovsky, V., Offenberger, A. A., Hammel, B. A., Perry, M. D., Josi, C., 1998. Experimental measurements of hot electrons generated by ultraintense (>10^{19} W/cm^2) laser-plasma interactions on solid-density targets , *Phys. Rev. Lett.* 81, 822.

Wilks, S. C. and Kruer W. L., 1997. Absorption of ultrashort, ultra-intense laser light by solids and overdense plasmas, *IEEE J. Quan. Electron.* 33(11) November, 1954–1967.

Yaakobi, B., Bourke, P., Conturie, Y., Delettrez, J., Forsyth, J. M., Frankel, R. D., Goldman, L. M., McCrory, R. L., Seka, W., and Soures, J. M., 1980. High X-ray conversion efficiency with target irradiation by a frequency tripled Nd:Glass laser, *Opt. Commun.* 38, 1981.

Zel'dovich Ya. B. and Raizer, Yu. P., 1966, *Physics of Shock Waves and High-Temperature Hydrodynamic Phenomena*, Academic Press, New York.

4

X-Ray Lasers

Hiroyuki Daido, Tetsuya Kawachi, Kengo Moribayashi,
and Alexander Pirozhkov

CONTENTS

4.1 Introduction

Saturated amplification x-ray lasers using collisional-excitation schemes
have been achieved by a laser whose pulse energy contains a few kilojoules
in the early stage of the development and then later more than 100 J in a few
hundred picosecond pulse duration. These historic achievements at present
and future directions of x-ray laser study are described in Ref. [1]. Although
the collisional schemes brought us many successful results on the x-ray laser
development itself as well as its applications such as probing a moving object,
the required large-scale pumping laser system restricts its applicability to the
wider aspects of the scientific field. In 1995, the transient collisional-excitation
(TCE) scheme, which requires a much smaller pumping system with a much

shorter pumping laser pulse width of a few ps, was demonstrated experimentally [2]. Now, this scheme has been modified into the grazing incidence pumping (GRIP) scheme, which has grown to be the most popular scheme among x-ray lasers. We can obtain gain-saturated x-ray lasers in the wavelength region down to 12 nm with a pumping energy of <1 J [3]. In this section, the details of the mechanism of these schemes and their applications to other research fields are reviewed. Research on the shorter wavelength x-ray laser, mainly devoted to inner-shell excitation schemes, is also reviewed. We briefly describe coherent soft x-ray generation through high-order harmonics from a laser-irradiated neutral gas as well as from a laser-irradiated solid thin relativistic plasma.

4.2 Electron Collision-Excitation Lasers

4.2.1 Development of X-Ray Lasers

Population kinetics of the collisional-excitation laser is in principle rather simple compared to that of a recombining plasma laser. The gain medium ions for the collisional-excitation laser are closed shell ions such as Ne- or Ni-like ions because of their large abundance in plasmas as well as their large excitation rate from the ground state to the excited states. Figure 4.1 shows the simplified energy level diagram of the Ne- and Ni-like ions. Consider the $(2p_{1/2}, 3s_{1/2})_1$ and $(2p_{1/2}, 3p_{1/2})_0$ excited levels of the Ne-like ions. In high-temperature and high-density plasmas, collisional excitation (so-called monopole excitation) from the ground state, $2p^6$ ($J = 0$), to these excited levels ($\Delta J = 0$, where ΔJ is the difference of the J number) becomes substantial. The $(2p_{1/2}, 3s_{1/2})_1$ level has fast radiative decay probability to the ground state, whereas the transition from the $(2p_{1/2}, 3p_{1/2})_0$ to the ground state is optically forbidden;

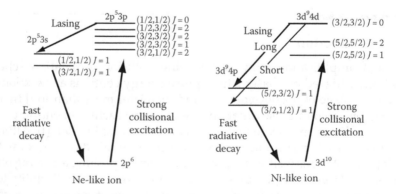

FIGURE 4.1
Energy level diagrams of Ne- and Ni-like ions.

therefore, population inversion is created between these two levels. In the same way, population inversion can be created between the $(3d_{3/2}, 4d_{3/2})_0$ and $(3d_{5/2}, 4p_{3/2})_1$ levels of the Ni-like ions.

The collisional scheme was verified experimentally in 1984 by the use of NOVA laser in Lawrence Livermore National Laboratory (LLNL) [4]. In this experiment, a frequency doubled ($\lambda = 532$ nm) single pulse with a duration of 600 ps and energy of 6 kJ was used to pump the Se foil target. Obvious amplification of the $(2p_{1/2}, 3s_{1/2})_1-(2p_{1/2}, 3p_{1/2})_0$ transition of the Ne-like Se ions was observed. The succeeding experimental and theoretical works allowed us to achieve saturated amplification in the Ne- and Ni-like ion lasers.

The large pumping energy required for collisional-excitation lasers in early experiments was reduced by the multiple pumping technique, in which the target was illuminated by two laser pulses temporally separated by 0.2–2 ns. The first laser pulse was used to create a preformed plasma with long-scale length, which had a small density gradient in the electron density region for the gain. Then, the energy of the following second (heating) pulse was absorbed more efficiently, resulting in a high-temperature plasma with appropriate abundance of Ne- or Ni-like ions, which worked as a gain medium. By adopting this technique, the pumping energy came down by more than an order of magnitude: substantial lasing in the wavelength region of 6–8 nm was obtained by the pumping energy of 200–300 J [5]. The gain saturation of the Ni-like Sm laser (7.3 nm) was obtained with a pumping energy of 70 J [6] and that of the Ni-like Pd and Ag lasers were obtained with 30 J [7].

In the experiments described above, the typical duration and intensity of the heating pulse were ~100 ps and ~10^{13} W/cm^2, respectively. Under such a heating condition, the plasma parameters, especially for T_e (the electron temperature), changed slowly compared with the relaxation time of the excited level populations of the ions, and quasisteady state (QSS) approximation was valid to describe the population distribution. Therefore, this scheme was called QSS collisional-excitation laser. More recently, the use of the heating pulse with ps duration made it possible to realize transiently high gain, resulting in the reduction of the required pumping energy for the saturated amplification. This scheme was called TCE laser [8]. TCE laser was first demonstrated by the Max-Born Institute in 1997 [2], and a large gain coefficient of ~35 cm^{-1} was obtained in the Ne-like Ti ion laser at 32.6 nm by the use of a chirped pulse amplification (CPA) Nd:glass laser, which delivered a ~ps duration pulse.

The TCE scheme makes it possible to reduce the size of the x-ray laser system from the large driver laser dedicated to the study of inertial confinement fusion to a laboratory-size tabletop system. The gain saturation of 10–30 nm x-ray lasers can be achieved by only 5–10 J of pumping energy, and the optimization of the pumping condition and the characterization of the x-ray laser beam have been intensively studied by many laboratories [9–11].

Figure 4.2 shows a typical TCE x-ray laser output as a function of the plasma length, L. In $L < 3$ mm, the x-ray laser intensity increases exponentially with an increase in the plasma length. For $L < 4$ mm, obviously the gradient

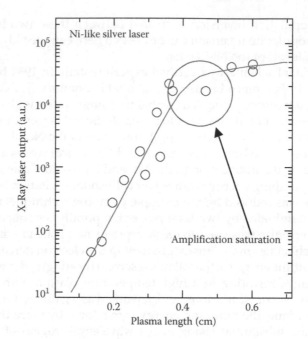

FIGURE 4.2
Typical amplification curve of TCE x-ray laser.

of the amplification curve becomes small. This is called saturated amplification. Under an assumption that the spectral line profile is determined by homogeneous (collision and natural) broadening, the saturation intensity, I_{sat}, at the central frequency of the line profile can be expressed as

$$I_{sat} = \frac{h\nu}{\sigma_{se}\tau_2} = \frac{h\nu}{\left[\lambda^2(\tau_1 + \tau_2)/4\pi^2 t_{spont}\right]\tau_2}, \tag{4.1}$$

where
  $h\nu$ is the photon energy of the x-ray
  σ_{se} is the stimulated emission cross section
  τ_1 and τ_2 are the collisional-radiative destruction time of the lower and upper lasing levels, respectively
  λ is the wavelength of the lasing line
  t_{spont} is the decay time of the spontaneous transition of the lasing line

In the case of Ni-like ion laser, typical I_{sat} is around 10^{10}–10^{11} W/cm^2 [6,11]; that is, in order to increase the output energy of the x-ray lasers, increasing the size of the gain region is indispensable.

It should be noted that the condition of transiently high gain proposed by Ref. [8] is realized by sufficiently fast collisional pumping compared with the collisional mixing between the excited levels. However, the kinetics code implies that the collisional destruction rates of the upper and lower lasing levels (~0.15 ps for the nickel-like silver laser) are much shorter than the ps-duration pumping pulse used in the experiments. Therefore, in the actual experiments, it may be realistic to consider that the maximum abundance of gain medium ions and the optimum electron temperature for the population inversion are simultaneously realized by the short-pulse pumping. It means that if the pumping duration is long, overionization occurs under such a high electron temperature. Indeed, recent investigations suggest that the lifetime of the population inversion density is comparable to the ionization time of the nickel-like ion in plasma.

Compared to the conventional transverse pumping, recent progress in pumping methods, that is, longitudinal [12] or oblique pumping, reduces the pumping energy by one order of magnitude compared with that of the TCE scheme. These are the optical-field ionization (OFI) scheme and GRIP scheme.

The OFI scheme was first demonstrated in the recombining scheme [13], and a strong amplification was obtained in the collisional-excitation scheme using Pd-like Xe ions [14] and Ni-like Kr ions [15]. The gas target was irradiated by the circularly polarized ultrashort laser pulse, and the OFI process instantaneously generated the gain medium ions, which collided with plasma electrons driven by the laser electric field to generate population inversion. This scheme was improved by employing sophisticated gas targets, such as capillary tube [16], and now an x-ray laser pulse with μJ energy and 10 Hz repetition rate is obtained with the pumping energy of 1 J in the wavelength region of 30–40 nm.

The effect of GRIP was firstly reported, to the author's knowledge, by Tommasini and coworkers. In their work, the Ni-like Mo laser line at 18.8 and 22.6 nm was observed by oblique incidence pumping with 150 fs, 300 mJ pulse [17]. The scheme is schematically shown in comparison with normal incidence pumping in Figure 4.3. The following analysis, involving the ray tracing of the pump laser in the plasma, gave us the optimum angle of incidence and the required pumping intensity. The gain-saturated amplification in the GRIP scheme was first reported by the group of Colorado state university and LLNL in the Ni-like Mo ions [18]. Now, a gain-saturated

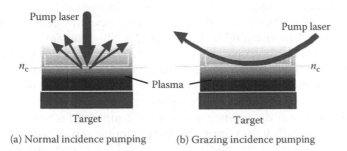

(a) Normal incidence pumping   (b) Grazing incidence pumping

FIGURE 4.3
Schematic views of GRIP compared with the normal incidence pumping.

x-ray laser with 5–10 Hz repetition rate in the wavelength region as short as 12.0 nm is operated by the commercial base tabletop laser system [19]. The average power reaches ~μW, and it is used as the light source for the x-ray laser microscope [20].

The maximum electron density achieved by the pump laser, $n_{e,max}$, is determined by the angle of incidence θ (degree), i.e., $n_{e,max} = n_{cr} \times \cos^2(90 - \theta)$. This implies that we can select the heating region in the plasma by adjusting the angle of incidence, which may become an important feature in improving the beam quality of the x-ray laser.

4.2.2 Improvement of Spatial Coherence of Collisional-Excitation Laser

Since the gain region of the collisional-excitation laser is produced in plasmas with macroscopic density gradient and microscopic density fluctuation, the refraction affects the propagation of amplified x-ray, which leads to the large beam divergence compared with that of diffraction limited and intensity fluctuation in the near-field pattern. Due to this distortion, the degree of intrinsic spatial coherence of the x-ray laser beam from the single flat target is not high enough; typically, the coherent length is around several 100 μm at the position of 1 m from the source in the case of the x-ray laser beam with the divergence angle of 5 mrad. To improve the beam quality, several experimental efforts have been made in these experiments. We take two examples, i.e., curved target and double target techniques.

The objective of the curved target is to compensate the influence of macroscopic density gradient. The propagation length of the amplified x-ray in the gain medium with the size d (~20 μm) under the density gradient of grad(n_e) is expressed as $l_p = (2n_{cr}d/\mathrm{grad}(n_e))^{1/2}$, where n_{cr} (~10^{25} cm^{-3} for 10 nm x-ray) is the critical electron density for the amplified x-ray. The refraction angle $\theta \sim d/l_p$ can be quantitatively estimated from the hydrodynamics simulation code, and the effect of the refraction can be compensated by the use of a curved target with the curvature of $R \sim \theta/l_p$. A group from Osaka University in collaboration with the Rutherford Appleton Laboratory conducted an experiment on curved target, and a beam divergence with 1 mrad at a wavelength of 19.6 nm of neon-like germanium ion laser was obtained [21].

The experiment, using double targets, could be seen in the QSS collisional-excitation laser in 1990s [22], and recently a fully spatially coherent x-ray laser at 13.9 nm was demonstrated in TCE laser [23]. The schematic diagram of the experimental setup is shown in Figure 4.4. The x-ray laser beam from the first gain medium worked as a seed x-ray generator, and a portion of the first x-ray laser beam was injected into the second gain medium (x-ray amplifier), in which only a spatial coherent component of the seed x-ray was amplified. The obtained beam divergence was only 0.2 mrad, which was comparable with that of the diffraction limit [23]. In the double target geometry, the first target can be replaced by another source such as higher-order harmonics. Recently, a French group and a Japanese group have demonstrated the injection of the high-order harmonics to the x-ray laser amplifier [24,25]. In both

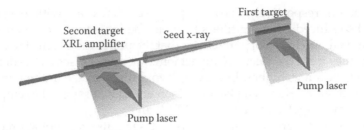

FIGURE 4.4
The experimental setup of double target experiment.

cases beam divergence and the output intensity of the x-ray laser beam were obviously improved. Such highly coherent x-ray lasers can provide unique applications such as soft x-ray speckles produced by domain structures in ferroelectric materials. In this case, dynamics of the domain structure can be taken with a picosecond time resolution [26,27]. Such unique applications using x-ray lasers might be encouraged by many scientists.

4.3 Toward Shorter Wavelength X-Ray Lasers

4.3.1 Introduction

Laser-plasma-based x-ray lasers have been developed worldwide and application experiments have been performed mainly with the electron collisional-excitation scheme [1,28]. However, the pumping energy for the scheme still needs to be relatively large if one desires short wavelength x-ray lasers in the water window spectral range (or kiloelectron volt x-ray lasers).

An x-ray pumped inner-shell ionization laser has been considered to be one of the candidates toward a shorter wavelength x-ray laser. Duguay and Rentzepis proposed this for the first time [29]. Silfvast et al. produced the laser at 441.6 and 325 nm of Cd vapor pumped by a 12 eV soft-x-ray source with an approximated blackbody distribution [30]. Kapteyn et al. and Sher et al. made a laser at 108.9 nm with the use of selective auger decay of the photo-inner-shell ionization state of Xe [31,32]. Walker et al. observed the laser at 127, 130.6, and 131 nm through super Coster-Kronig decay of Zn [33]. However, there has yet been no experimental demonstration of x-ray lasing using inner-shell processes, while several models have been proposed. Kapteyn [34] theoretically showed that a lasing gain of 10 cm^{-1} frequency regime associated with the Kα transitions is possible by irradiating x-rays with a temperature of 600 eV and 50 fs duration on Ne atoms with 10^{20} cm^{-3} density. On the other hand, Moon et al. [35] obtained the same results by using x-rays with temperatures of 150 and 500 eV irradiated on the C atoms and Ne atoms, respectively. They also proposed that x-rays with low energies should be reduced by using filters to avoid the ionization of the outer-shell electrons [35]. Axelrod [36] pointed out that the population of the lower states, which comes from the electron impact ionization, increases as the density of the targets becomes higher. In order to avoid the increase in the population of the lower states, Kapteyn [34] and Moon et al. [35] proposed to employ the relatively low target atom density of 10^{19-20} cm^{-3}.

Recent rapid progress in high-intensity laser pulse technology has made possible new high-intensity short-pulse (~fs) x-ray sources such as those from Larmor radiation [37] due to the high-intensity laser field. Larmor radiation comes from the interaction of electrons in plasmas with the electromagnetic field of high-intensity lasers. In the relativistic region for laser fields, the eight figure motion of the electrons, which corresponds to very high acceleration, produces high brightness x-rays. Moribayashi et al. [38,39] proposed an x-ray inner-shell ionization laser pumped by Larmor radiation. They also proposed that hollow atom x-ray lasers are good for short wavelength and that Na and Mg vapor can reduce the necessary x-ray intensity in an inner-shell ionization x-ray laser.

High-energy electrons are also proposed as a pumping source for ionization of an inner-shell electron. Fill et al. [40] proposed as follows: high-energy electrons accelerated by high-intensity lasers produce a hole in an electron in K-shell and x-rays are emitted through the radiation transition from the 2p to the 1s state. These x-rays are employed as pumping sources. On the other hand, Kim et al. [41] found a new physical scheme for femtosecond x-ray lasers, in which the upper lasing level is pumped by fast electron impact and the lower lasing level is depopulated via Coster-Kronig transitions. Ivanova and Knight [42] proposed that the electron impact x-ray laser scheme extends inner-shell ionization x-ray lasers. They considered the inner-shell electron excitation produced by the electron impact. For example, in the case of a Ne target, an electron in the 2s state is excited and the x-rays emitted through radiative transition 2p → 2s are employed as an inner-shell ionization x-ray laser.

The hollow atom production has been considered to be a fundamental process for inner-shell ionization x-ray lasers. Hollow atoms have also attracted attention to the spectroscopic measurement of high-density plasmas made by high-intensity laser irradiation [43,44]. In this case, the hollow atoms may be produced from inner-shell excitation and ionization processes by the fast electron impact or bremsstrahlung x-ray or recombination [44,45]. The hollow atom productions from the collision of ions with a solid have been investigated by spectroscopy of x-rays irradiated from hollow atoms both experimentally [46] and theoretically [47]. Briand et al. measured Kα x-ray spectra from hollow atoms produced by collisions of Ar^{17+} ions with a solid at various ion energies [46]. Suto and Kagawa theoretically reproduced this experimental result with the use of the multistep-capture-and-loss model [47]. On the other hand, Moribayashi et al. [48] proposed the following: In a laser-produced plasma, atoms with a relatively low atomic number such as Mg and Al can be easily ionized to hydrogen-like and fully ionized states. If a cold surface is located close to the plasma, the highly charged ions interact with the solid surface. They capture many electrons at once to form the hollow atom state.

By using the hollow atom production by x-rays, the following applications have been proposed. Moribayashi et al. have showed that x-ray emissions from hollow atoms inform us of the x-ray intensities of high irradiance x-ray sources [49]. Furthermore, they have proposed the measurement method of the pulses of femtosecond pulse x-rays by using fluorescence x-rays from multi-inner-shell excited states and an autocorrelation method [50].

In the next section, the atomic processes in inner-shell ionization and hollow atom x-ray lasers are described. We discuss inner-shell ionization and hollow atom x-ray lasers for necessary intensities of x-rays and show an example of experimental setup.

4.3.2 Atomic Processes in Inner-Shell Ionization and Hollow Atom X-Ray Lasers

Here, we show atomic processes used in this study. We employ x-ray absorption ionization (A_1), radiative transition (A_r), autoionization (A_a), and electron impact ionization (E_{II}) processes. Here the electrons used in Equation 4.7 are produced through the processes of photoionization absorptions and autoionization. Their energies are decided by the processes. As mentioned in Ref. [36], this process plays a role in the increase in population of the lower states, that is, decrease in the gain values. The atomic data for A_1, A_r, and A_a employed here have often been calculated using an atomic data code such as Cowan's code [51]. On the other hand, those for E_{II} have been calculated using Lotz formula [52] or BEB theory [53].

With these atomic processes, the population dynamics of the various atomic states as illustrated in Figures 4.5 and 4.6 can be investigated by the rate equations. For the inner-shell ionization x-ray laser, three states are considered, that is, the ground state, the upper state, and the lower state of the

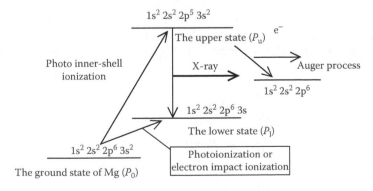

FIGURE 4.5
Atomic processes in x-ray emission from inner-shell ionization x-ray laser of Mg.

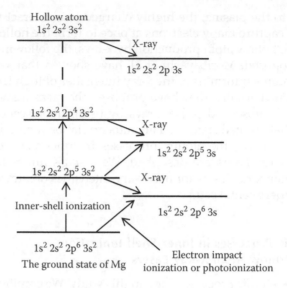

FIGURE 4.6
Atomic processes in x-ray emission from hollow atom x-ray lasers of Mg atom.

laser. Suppose that the population of these three states is P_0, P_u, P_l, then the rate equation becomes

$$\frac{dP_0}{dt} = -\beta_0 P_0,$$

$$\frac{dP_u}{dt} = \alpha_{0u} P_0 - \beta_u P_u,$$

$$\frac{dP_l}{dt} = \alpha_{0l} P_0 + \alpha_{ul} P_l - \beta_l P_l,$$

(4.2)

where $\alpha_{m,k}$ is the transition rate from the m to the k state and β_k is the decay rate via A_r, A_a, E_{II}, and P_I processes in the kth state (see Figure 4.5). On the other hand, in the case of hollow atom x-ray lasers, we must consider larger number of states as shown in Figure 4.6. Then, the rate equation becomes

$$\frac{dP_0}{dt} = -\beta_0 P_0,$$

$$\frac{dP_1}{dt} = \alpha_{0,1} P_0 - \beta_1 P_1,$$

$$\vdots$$

$$\frac{dP_n}{dt} = \sum_k \alpha_{k,n} P_n - \beta_n P_n,$$

(4.3)

where P_0, P_1, P_2, ..., P_n are the populations of the ground state, upper, and lower state for inner-shell ionization, double inner-shell ionization, and hollow atom x-ray lasers of Mg shown in Figure 4.6, respectively.

For analysis of the x-ray laser from the lasing process, the gain Γ is given by

$$\Gamma = 2.7 \times 10^{-2} \phi_i \frac{Q}{g} f_{ul}, \tag{4.4}$$

with $Q = P_{up} - g P_{low}$ and $g = g_{low}/g_{up}$, where $g_{up(low)}$ and f_{ul} are the statistical weights of the upper (lower) state and the oscillator strengths, and ϕ_i is the lifetime of the upper state.

4.3.3 Inner-Shell Ionization and Hollow Atom X-Ray Laser

Figure 4.7a and b shows the gain as a function of time for various x-ray intensities in inner-shell ionization and hollow atom x-ray lasers, respectively. The intensities of 10^{13} and 3×10^{13} W/cm² are required in inner-shell ionization and hollow atom x-ray lasers, respectively. For $I > 10^{14}$ and 10^{15} W/cm², the time duration of x-ray laser decreases as the intensity of x-ray becomes larger for the inner-shell and hollow atom x-ray lasers, respectively. Figure 4.8 shows an example of details required for an experiment of the inner-shell x-ray laser. As soon as the high-intensity (10^{20} W/cm²) short-pulse (100 fs) laser irradiated a plasma with an electron density of 3×10^{20} cm⁻³, $R_s = 5\,\mu$m, and $R_L = 90\,\mu$m, high-brightness (10^{13} W/cm²), short-pulse (100 fs) Larmor x-rays are emitted. By using the Larmor x-rays for an x-ray pump source and Mg vapor with a density of 10^{17} cm⁻³ for a target, the inner-shell ionization x-ray laser may be realized. Here R_s and R_L are the diameters of the spot size and the propagation length of the laser.

A hollow atom x-ray laser has a clear advantage over the conventional inner-shell ionization method, that is (1) a much larger population of the upper states can be obtained, (2) the lower states decay much more rapidly

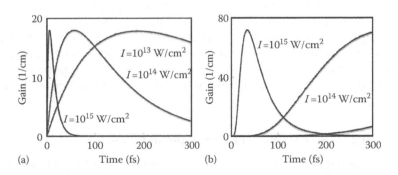

(a)   (b)

FIGURE 4.7
Gain versus time for various x-ray intensities (I): (a) inner-shell ionization x-ray laser and (b) hollow atom x-ray laser.

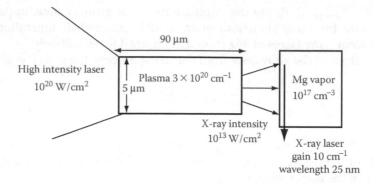

FIGURE 4.8
An example of detailed requirements for the experiment.

because of the ultrafast inner-shell ionization process: the effect of secondary electron impact ionization process may be ignored, (3) with the use of a high initial density of atom ($\sim 10^{22}$ cm^{-3}), high gain x-ray laser emission occurs for a longer time. This arises from the properties (1) and (2), and (4) the wavelength becomes shorter.

We have reported the atomic processes in inner-shell ionization and hollow atom x-ray lasers. For the inner-shell ionization x-ray laser, an example of detailed parameters for an experiment is shown by using a 30 TW laser, a plasma with a density of 3×10^{20} cm^{-3}, and Mg vapor with a density 10^{17} cm^{-3}. In this case, it is estimated that the gain of the inner-shell ionization x-ray laser is about 10 cm^{-1}. Further, the advantage of a hollow atom x-ray laser is explained.

4.4 Other Mechanisms of Coherent X-Ray Generation

Coherent x-ray radiation can also be generated by other mechanisms not related to lasing in plasma. These mechanisms are based on (1) the generation of high-order harmonics of lasers operated in the visible or near-infrared spectral regions and (2) reflection of laser pulses from a relativistic mirror. For completeness, we briefly introduce here the main experimental and theoretical results achieved in these fields. For a detailed description, see the reviews [54–57].

4.4.1 High-Order Harmonics Generated in Gases

The spectral region of x-ray lasers implies that this high-frequency radiation cannot propagate in nonlinear crystals. However, this radiation can propagate in gases. High-order harmonic generation in gases was first observed experimentally in Ref. [58] and later by many other groups. Under the typical

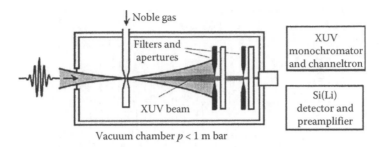

Vacuum chamber $p < 1$ m bar

FIGURE 4.9
Schematic of the experimental setup of a typical high harmonic generation experiment. The laser beam is focused into the target chamber. The target is formed by a tube, where walls are holes bored by the laser itself. The tube is continuously backed with some noble gas. The generated high harmonic radiation is measured either by wavelength-dispersive spectral analysis or by energy-dispersive x-ray spectrometry. (From Brabec, T. and Krausz, F., *Rev. Mod. Phys.*, 72, 545, 2000. With permission.)

experimental conditions, intense (10^{13}–10^{15} W/cm^2) short (femtosecond or picoseconds) laser pulses are focused into a gas jet, cell, or gas-filled fiber (Figures 4.9 and 4.10). The observed spectra have a characteristic shape (Figure 4.11): they consist of a large number of odd harmonics (typically, from few tens to up to a few hundred); at low orders, the conversion efficiency quickly decreases with harmonic number, but at higher orders the conversion efficiency changes relatively slowly (so-called harmonic plateau). Finally, at highest orders, the conversion efficiency quickly drops to zero (harmonic cutoff). The highest harmonic photon energy ever reported is 1.3 keV ($\lambda_x \approx 1$ nm) [60].

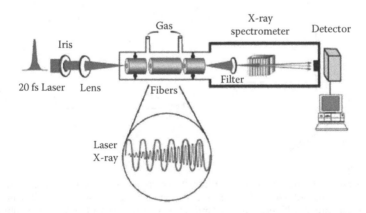

FIGURE 4.10
Experimental setup for phase-matching of soft x-rays in a capillary waveguide. The inset shows the growth of the x-ray wave when phase-matched. Note that the laser and x-ray wavelengths are not to scale. (From Rundquist, A., et al., *Science*, 280, 1412, 1998. With permission.)

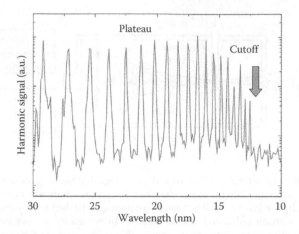

FIGURE 4.11
A partial high harmonic spectrum from neon irradiated by 50 fs, 800 nm pulses versus wavelength. In energy the corresponding peaks are equidistant and separated by 3.1 eV twice the Ti–sapphire photon energy. The graph clearly shows the plateau and cutoff characteristics of this type of emission. The cutoff occurs in this case for orders higher than 61. The lowest order on the left of the picture is the 27th. (From Agostini, P. and DiMauro, L.F., *Rep. Prog. Phys.*, 67, 813, 2004. With permission.)

The process of harmonic generation depends on single-atom and propagation effects. On the single-atom level, it can be understood using a so-called three-step model [61,62] (Figure 4.12, left part). At the first step, the atom is ionized by a strong laser field (under the typical experimental parameters, ionization is in the tunneling regime; the moment of ionization is therefore

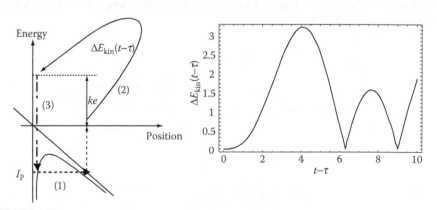

FIGURE 4.12
The left part of the figure illustrates the three-step recollision model: (1) tunneling; (2) classical trajectory under the influence of the electromagnetic field practically alone. During this part the electron acquires a kinetic energy (and its wavefunction a phase); (3) recombination in which the kinetic energy acquired in (2) plus the ionization energy I_p overcome in (1) are converted into a high-energy photon. The right part of the figure shows the classical kinetic energy (in units of the ponderomotive energy U_p) versus the return time, showing in particular the maximum at 3.2, which accounts for the harmonic cutoff $3.2U_p + I_p$ (Equation 4.5). (From Agostini, P. and DiMauro, L.F., *Rep. Prog. Phys.*, 67, 813, 2004. With permission.)

close to the electric field maximum). After the ionization, the electron is pulled from the atom (ion) by the laser field, gaining substantial energy. After a quarter of the optical period, the electron is pulled back. If the laser polarization is linear, the electron passes through the atom and can eventually recombine, radiating the acquired energy in the form of a high-energy photon. This process is repeated twice in every laser period as the ionization happens both at positive and negative peaks of the electric field. This quasi-periodical emission means that the spectrum of high-frequency radiation consists of odd harmonics. The energy acquired by the electron during the time from the ionization to recombination depends on the ionization time; the maximum energy is equal to $3.17U_p$ (Figure 4.12, right part), where $U_p = e^2E^2/(4m\omega^2)$ is the ponderomotive potential ($U_p[\text{eV}] \approx 9.3 \times 10^{-14} (\lambda[\mu\text{m}])^2 I[\text{W}/\text{cm}^2]$). Here λ, ω, and E are the laser wavelength, frequency, and electric field, and e and m are the electron charge and mass. The maximum photon energy (harmonic cutoff) is then

$$\hbar\omega_x = 3.17U_p + I_p, \tag{4.5}$$

where I_p is the atomic ionization potential. As follows from the three-step model, the driving laser's intensity should be high enough to cause the ionization, and the pulse duration should be short enough so that the gas is not fully ionized before the pulse peak comes (low ionization degree is also favorable for propagation, see below). The highest laser intensity and thus the maximum frequencies that can be generated are limited by the ionization of atoms in gas media [54–56,60].

Propagation effects (defocusing, absorption, and phase mismatch) strongly influence the harmonic generation efficiency. Defocusing arises due to the free electron density profile, which has the maximum near the laser beam center, where the intensity is highest and the ionization is strongest. Absorption of high-frequency harmonic radiation (extreme ultraviolet [XUV] and soft x-ray spectral ranges) is mainly due to the single-photon ionization. Phase mismatch arises due to the different phase velocities of the laser and harmonic waves in the partly ionized gas, geometrical effects (the Gouy phase), and the intensity-dependent phase of the emitted harmonics. By optimizing the experimental parameters, it is possible to achieve the phase matching at substantial interaction length, which results in relatively high conversion efficiencies. In this way, the single harmonic energy of $0.3\,\mu\text{J}$ with a conversion efficiency of $\eta = 1.5 \times 10^{-5}$ has been achieved at $\lambda_x = 30\,\text{nm}$ (27th harmonics of Ti:sapphire laser) [63]. At longer wavelength, the achieved efficiency is higher ($\eta = 4 \times 10^{-5}$ at $\lambda_x = 53\,\text{nm}$ [64]; $7\,\mu\text{J}$ and $\eta = 4 \times 10^{-4}$ at $\lambda_x = 73\,\text{nm}$ [65]); at shorter wavelength, the efficiency decreases ($25\,\text{nJ}$ and $\eta = 5 \times 10^{-7}$ at $\lambda_x = 13.5\,\text{nm}$ [66]). Another approach is the phase matching [59,67] or quasiphase matching [68] in a gas-filled hollow waveguide. In the latter case, a harmonic source driven by a 1–3 mJ, 22 fs, 1 kHz Ti:sapphire laser system has produced ~10^6–10^8 photons/s at carbon K edge ($\lambda_x = 4.4\,\text{nm}$) in 10% bandwidth [68].

For sufficiently short driving laser pulses (typically, shorter than 100 fs), high-order harmonics generated in gases are emitted in a form of a narrow beam with the divergence decreasing with harmonic order. The typical divergence is ~mrad or smaller near the cutoff. Near-diffraction-limited divergence was demonstrated (e.g., Refs. [69,70]), implying a high degree of coherence of harmonic radiation. An interference experiment employing two pinholes [71] gave the fringe visibility value approaching unity for harmonics generated in a gas-filled hollow fiber (23rd to 29th harmonics of Ti:sapphire laser). The pinhole diameters were 50 μm and separation was 400 μm; the distance from the exit of the hollow fiber to the pinholes was 60 cm (the harmonic beam diameter was ~1 mm at this point); and the distance from the pinholes to the detector (charge-coupled device [CCD] camera) was ~3 m. High-order harmonic radiation can be focused to a near-diffraction limited spot. A spot size of 1 μm ($M^2 = 1.4$) and an estimated intensity of 10^{14} W/cm² were achieved at $\lambda_x = 30$ nm in Ref. [72]. The M^2 indicates that the spot size is M^2 times larger than that of an ideal Gaussian beam; more rigorous definitions can be found in Refs. [73,74].

It is worth noting that the harmonics are emitted in the form of pulse trains, with the individual pulse duration in the attosecond domain. The proposal of attosecond pulse train generation was formulated in Refs. [75,76] and first implemented in Ref. [77], where the train of 250 as pulses consisting of harmonics 13th to 19th of Ti:sapphire laser was observed. For sufficiently short driver pulses, single attosecond pulses can be generated [78]. Using 5 fs driver pulses with the modulated polarization, single 130 as pulses near $\hbar\omega_x = 36$ eV were reported in Ref. [79]; this pulse duration equals less than 1.2 optical cycles at this wavelength.

4.4.2 High-Order Harmonics in the Relativistic Regime

Other mechanisms of harmonic generation result from the relativistic motion of electrons driven by relativistic-irradiance laser pulses in plasmas. Though less studied experimentally than harmonics generated in gases, relativistic harmonics attract attention because higher energy and higher frequencies are expected. Relativistic harmonics have the potential to generate higher frequencies than the gas harmonics, because in the former case the plasma is already ionized, and the ionization does not limit the maximum laser intensity (cf. the cutoff in the latter case, Equation 4.5). The electron motion becomes relativistic when the dimensionless laser field amplitude a approaches or exceeds unity:

$$a = eE/(mc\omega), I \approx a^2 \times 1.37 \times 10^{18}\ \text{W/cm}^2 \times (\mu m/\lambda)^2 \tag{4.6}$$

where c is the velocity of light. Under these conditions, the matter is inevitably ionized, and laser pulses interact with plasma. Due to the large mass difference between electrons and ions, the interaction with electrons is much more efficient.

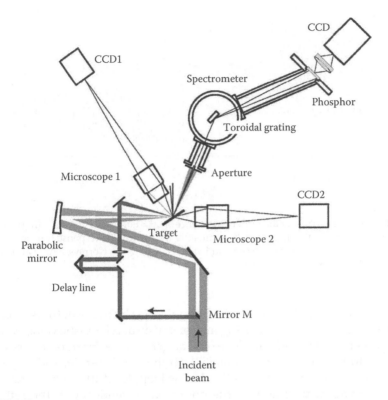

FIGURE 4.13
Experimental setup for high-order harmonic generation from solid target. The mirror M in the center of the incident beam together with the aperture in front of the spectrometer, placed at the image position of the mirror M, prevents the laser light from entering into the spectrometer. CCD1 and CCD2 are for precise alignment of the target. (From Tarasevitch, A., et al., *Phys. Rev. A*, 62, 023816, 2000. With permission.)

High-order harmonics from laser-driven plasmas were observed in the early 1980s using infrared nanosecond CO_2 lasers [80,81] whose wavelength was 10 μm and later using near-IR pico- and femtosecond lasers (see, Refs. [82–87]) focused on solid targets. A typical experimental setup is shown in Figure 4.13. Both odd and even harmonics were observed (Figure 4.14), which is an indication of the absence of the inversion symmetry near the surface at oblique incidence. Harmonic emissions in the water window with a wavelength of $\lambda_x = 3.6$ nm and a conversion efficiency of $\eta > 10^{-6}$ were observed, with the indication of harmonics at 1.2 nm [86].

Harmonic generation at the surface of a solid target is attributed to the driving laser light reflection from an oscillating mirror formed by the electrons at the sharp plasma–vacuum interface or at the critical density surface, if the density profile is gradual. The explanation in terms of oscillating mirror driven at twice the laser frequency by a ponderomotive force was proposed in Ref. [88]. The oscillating mirror was studied theoretically and

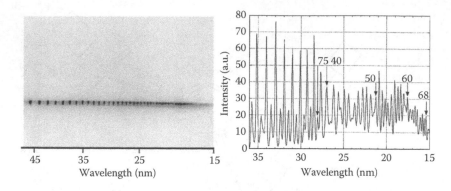

FIGURE 4.14
(Left) Harmonic spectrum for an incident irradiance of $I\lambda^2 = 1.0 \times 10^{19}\,W\,\mu m^2\,cm^{-2}$. (Right) Lineout from the spectrum (left) in the range 36–15 nm after the instrument response function has been removed by a maximum entropy deconvolution procedure. (From Norreys, P.A., et al., *Phys. Rev. Lett.*, 76, 1832, 1996. With permission.)

with the help of particle-in-cell (PIC) simulations by many authors (see, Refs. [89–94]). In general, the oscillating mirror driven at relativistic velocity moves both perpendicular and parallel to the target surface. Harmonics are generated due to the following reasons [57]: (1) the nonlinear dependency of the velocity on the electric field; (2) the double Doppler shift upon the reflection at the moving mirror; and (3) the nonlinear dependency of the reflection instant on time (retardation effect).

Harmonics generated by the oscillating mirror have nontrivial polarization properties [88–93]. At normal incidence and linear driver pulse polarization, odd harmonics with the same polarization are generated. At oblique incidence, p-polarized driver pulses generate p-polarized odd and even harmonics (Figure 4.15a), while s-polarized driver pulses generate s-polarized odd (Figure 4.15b) and p-polarized even harmonics (Figure 4.15c).

The oscillating mirror model assumes a sufficiently steep electron density gradient; otherwise, the harmonic generation efficiency is exponentially small. In the experiments employing pico- and nanosecond pulses [80,81,84], the steep density gradient near the critical plasma density was formed because the driver pulse ponderomotive pressure exceeded the thermal pressure, and the pulse was sufficiently long. In this case, the oscillating mirror surface was rippled, which resulted in decreased coherence, broad angular distribution of harmonic emission, and lack of polarization properties. To achieve a flat reflective plasma surface, it is necessary to use short (femtosecond) driver pulses with a high contrast. The latter means that there should be neither prepulses nor pedestal, which create plasma before the main pulse. For example, the highest harmonic orders [86] were generated employing two plasma mirrors that improved the contrast ratio from $10^7{:}1$ to $> 10^{11}{:}1$. Other techniques based on frequency doubling [87], polarization rotation, and optical parametric chirped pulse amplification (OPCPA) are also employed to achieve high-contrast pulses.

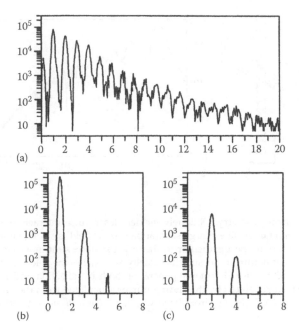

FIGURE 4.15
Results of the PIC simulations. High harmonic and low-frequency generation for a sharp bound-ary plasma and a relativistic-irradiance laser pulse with $a_0 = 3$, $\tau_0 = 16\pi/\omega_0$, incidence angle $\theta = 60°$, and plasma density $n_0 = 6n_{cr}$. P-polarized incident pulse generates odd and even p-polarized har-monics (a). S-polarized incident pulse generates s-polarized odd and p-polarized even harmonics (b) and (c), respectively. (From Bulanov, S.V., et al., *Phys. Scr.*, T63, 258, 1996. With permission.)

Relativistic harmonics have a broad spectrum; furthermore, PIC simula-tions show that high-order harmonics are phase locked. Using the spectral filter which selects high harmonic orders, one should be able to obtain very short high-frequency electromagnetic pulses in the attosecond or even zeptosecond range [95–98].

In the case of a thin overdense target, an analytical theory describes self-consistently the dynamics of the driving electric field and the target, the so-called sliding mirror model [93,95,96]. In this model, it is assumed that the electrons move collectively as a whole, while the ions remain immobile. The charge separation electric field is strong enough to suppress the elec-tron layer motion perpendicular to the target surface, so it moves (slides) along the target (Figure 4.16a). The harmonics are generated by the nonlin-ear surface current (Figure 4.16b). This mechanism promises attosecond pulse generation without the spectral filtering, with the efficiency of a few percent [95,96]. For efficient attosecond pulse generation, the laser pulse amplitude should be approximately equal to the dimensionless areal electron density ε:

$$a \approx \varepsilon = (\pi n_0 l)/(n_{cr}\lambda). \tag{4.7}$$

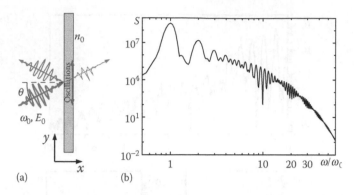

(a)   (b)

FIGURE 4.16
(a) The sliding mirror geometry. Obliquely incident laser pulse interacts with thin ($l \ll \lambda_0$) overdense ($n_0 \gg n_{cr}$) target. The charge separation electric field is strong enough to suppress the electron layer motion perpendicular to the target surface, so it moves (slides) along the target. This motion is nonlinear because its velocity approaches the velocity of light. The nonlinear surface current causes harmonic generation both in reflection and transmission. (b) Harmonic spectrum generated by the sliding mirror (calculated from the analytical model); $a_0 = 15$, $\varepsilon_0 = 13.5$, $\theta = 42°$. (From Pirozhkov, A.S., et al., *Phys. Plasmas*, 13, 013107, 2006. With permission.)

where
　n_0 is the electron density
　l is the target thickness

Polarization properties of the sliding mirror are the same as those of the oscillation mirror.

Another mechanism of efficient attosecond pulse generation in the so-called λ^3 (lambda-cube) regime is studied in Refs. [99–101] using the PIC simulations. In this regime, relativistic-irradiance few cycle laser pulses, focused to the focal spot with the diameter $\sim\lambda$, interact with overdense plasma. The optimum plasma density is $n = n_{cr}a/2$ (this relation is derived from PIC simulations). Driver pulses create an oscillating concave plasma–vacuum boundary with sufficiently steep gradient, from which the driver light itself is reflected in the form of several attosecond pulses. These pulses are reflected in a few different directions, allowing the selection of a single attosecond pulse. The pulse duration scales as $\tau_x \sim a^{-1}$.

4.4.3　Relativistic Mirror

Coherent high-frequency radiation can be generated upon laser pulse reflection from a mirror moving with relativistic velocity [102]. The frequency of the reflected laser pulse is upshifted and the duration is reduced by the factor $\approx 4\gamma^2\cos^2(\theta/2)$, where γ is the relativistic gamma-factor of the mirror and θ is the incidence angle. Furthermore, the reflected high-frequency pulse inherits the coherence, polarization, and temporal shape from the original laser pulse, which makes pulse-shaping techniques from the visible and near-IR spectral regions available in the XUV and x-ray regions. A partly reflecting relativistic mirror (flying mirror) can be formed by a breaking wake wave created by a strong driver pulse propagating in underdense plasma, as proposed in Ref. [103] and later studied using PIC simulations in Ref. [104] (Figure 4.17). Depending on the electron density distribution in the flying mirror, the reflectivity of the counterpropagating source pulse in terms of photon number is expected to be $R = 1/(2\gamma^3)$ or $R \approx 0.11/\gamma^4$ (for simplicity it is assumed that $\theta = 0$ (normal incidence) and the frequencies of the driver and source pulses are equal). Here, γ is the gamma-factor corresponding to the

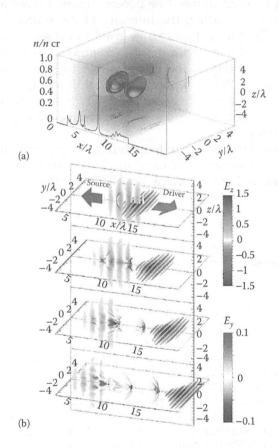

FIGURE 4.17
Relativistic flying mirror in 3D PIC simulations. (a) The electron density in the wake of the driver laser pulse at $t = 14 \times 2\pi/\omega_d$. The $(x; y = -6\lambda; z)$ plane: density profile along the symmetry axis. Thin curves are for density values $n = 0.12, 0.24$, and $0.36n_{cr}$ on the corresponding perpendicular planes of symmetry; isosurfaces for value $n = 0.15n_{cr}$; greyscale for lower values. (b) The cross sections of the electric field components. The $(x; y; z = 0)$ plane: $E_z(x; y; z = 0)$ (upper color scale, the driver pulse); the plane $(x; y = 0; z)$: $E_y(x; y = 0; z)$ (lower color scale, the source pulse and reflected radiation) at $t = 16, 18, 20$, and $22 \times 2\pi/\omega_d$ (top–down). (From Bulanov, S.V., et al., *Phys. Rev. Lett.*, 91, 085001, 2003. With permission.)

phase velocity of the flying mirror. This velocity equals the group velocity of the driver pulse, which depends on the pulse shape, focal spot size, etc. For estimations, a simple relation $\gamma \approx \omega_d / \omega_p$ can be used, where ω_d is the driver pulse frequency and ω_p is the plasma frequency. For sufficiently low plasma density, the gamma-factor can be quite high. Because each reflecting photon has $\approx 4\gamma^2$ times larger energy, the reflectivity in terms of energy is either $2/\gamma$ or $\approx 0.44/\gamma^2$. Due to the pulse compression, the power of the source pulse upon reflection is multiplied by either 8γ or ≈ 1.8. An interesting property of the flying mirror is a paraboloidal shape of the surface, which can cause the reflected pulse focusing (Figure 4.17a). The focus size can be as small as $\lambda'_s = \lambda_s/(2\gamma)$, where λ_s and λ'_s are the wavelengths of the source pulse in the laboratory and boosted frames (the boosted frame moves with the flying mirror). Due to the focusing, the intensity of the source pulse increases either $32\gamma^3$ or $\approx 7\gamma^2$ times. This can be used to achieve record intensities approaching the Schwinger limit [104]. A proof-of-principle experiment demonstrating the ability of the flying mirror to reflect laser pulses was reported in Refs. [105–106]. The frequency upshifting factors from 55 to 114 were observed at the incidence angle of $\theta = 45°$ (the gamma-factors $\gamma \approx 4$–6). The reflected radiation was in the XUV spectral region (the wavelengths from $\lambda_x = 7$ to $14\,\text{nm}$).

The wake wave can also generate high-frequency radiation upon interaction with regular coherent structures in plasma, such as electromagnetic solitons, vortices, and other wake waves [107]. In this case, the frequency of the radiation is $\approx 4\gamma^2\omega_p$.

References

1. H. Daido, *Rep. Prog. Phys.*, **65**(10), 1513, 2002.
2. P. V. Nickles, V. N. Shlyaptsev, M. Kalachinikov, M. Schnürer, I. Will, and W. Sandner, *Phys. Rev. Lett.*, **78**, 2748, 1997.
3. B. M. Luther, Y. Wang, M. A. Larotonda, D. Alessi, M. Berrill, M. C. Marconi, and J. J. Rocca, *Opt. Lett.*, **30**, 165, 2005.
4. D. L. Matthews et al., *Phys. Rev. Lett.*, **54**, 110, 1985.
5. H. Daido, Y. Kato, K. Murai, S. Ninomiya, R. Kodama, G. Yuan, Y. Oshikane, M. Takagi, H. Takabe, and F. Koike, *Phys. Rev. Lett.*, **75**, 1074, 1995.
6. J. Zhang, A. G. MacPhee, J. Lin, E. Wolfrum, R. Smith, C. Danson, M. H. Key, C. L. S. Lewis, D. Neely, J. Nilsen, G. J. Pert, G. J. Tallents, and J. S. Wark, *Science*, **276**, 1097, 1997.
7. J. E. Balmer, R. Tommasini, and F. Lowenthal, *IEEE J. Select. Top. Quantum Electron.*, **5**(6), 1435–1440, 1999.
8. Y. A. Afanasiev and V. N. Shlyaptsev, *Sov. J. Quantum Electron.*, **19**, 1606, 1989.
9. J. Dunn, A. L. Osterheld, R. Shepherd, W. E. White, V. N. Shlyaptsev, and R. E. Stewart, *Phys. Rev. Lett.*, **80**, 2825, 1998.
10. A. Klisnick et al., *J. Opt. Soc. Am. B*, **17**, 1093, 2000.

11. T. Kawachi et al., *Phys. Rev. A*, **66**, 033815, 2002.
12. T. Ozaki, R. A. Ganeev, A. Ishizawa, T. Kanai, and H. Kuroda, *Phys. Rev. Lett.*, **89**, 253902, 2002.
13. Y. Nagata, K. Midorikawa, K. Obara, M. Tashiro, and K. Toyoda, *Phys. Rev. Lett.*, **71**, 3774, 1993.
14. B. E. Lemoff, G. Y. Yin, C. L. Gordon III, C. P. J. Barty, and S. E. Harris, *Phys. Rev. Lett.*, **74**, 1574, 1995.
15. S. Sebban et al., *Phys. Rev. Lett.*, **89**, 253901, 2002.
16. A. Butler, A. J. Gonsalves, C. M. Mckenna, D. J. Spence, S. M. Hooker, S. Sebban, T. Mocek, I. Bettaibi, and B. Cros, *Phys. Rev. Lett.*, **91**, 205001, 2003.
17. R. Tommasini, J. Nilsen, and E. Fill, *Proc. SPIE*, **4505**, 85, 2001.
18. B. M. Luther, Y. Wang, M. A. Larotonda, D. Alessi, M. Berrill, M. C. Marconi, J. J. Rocca, and V. N. Shlyaptsev, *Opt. Lett.*, **30**, 165, 2005.
19. Y. Wang, M. A. Larotonda, B. M. Luther, D. Alessi, M. Berrill, V. N. Shlyaptsev, and J. J. Rocca, *Phys. Rev. A*, **72**, 053807, 2005.
20. G. Vaschenko et al., *Opt. Lett.*, **31**, 1214, 2006.
21. R. Kodama, D. Neely, Y. Kato, H. Daido, K. Murai, G. Yuan, A. MacPhee, and C. L. Lewis, *Phys. Rev. Lett.*, **73**, 3215, 1994.
22. H. Daido, H. Tang, Y. Kato, and K. Murai, *Comptes Rendus*, Special Issue on X-Ray Lasers, Serie **IV**, No.8, p.999 2000.
23. M. Nishikino, M. Tanaka, K. Nagashima, M. Kishimoto, M. Kado, T. Kawachi, K. Sukegawa, Y. Ochi, N. Hasegawa, and Y. Kato, *Phys. Rev. A*, **68**, 061802, 2003.
24. Ph. Zeitoun et al., *Nature*, **431**, 426–429, 2004.
25. N. Hasegawa, A. V. Kilpio, K. Nagashima, T. Kawachi, M. Kado, M. Tanaka, S. Namba, K. Takahashi, K. Sukegawa, P. Lu, H. Tang, M. Kishimoto, R. Tai, H. Daido, and Y. Kato, *Proc. SPIE*, **4505**, 204, 2001.
26. R. Z. Tai et al., *Phys. Rev. Lett.*, **89**, 257602, 2002.
27. R. Tai, K. Namikawa, A. Sawada, M. Kishimoto, M. Tanaka, P. Lu, K. Nagashima, H. Maruyama, and M. Ando, *Phys. Rev. Lett.*, **90**, 087601, 2004.
28. P. V. Nickles, K. A. Janulewicz, editors, X-Ray Lasers 2006, *Springer Proceedings in Physics* 115, Dordrecht, the Netherlands, 2006.
29. M. A. Duguay and M. Rentzepis, *Appl. Phys. Lett.*, **10**, 350–352, 1967.
30. W. T. Silfvast, J. J. Macklin, and O. R. Wood II, *Opt. Lett.*, **8**, 551–553, 1983.
31. H. C. Kapteyn, R. W. Lee, and R. W. Falcone, *Phys. Rev. Lett.*, **57**, 2939–2942, 1986.
32. M. H. Sher, J. J. Macklin, J. F. Young, and S. E. Harris, *Opt. Lett.*, **12**, 891–893, 1987.
33. D. J. Walker, C. P. J. Barty, G. Y. Yin, J. F. Young, and S. E. Harris, *Opt. Lett.*, **12**, 894–896, 1987.
34. H. C. Kapteyn, *Appl. Opt.*, **31**, 4931–4939, 1992.
35. S. J. Moon, D. C. Eder, and G. L. Strobel, *AIP Conf. Proc.*, **332**, 262–266, 1994.
36. T. S. Axelrod, *Phys. Rev. A*, **13**(1), 376, 1976.
37. Y. Ueshima, Y. Kishimoto, A. Sasaki, and T. Tajima, *Laser Part. Beam*, **17**, 45–58, 1999.
38. K. Moribayashi, A. Sasaki, and T. Tajima, *Phys. Rev. A*, **58**, 2007–2015, 1998.
39. K. Moribayashi, A. Sasaki, and T. Tajima, *Phys. Rev. A*, **59**, 2732–2737, 1999.
40. E. Fill, D. Eder, K. Eidmann, J. Meyer-ter-Vehn, G. Pretzler, A. Pukhov, and A. Saemann, *Proceedings of 6th International Conference on X-Ray Lasers*, Vol. 159 (159), 301, 1999.
41. D. Kim, C. Toth, and C. P. J. Barty, *Phys. Rev. A*, **59**(6), 4129 1999.

42. E. Ivanova and L. Knight, *Proc. SPIE Soft x-ray Laser and Applications III*, **3776**, 263, 1999.
43. F. B. Rosmej, A. Y. Faenov, T. A. Pikuz, A. I. Magunov, I. Y. Skobelev, T. Auguste, P. D'Oliveira, S. Hulin, P. Monot, N. E. Andreev, M. V. Chegotov, and M. E. Veisman, *J. Phys. B*, **32**, L107–L117, 1999.
44. A. Y. Faenov, J. Abdallah Jr., R. E. H. Clark, J. Cohen, R. P. Johnson, G. A. Kyrala, A. I. Magunov, T. A. Pikuz, I. Y. Skobelev, and M. D. Wilke, *Proc. SPIE*, **3157**, 10–14, 1997.
45. K. Moribayashi, A. Sasaki, and A. Zhidkov, *Phys. Scr.*, **T92**, 185–187, 2001.
46. J. P. Briand, L. de Billy, P. Charles, S. Essabaa, P. Briand, R. Geller, J. P. Desclaux, S. Bliman, and C. Rastori, *Phys. Rev. Lett.*, **65**, 159–162, 1990.
47. K. Suto and T. Kagawa, *Phys. Rev. A*, **58**, 5004–5007, 1998.
48. K. Moribayashi, K. Suto, A. Zhidkov, A. Sasaki, and T. Kagawa, *Laser Part. Beams*, **19**, 643–646, 2001.
49. K. Moribayashi, T. Kagawa, and D. E. Kim, *J. Phys. B*, **37**, 4119, 2004.
50. K. Moribayashi, T. Kagawa, and D. E. Kim, *J. Phys. B*, **38**, 2187, 2005.
51. R. D. Cowan, *J. Opt. Soc. Am.*, **58**, 808, 1968.
52. W. Lotz, *Z. Phys.*, **232**, 101, 1970.
53. Y. K. Kim and M. E. Rudd, *Phys. Rev. A*, **50**, 3954, 1994.
54. T. Brabec and F. Krausz, Intense few-cycle laser fields: Frontiers of nonlinear optics, *Rev. Mod. Phys.*, **72**, 545, 2000.
55. J. G. Eden, High-order harmonic generation and other intense optical field-matter interactions: review of recent experimental and theoretical advances, *Prog. Quantum Electron.*, **28**, 197, 2004.
56. P. Agostini and L. F. DiMauro, The physics of attosecond light pulses, *Rep. Prog. Phys.*, **67**, 813, 2004.
57. G. A. Mourou, T. Tajima, and S. V. Bulanov, Optics in the relativistic regime, *Rev. Mod. Phys.*, **78**, 309, 2007.
58. A. McPherson, G. Gibson, H. Jara, U. Johann, T. S. Luk, I. A. McIntyre, K. Boyer, and C. K. Rhodes, Studies of multiphoton production of vacuum-ultraviolet radiation in the rare gases, *J. Opt. Soc. Am. B*, **4**, 595, 1987.
59. A. Rundquist, C. G. Durfee III, Z. Chang, C. Herne, S. Backus, M. M. Murnane, and H. C. Kapteyn, Phase-matched generation of coherent soft X-rays, *Science*, **280**, 1412, 1998.
60. J. Seres, E. Seres, A. J. Verhoef, G. Tempea, C. Streli, P. Wobrauschek, V. Yakovlev, A. Scrinzi, C. Spielmann, and F. Krausz, Source of coherent kiloelectronvolt X-rays, *Nature*, **433**, 596, 2005.
61. P. B. Corkum, Plasma perspective on strong-field multiphoton ionization, *Phys. Rev. Lett.*, **71**, 1994, 1993.
62. M. Lewenstein, Ph. Balcou, M. Yu. Ivanov, A. L'Huillier, and P. B. Corkum, Theory of high-harmonic generation by low-frequency laser fields, *Phys. Rev. A*, **49**, 2117, 1994.
63. E. Takahashi, Y. Nabekawa, T. Otsuka, M. Obara, and K. Midorikawa, Generation of highly coherent submicrojoule soft x-rays by high-order harmonics, *Phys. Rev. A*, **66**, 021802R, 2002.
64. E. Constant, D. Garzella, P. Breger, M. Mevel, Ch. Dorrer, C. Le Blanc, F. Salin, and P. Agostini, Optimizing high harmonic generation in absorbing gases: Model and experiment, *Phys. Rev. Lett.*, **82**, 1668, 1999.
65. E. Takahashi, Y. Nabekawa, and K. Midorikawa, Generation of 10-µJ coherent extreme-ultraviolet light by use of high-order harmonics, *Opt. Lett.*, **27**, 1920, 2002.

66. E. J. Takahashi, Y. Nabekawa, and K. Midorikawa, Low-divergence coherent soft x-ray source at 13 nm by high-order harmonics, *Appl. Phys. Lett.*, **84**, 4, 2004.
67. Y. Tamaki, Y. Nagata, M. Obara, and K. Midorikawa, Phase-matched high-order-harmonic generation in a gas-filled hollow fiber, *Phys. Rev. A*, **59**, 4041, 1999.
68. E. A. Gibson, A. Paul, N. Wagner, R. Tobey, D. Gaudiosi, S. Backus, I. P. Christov, A. Aquila, E. M. Gullikson, D. T. Attwood, M. M. Murnane, and H. C. Kapteyn, Coherent soft X-ray generation in the water window with quasi-phase matching, *Science*, **302**, 95, 2003.
69. Ch. Spielmann, N. H. Burnett, S. Sartania, R. Koppitsch, M. Schnürer, C. Kan, M. Lenzner, P. Wobrauschek, and F. Krausz, Generation of coherent X-rays in the water window using 5-femtosecond laser pulses, *Science*, **278**, 661, 1997.
70. E. Takahashi, Y. Nabekawa, M. Nurhuda, and K. Midorikawa, Generation of high-energy high-order harmonics by use of a long interaction medium, *J. Opt. Soc. Am. B*, **20**, 158, 2003.
71. A. R. Libertun, X. Zhang, A. Paul, E. Gagnon, T. Popmintchev, S. Backus, M. M. Murnane, and H. C. Kapteyn, Design of fully spatially coherent extreme-ultraviolet light sources, *Appl. Phys. Lett.*, **84**, 3903, 2004.
72. H. Mashiko, A. Suda, and K. Midorikawa, Focusing coherent soft-x-ray radiation to a micrometer spot size with an intensity of 10^{14} W/cm^2, *Opt. Lett.*, **29**, 1927, 2004.
73. A. Siegman, New developments in laser resonators, in *Optical Resonators*, D. A. Holmes, ed., *Proc. SPIE* **1224**, 2 1990.
74. Lasers and laser-related equipment – test methods for laser beam parameters – beam widths, divergence angle, and beam propagation factor, ISO 11146: 1999(E) (International Organization for Standardization, Geneva, Switzerland, 1999).
75. Gy. Farkas and Cs. Tóth, Proposal for attosecond light pulse generation using laser induced multiple-harmonic conversion processes in rare gases, *Phys. Lett. A*, **168**, 447, 1992.
76. S. E. Harris, J. J. Macklin, and T. W. Hänsch, Atomic scale temporal structure inherent to high-order harmonic generation, *Opt. Commun.*, **100**, 487, 1993.
77. P. M. Paul, E. S. Toma, P. Breger, G. Mullot, F. Augé, Ph. Balcou, H. G. Muller, and P. Agostini, Observation of a train of attosecond pulses from high harmonic generation, *Science*, **292**, 1689, 2001.
78. M. Hentschel, R. Kienberger, Ch. Spielmann, G. A. Reider, N. Milosevic, T. Brabec, P. Corkum, U. Heinzmann, M. Drescher, and F. Krausz, Attosecond metrology, *Nature*, **414**, 509, 2001.
79. G. Sansone, E. Benedetti, F. Calegari, et al., Isolated single-cycle attosecond pulses, *Science*, **314**, 443, 2006.
80. R. L. Carman, D. W. Forslund, and J. M. Kindel, Visible harmonic emission as a way of measuring profile steepening, *Phys. Rev. Lett.*, **46**, 29, 1981.
81. R. L. Carman, C. K. Rhodes, and R. F. Benjamin, Observation of harmonics in the visible and ultraviolet created in CO_2-laser-produced plasmas, *Phys. Rev. A*, **24**, 2649, 1981.
82. D. von der Linde, T. Engers, G. Jenke, P. Agostini, G. Grillon, E. Nibbering, A. Mysyrowics, and A. Antonetti, Generation of high-order harmonics from solid surfaces by intense femtosecond laser pulses, *Phys. Rev. A*, **52**, R25, 1995.
83. S. Kohlweyer, G. D. Tsakiris, C. G. Wahlström, C. Tillman, and I. Mercer, Harmonic generation from solid-vacuum interface irradiated at high laser intensities, *Opt. Comm.*, **117**, 431, 1995.

84. P. A. Norreys, M. Zepf, S. Moustaizis, A. P. Fews, J. Zhang, P. Lee, M. Bakarezos, C. N. Danson, A. Dyson, P. Gibbon, P. Loukakos, D. Neely, F. N. Walsh, J. S. Wark, and A. E. Dangor, Efficient extreme UV harmonics generated from picosecond laser pulse interactions with solid targets, *Phys. Rev. Lett.*, **76**, 1832, 1996.

85. A. Tarasevitch, A. Orisch, D. von der Linde, Ph. Balcou, G. Rey, J. P. Chambaret, U. Teubner, D. Klöpfel, and W. Theobald, Generation of high-order spatially coherent harmonics from solid targets by femtosecond laser pulses, *Phys. Rev. A*, **62**, 023816, 2000.

86. B. Dromey, M. Zepf, A. Gopal, K. Lancaster, M. S. Wei, K. Krushelnick, M. Tatarakis, N. Vakakis, S. Moustaizis, R. Kodama, M. Tampo, C. Stoeckl, R. Clarke, H. Habara, D. Neely, S. Karsch, and P. Norreys, High harmonic generation in the relativistic limit, *Nat. Phys.*, **2**, 456, 2006.

87. A. Tarasevitch, K. Lobov, C. Wünsche, and D. von der Linde, Transition to the relativistic regime in high order harmonic generation, *Phys. Rev. Lett.*, **98**, 103902, 2007.

88. S. V. Bulanov, N. M. Naumova, and F. Pegoraro, Interaction of an ultrashort, relativistically strong laser pulse with an overdense plasma, *Phys. Plasmas*, **1**, 745, 1994.

89. S. V. Bulanov, N. M. Naumova, T. Zh. Esirkepov, and F. Pegoraro, Evolution of the frequency spectrum of a relativistically strong laser pulse in a plasma, *Phys. Scr.*, **T63**, 258, 1996.

90. R. Lichters, J. Meyer-ter-Vehn, and A. Pukhov, Short-pulse laser harmonics from oscillating plasma surfaces driven at relativistic intensity, *Phys. Plasmas*, **3**, 3425, 1996.

91. P. Gibbon, Harmonic generation by femtosecond laser-solid interaction: A coherent "water-window" light source? *Phys. Rev. Lett.*, **76**, 50, 1996.

92. D. von der Linde and K. Rzàzewski, High-order optical harmonic generation from solid surfaces, *Appl. Phys. B: Lasers Opt.*, **63**, 499, 1996.

93. V. A. Vshivkov, N. M. Naumova, F. Pegoraro, and S. V. Bulanov, Nonlinear electrodynamics of the interaction of ultra-intense laser pulses with a thin foil, *Phys. Plasmas*, **5**, 2727, 1998.

94. S. V. Bulanov, T. Zh. Esirkepov, N. M. Naumova, and I. V. Sokolov, High-order harmonics from an ultraintense laser pulse propagating inside a fiber, *Phys. Rev. E*, **67**, 016405, 2003.

95. A. S. Pirozhkov, S. V. Bulanov, T. Zh. Esirkepov, M. Mori, A. Sagisaka, and H. Daido, Generation of high-energy attosecond pulses by the relativistic-irradiance short laser pulse interacting with a thin foil, *Phys. Lett. A*, **349**, 256, 2006.

96. A. S. Pirozhkov, S. V. Bulanov, T. Zh. Esirkepov, M. Mori, A. Sagisaka, and H. Daido, Attosecond pulse generation in the relativistic regime of the laser-foil interaction: The sliding mirror model, *Phys. Plasmas*, **13**, 013107, 2006.

97. G. D. Tsakiris, K. Eidmann, J. Meyer-ter-Vehn, and F. Krausz, Route to intense single attosecond pulses, *New J. Phys.*, **8**, 19, 2006.

98. S. Gordienko, A. Pukhov, O. Shorokhov, and T. Baeva, Relativistic Doppler effect: Universal spectra and zeptosecond pulses, *Phys. Rev. Lett.*, **93**, 115002, 2004.

99. N. M. Naumova, J. A. Nees, I. V. Sokolov, B. Hou, and G. A. Mourou, Relativistic generation of isolated attosecond pulses in a λ^3 focal volume, *Phys. Rev. Lett.*, **92**, 063902, 2004.

100. J. Nees, N. Naumova, E. Power, V. Yanovsky, I. Sokolov, A. Maksimchuk, S. W. Bahk, V. Chvykov, G. Kalintchenko, B. Hou, and G. Mourou, Relativistic generation of isolated attosecond pulses: a different route to extreme intensity, *J. Mod. Opt.*, **52**, 305, 2005.

101. N. M. Naumova, J. Nees, and G. A. Mourou, Relativistic attosecond physics, *Phys. Plasmas*, **12**, 056707, 2005.

102. A. Einstein, Zur Elektrodynamik bewegter Körper, *Ann. Phys. (Leipzig)*, **17**, 891, 1905.

103. S. V. Bulanov, I. N. Inovenkov, V. I. Kirsanov, N. M. Naumova, and A. S. Sakharov, Electromagnetic wave frequency upshifting upon interaction with nonlinear plasma waves, *Kratk. Soobshch. Fiz.*, 6, 9. 1991 [Bull. Lebedev Phys. Inst.].

104. S. V. Bulanov, T. Zh. Esirkepov, and T. Tajima, Light intensification towards the Schwinger limit, *Phys. Rev. Lett.*, **91**, 085001, 2003.

105. M. Kando et al., Demonstration of laser-frequency upshift by electron-density modulations in a plasma wakefield, *Phys. Rev. Lett.*, 99, 135001, 2007.

106. A. S. Pirozhkov et al., Frequency multiplication of light back-reflected from a relativistic wake wave, *Phys. Plasmas*, 14, 123106, 2007.

107. S. S. Bulanov, T. Zh. Esirkepov, F. F. Kamenets, and F. Pegoraro, Single-cycle high-intensity electromagnetic pulse generation in the interaction of a plasma wakefield with regular nonlinear structures, *Phys. Rev. E*, **73**, 036408, 2006.

5

Nuclear and Particle Physics with Ultraintense Lasers

José Tito Mendonça and Shalom Eliezer

CONTENTS

5.1 Introduction

With the advent of ultraintense lasers, a new area was opened for nuclear and high-energy physics research (Ledingham et al., 2003). A wide variety of phenomena, such as heavy ion-induced nuclear reactions, fusion evaporation, isotope production for positron emission tomography (PET), or photon-transmutation of long-lived nuclear waste, have been studied in recent years mainly using 100 TW laser intensities (RAL, 2003). These nuclear phenomena result from the efficient acceleration processes taking place in intense laser–plasma interaction experiments, which lead to the generation of fast electron and ion beams, and the subsequent emission of high-energy photons and neutron beams (Mendonça et al., 2001). Intense laser–plasma interactions can therefore provide a new environment for nuclear physics research, but no new nuclear processes have been considered so far.

In the first part of this chapter we summarize the past achievements of the subject of nuclear and particle physics induced by ultraintense lasers. In particular, in Sections 5.2 through 5.5, the following subjects are summarized: particle production, photonuclear reactions, nuclear fission, and nuclear medicine with very intense laser beams. The important subject of nuclear fusion of hydrogen isotopes is discussed in Chapter 1 on inertial fusion energy, and therefore it is out of the scope of this chapter. In this context the neutron production was also analyzed (Norreys et al., 1998; Disdier et al., 1999; Hilscher et al., 2001).

In the second part of this chapter we review possible new nuclear processes using ultraintense lasers. We assume a double laser beam configuration, which can be of considerable interest in this area, because it allows for the possible enhancement of several types of nuclear reactions. Two counter-propagating PW (Peta-Watt) laser pulses, with a very short duration, interacting with a gas jet or a thin solid target will be considered as a typical experimental configuration. Relativistic particle-in-cell code simulations, using the Osiris code (Hemker, 2000), can help to illustrate the existence of two different interaction regions, pertinent to nuclear physics studies. In the central region of the target, the electron plasma density is considerably increased due to target compression, which leads to a good space and time resolution of the nuclear processes. Furthermore, the absence of ponderomotive forces in this central region and the increase in the electron density will eventually allow for the enhancement of the cross section of electron–nuclear interactions by Coulomb focusing, as recently discussed (Milosevic et al., 2004). This leads to the enhancement of nuclear processes in the dense plasma, as illustrated by using four different examples.

We discuss here laser-induced nuclear fission (Boyer et al., 1988), stimulated muon catalyzed fusion (Eliezer and Henis, 1988), and electron–positron pair production by electron–ion collisions (Liang et al., 1998). These three processes have already been considered before, but here we specify the appropriate plasma environment and the influence of the laser field enhancement of the nuclear cross sections. The occurrence of stimulated emission in the gamma-ray energy range by isomeric nuclei in the presence of the intense laser beam is also considered. An intense and short gamma ray pulse can eventually be generated. This is an important process, which can lead to progress in the area of gamma-ray lasers.

Another area of future application for ultraintense lasers is the exploration of high-energy physics and of the quantum vacuum properties. This will be discussed in the second part of this chapter. Vacuum is one of the most fascinating physical concepts. Several aspects of nonlinear vacuum have been studied in recent years, in the frame of quantum electrodynamics (QED). It is well known from QED that photon–photon interactions in vacuum are possible due to the existence of virtual electron–positron pairs (Heisenberg and Euler, 1936; Schwinger, 1951). This leads to nonlinear corrections of the photon dispersion relation in vacuum, and vacuum becomes a nonlinear optical medium. However, such QED corrections are extremely small, and only recently, with the advent of very intense laser systems, in the PW regime, the prospect for experimental observation of such nonlinearities started to be seriously considered. A large variety of different effects has been discussed, including photon splitting (Bialynicka-Birula and Bialynicka-Birula, 1970; Adler, 1971), harmonic generation (Kaplan and Ding, 2000), self-focusing (Soljacic and Segev, 2000), nonlinear wave (Brodin et al., 2001), and sideband generation by rotating magnetic fields. In addition, attention has been paid to collective photon phenomena (Marklund and Shukla, 2006), such as the electromagnetic wave collapse (Shukla and Eliasson, 2004; Marklund et al., 2005), as well as the formation of photon bullets (Marklund et al., 2004) and light wedges (Shukla et al., 2004). Recently, we have explored another process associated with the collective photon interactions, related to possible frequency shift of test photons immersed in a modulated radiation background (Mendonça et al., 2006a). This new process can be called photon acceleration in vacuum, because of its obvious analogies with the well-known photon acceleration processes that can occur in a plasma or in other optical media (Mendonça, 2001). Photon acceleration in plasmas is well documented by different kinds of experiments (Dias et al., 1997; Murphy et al., 2006). Another kind of nonlinear optical QED effects is associated with magnetized vacuum. Recent work (Adler, 2007; Mendonça et al., 2006b; Mendonça, 2007b) considers a rotating magnetic field. In this case birefringence is inhibited, and optical sidebands are excited, providing a possible signature of the vacuum nonlinearity.

We also consider vacuum detuning of an optical cavity, produced by an intense laser pulse. We assume that this intense laser beam crosses the empty cavity in the perpendicular direction. The optical length of the cavity is then slightly changed, due to the nonlinear QED perturbation of vacuum, and the cavity modes are frequency shifted (or frequency modulated) and detuned. This sudden change of the refractive index of a cavity can be seen as time refraction (Mendonça and Guerreiro, 2005), or alternatively as dynamical Casimir effect (Dodonov et al., 1993). What is new and relevant here is that such processes do not take place in a variable optical medium, or in a variable cavity, but in pure vacuum. The basic principles behind the proposed optical configuration rely on the phase shift amplification associated with resonant optical cavities.

Another interesting aspect of laser experiments is related to the search for pseudoscalar particles or axions. Although doubtful, the existence of axions has stimulated the construction of different kinds of experiments as well as campaigns of solar and astrophysical observations (van Bibber and Rosenberg, 2006). Recently, we have proposed the use of ultraintense lasers and have shown that the axion field, if it exists, will be parametrically unstable to the intense laser fields, which will eventually lead to an experimental proof or disproof of these particles (Mendonça, 2007a). As in most of the present review, such experimental proposals are simultaneously risky and extremely exciting. We mainly discuss highly relevant physical problems, which can now start to be explored with the present and near future laser systems.

5.2   Particle Production

The kinetic energy that the electron acquires in the laser field is of the order of the ponderomotive potential energy $U_p \sim (\gamma - 1)m_e c^2$, where $m_e c^2$ is the electron mass energy. In the relativistic regime, the relativistic factor γ is related to the quiver energy E_q and the quiver velocity v_q of the electrons in the electromagnetic field of the laser with an electric field E_L, irradiance I_L, and wavelength λ_L by

$$E_q = \gamma m_e c^2 = [(\gamma m_e v_q)^2 c^2 + m_e^2 c^4]^{1/2}, \quad v_q = \frac{2\pi \lambda_L e E_L}{\gamma m_e c} \tag{5.1}$$

$$\gamma = \left(1 + \frac{2e^2 I_L \lambda_L^2}{\pi m_e^2 c^5}\right)^{1/2} \simeq \left[1 + \frac{I_L \lambda_L^2}{1.37 \times 10^{18}\,(\text{W/cm}^2)\mu\text{m}^2}\right]^{1/2} \tag{5.2}$$

If $I_L < 10^{18}$ W/cm^2, $\lambda_L = 1\,\mu$m, then one has the nonrelativistic regime with $\gamma \sim 1$. At irradiances about 10^{21} W/cm^2 one gets $\gamma = 27$ for $1\,\mu$m laser wavelength, namely a kinetic energy of about 13 MeV.

Laser photons are absorbed by electrons (Eliezer, 2002). The origin of high-energy electrons is a subject of intensive research (Umstadter, 2001). Besides the ponderomotive force described here, there are other electron acceleration mechanisms such as wakefield acceleration, plasma wave breaking, azimuthal magnetic field, and resonance absorption. For example, it was suggested (Bertrand et al., 1994) that electron acceleration is caused by wave breaking of

Raman forward scattering of a plasma wave to achieve relativistic energies and wave breaking of Raman backscattered waves for slower velocities. Relativistic self-channeling of a picosecond laser pulse in preformed plasma near critical density has been observed both experimentally and in 3D particle-in-cell simulations (Borghesi et al., 1997). Furthermore, the electron energy distribution can be described by two temperatures (Malka et al., 1997), implying two different schemes of acceleration such as wave breaking and ponderomotive forces.

The high-energetic photons are created by bremsstrahlung of the laser-induced electrons in plasma. The relativistic Maxwellian electron energy E_e spectrum in d dimensions is given by (Davies, 2002)

$$N(E_e) \propto \gamma(\gamma^2 - 1)^{(d/2)-1} e^{-E_e/k_B T}, \quad \gamma = 1 + \frac{E_e}{m_e c^2} \tag{5.3}$$

For $\gamma \gg 1$ and $d = 3$ this leads to

$$N(E_e) = N_0 E_e^2 e^{-E_e/k_B T} \tag{5.4}$$

These electrons produce a photon spectrum via bremsstrahlung.

The protons are produced in laser–plasma interaction from contamination layers like water or hydrocarbons on the surface of a target. The very energetic electrons produced from the laser–plasma interaction ionize the impurity layer creating protons. These protons are pulled off the surface of the target by the cloud of electrons and accelerated to multi-MeV energies (Snavely et al., 2000; Mackinnon et al., 2001; Santala et al., 2001; Sakabe et al., 2004). In experiments where the target was irradiated by 1 PW laser with a peak intensity of 3×10^{20} W/cm^2, about 2×10^{13} protons were obtained with energy larger than 10 MeV and a current density of 10^8 A/cm^2, with transverse emittance smaller than 1.0π mm mrad, comparable to the CERN-Linac-2 50 MeV proton beam. The energy spectrum of the protons is a continuum with a sharp cutoff at about 60 MeV. These protons induce nuclear processes that have also been used to analyze the characteristics of the proton beam.

In general, the ions in a plasma are accelerated by a double layer (electrostatic sheath) created by the charge displacement. While for long laser pulses (~ns) the double layers are created by thermal expansion, in the case of short pulses (<ps) the ponderomotive force is responsible for the sheath formation (Eliezer and Hora, 1989; Sarkisov et al., 1999; Krushelnick et al., 1999). These ions are accelerated in the direction of maximum intensity gradient; particularly for a planar target the acceleration is perpendicular to the target.

To conclude this section we would like to mention that with a laser of 1 μm wavelength and a pulse duration of about 1 ps, one can create (Ledingham et al., 2002) at 10^{21} W/cm^2 the pion elementary particles, and at 10^{28} W/cm^2 the electron–positron pairs can be created from the vacuum.

5.3   Photonuclear Reaction

The binding energy of a neutron in a nucleus is of the order of a few MeV. In particular, the binding energies of a neutron and a proton in copper nuclei are about 10.8 and 6 MeV respectively. Therefore, a 10.8 MeV photon is required to liberate a neutron from copper; namely, the threshold for (γ, n) reaction in copper is 10.8 MeV. However, the threshold for (γ, p) is much larger than 6 MeV because of the Coulomb barrier during the proton release. The Coulomb barrier also reduces significantly the proton emission probability from the nucleus in comparison to the neutron emission.

When a large nucleus absorbs a photon, the electric field of this photon separates the protons and the neutrons inside the nucleus creating a "giant" dipole. If the laser-induced photon has the appropriate energy, a giant dipole resonance transition occurs. The lifetime of the dipole oscillation ($\sim 10^{-21}$ s) fixes the width of this resonance ($\sim$MeV).

In the first nuclear experiment at Rutherford Appleton Laboratory with the Vulcan TW laser, with a pulse energy of 100 J and intensity 5×10^{19} W/cm^2 (Ledingham et al., 2000), the following (γ, n) reaction was measured: $^{63}_{29}$Cu $+ \gamma \rightarrow {}^{62}_{29}$Cu $+ n$. The ^{62}Cu is a positron emitter with a half-life of 10 min. The emitted positron interacts with an electron and this annihilation simultaneously creates two 0.511 MeV photons that are detected in coincidence. The advantage of the coincidence measurement is the fact that the background is negligible.

Other photonuclear reactions detected in laser–plasma interaction are $(\gamma, 2n)$ and $(\gamma, 3n)$, etc, and even $(\gamma, 6n)$ in gold target was measured with PW lasers. In particular, the ratio of the nuclear activity of (γ, n) to $(\gamma, 3n)$ has been used to calculate the plasma temperature. The induced activity A of a target irradiated by high-energy photons is (Spencer et al., 2002)

$$A = \left(\frac{\ln 2}{t_{1/2}} \right) \sum_{E_\gamma} N_\gamma(E_\gamma) \sigma_\gamma(E_\gamma) N(t) \tag{5.5}$$

where

$N_\gamma(E_\gamma)$ is the number of photons in the considered energy bin, at energy E_γ
$\sigma_\gamma(E_\gamma)$ is the photonuclear reaction cross section at energy E_γ
$N(t)$ is the number of nuclei per cm^3 at time t
$t_{1/2}$ is the half-life of the radioisotope

The photons are created by bremsstrahlung of the laser-induced electrons in plasma. These electrons produce a photon spectrum via bremsstrahlung; in this way $N_\gamma(E_\gamma)$ is known for a given temperature. Comparing the ratio of theoretical activity A of the nuclei obtained from (γ, n) and $(\gamma, 3n)$ with the experimental measurement, one can fit the temperature T. In particular, for the stable tantalum isotope ^{181}Ta target (Spencer et al., 2002) irradiated by 120 J,

1 ps, 10^{20} W/cm^2 laser, a temperature of about 2.5 MeV was deduced from the activity ratio of ^{180}Ta and ^{178m}Ta, obtained from ^{181}Ta(γ, n)^{180}Ta and ^{181}Ta(γ, 3n)^{178m}Ta.

A very important nuclear photoreaction is the transmutation of the nuclear waste product ^{129}I: ^{129}I + $\gamma \rightarrow$ n +^{128}I. The half-life of ^{129}I is 15.7 million years while the half-life of ^{128}I is only 25 min. Therefore, it is evident that if we are able to transmute with laser the very long-lived nuclear waste products to a very short half-life then the nuclear waste problem would not be a burden of nuclear energy reactors. At present, about 10^6 nuclei are transmuted per PW laser shot.

5.4 Nuclear Fission

Nuclear fission induced by ultraintense laser matter interaction was suggested (Boyer et al., 1988) and detected experimentally (Cowan et al., 2000; Ledingham et al., 2000). These fission processes can be induced by high-energetic photons (photofission)

$$\gamma + A \rightarrow f_1 + f_2 + xn$$

or by the high-energetic electrons (electrofission) created in the laser-produced plasma

$$e^- + A \rightarrow f_1 + f_2 + xn + e^-.$$

where
  A is the nucleus that undergoes the fission process into nuclei f_1 and f_2
  x(= integer) neutrons n

For the E_1 giant dipole resonance matrix element, the photofission and electrofission appropriate cross sections are of the order of $\sigma_{\gamma f} \approx 10^{-25}$ cm^2 and $\sigma_{ef} \approx 10^{-27}$ cm^2.

The threshold energy of these processes is of the order of 10 MeV. The kinetic energy that the electron acquires in the laser field is of the order of the ponderomotive potential energy $U_p \sim (\gamma - 1)m_e c^2$, where $m_e c^2$ is the electron mass energy. Therefore, the fission processes will take place if $\gamma \geq 20$. The desired threshold energies for the fission processes are obtained at irradiances about 10^{21} W/cm^2 where $\gamma = 27$ (for 1 μm laser wavelength). However, the laser irradiance may be reduced if the electron acceleration in the plasma is caused by other mechanisms such as wavebreaking or wakefield acceleration.

One can estimate the number of electrofission events from the following simple consideration. The transition rate for relativistic electrons with a density

n_e is given by the transition probability P_{ef}; for a laser with a pulse duration τ_L, it is $P_{ef} = n_e \sigma_{ef} c \tau_L$. The plasma volume that the interaction occurs in is $V = \pi R_L^2 \delta = (P_L/I_L)\delta$, where πR_L^2 is the laser effective area irradiating the target, P_L is the laser power related to the laser energy W_L, $P_L = W_L/\tau_L$, and δ is the skin depth of the solid target. The number of nuclei in this volume is $V\rho_0 N_A/A$, where ρ_0 is the solid density and N_A is Avogadro's number. Furthermore, n_e is related to ρ_0 by $n_e = (Z\rho_0)/(Am_p)$, where m_p is the proton mass and Z is the number of free electrons per nucleus, assumed to be uniformly distributed within the solid plasma. Therefore, the upper-limit number of electrofission events N_{ef} is estimated by

$$N_{ef} = (n_e \sigma_{ef} c \tau_L)(\rho_0 \pi R_L^2 \delta)(N_a/A) \simeq \left(\frac{Z\rho_0^3}{2\pi^2 c m_e m_p A^3} \right)^{1/2} N_a \sigma_{ef} \qquad (5.6)$$

for $I_l \geq I_{th}(E_e \simeq 10\,\text{MeV})$, where I_{th} is the threshold irradiance to create electrons with an energy $E_e \sim 10\,\text{MeV}$. Using the $1\,\mu\text{m}$ wavelength, $W_L = 500\,\text{J}$ laser with a PW power ($10^{15}\,\text{W}$) with an intensity of $10^{21}\,\text{W/cm}^2$; this equation predicts for ^{238}U fully ionized (i.e., $Z = 92$) about 3×10^7 electrofissions per pulse and 100 times more photofissions per pulse. Using these equations one needs energy larger than $3 \times 10^9 \times 10\,\text{MeV} = 3 \times 10^{-3}\,\text{J}$ into the $10\,\text{MeV}$ (or more) energetic photons. This requirement implies a conversion ratio $E_\gamma(>10\,\text{MeV})/W_L \sim 10^{-5}$. In the uranium fission experiments (Cowan et al., 2000), a total yield of about 2×10^7 fission events per laser pulse was estimated: $260\,\text{J}$, $1.05\,\mu\text{m}$, $10^{20}\,\text{W/cm}^2$. This number is consistent with $E_\gamma(>10\,\text{MeV})/W_L \sim 10^{-7}$.

The fission events have been observed through the detection of fast fission fragments, fission neutrons, and gamma radiation from excited fission products. During the uranium fission, the nucleus splits into two unequal masses, typically around the strontium (light) and iodine (heavy) masses, as well as a few neutrons are emitted. The fission fragments are created in an excited state, and therefore they emit gamma photons, which are used to identify the fission products. In particular, the fission of ^{238}U characteristic gamma rays from the fission fragments ^{146}Ce, ^{141}Ba, ^{138}Cs, ^{138}Xe, ^{134}I, ^{134}Te, ^{128}Sn, ^{128}Sb, ^{107}Rh, ^{105}Ru, ^{104}Tc, ^{101}Tc, ^{95}Y, ^{94}Y, ^{93}Y, ^{93}Sr, ^{92}Sr, ^{88}Kr, and ^{87}Kr were observed (Cowan et al., 2000). For example, the energies of the gamma photons are 847 and 884 keV for ^{134}I, 1436 keV for ^{138}Cs, and 1384 keV for ^{92}Sr. A germanium-cooled detector is needed with a resolution of about 2 keV for the measurements of the gammas from the uranium fission products.

5.5   Nuclear Medicine

Using PW laser (at Livermore, California; RAL, United Kingdom; ILE, Osaka; and LULI, Paris), intensities of $10^{20}\,\text{W/cm}^2$ and more have been achieved.

Aluminum, gold, mylar, and other foils were used as targets with various thicknesses. From the contamination layers of water and hydrocarbons on the target surface, high-energetic protons with energies larger than 10 MeV have been produced. Beam of protons originated from the front side (where the laser irradiates the target) of the target (Clark et al., 2000; Maksimchuk et al., 2000) as well as from the back side (Snavely et al., 2000). Using an appropriate magnetic field, the protons are separated from the electrons and directed into a secondary target.

The proton beam was analyzed by stacks of copper filters. If the protons have an energy of 4.15 MeV or more, then the copper undergoes a nuclear reaction $^{63}Cu(p, n)^{63}Zn$. The created ^{63}Zn isotope is radioactive and decays with a half-life of 38.1 min via emission of a positron. The positron slows down in the solid and annihilates at rest with a free electron producing the photons of 0.511 MeV detected easily by coincidence measurement. The absolute activity A was calibrated with a known source (^{22}Na). Using the knowledge of the $^{63}Cu(p, n)^{63}Zn$ cross section and the fact that the protons slow down in the copper, the proton energy spectrum was calculated from the activity data on each copper stack (a "bin" of energy). The reaction $^{63}Cu(p, p + n)^{62}Cu$ was also detected for a proton threshold energy of 10.9 MeV.

In order to use the proton beam for further experiments, the calibration of the proton energy spectrum is done with one single foil of Cu (the natural stable copper contains 69.1% of ^{63}Cu and 30.9% of ^{65}Cu) using a germanium detector to measure the gamma spectrum of the generated isomers. In particular, the following isomers are created ^{62}Zn, ^{63}Zn, ^{65}Zn, ^{61}Cu, and ^{64}Cu via the following reactions respectively: $^{63}Cu(p, 2n)^{62}Zn$, $^{63}Cu(p, n)^{63}Zn$, $^{65}Cu(p, n)^{65}Zn$, $^{63}Cu(p, p + 2n)^{61}Cu$, and $^{65}Cu(p, n)^{64}Cu$. These reactions have different threshold energies, and their cross sections do not peak at the same proton energy. Therefore using the gamma line intensities and the half-life of the isomers, the decay branching ratios and the efficiency of the germanium detector, the energy spectrum of the protons can be calculated. This calculated spectrum is consistent with the previous measurements by the copper stack technique.

One of the suggested applications (Santala et al., 2001; Norreys, 2002; Ledingham et al., 2002) of laser-induced proton beams is the production of short-lived isotopes for PET. The positrons are used for PET diagnostics by using a pharmaceutical injection labeled with a short-lived positron emitter. In particular, the most widely used PET today is the ^{18}F-deoxyglucose (FDG), which is produced in only a few sites around the world, due to the expensive nuclear installation required to produce the F18 positron emitter. The "isotope molecules" metabolize at specific sites in the body and using the reaction $e^+ + e^- \rightarrow 2\gamma$, the gammas are imaged with an appropriate camera. Examples of appropriate nuclear reactions that create positrons are given in the table of Figure 5.1 (Ledingham et al., 2002).

The idea is to replace the expensive accelerator with a portable laser system that can produce suitable quantities of positron emitters like the ones given in Figure 5.1. If the PET pharmaceutical injection can be produced on site,

Nuclear Reaction	Positron Emitter	Lifetime (min)
$^{14}N + p \rightarrow ^{11}C + ^4He$	^{11}C	20
$^{13}C + p \rightarrow ^{13}N + n$	^{13}N	10
$^{15}N + p \rightarrow ^{15}O + n$	^{15}O	2
$^{18}O + p \rightarrow ^{18}F + n$	^{18}F	110

FIGURE 5.1
Positron-emitting nuclear reactions.

then the positron emitter with shorter lifetime than ^{18}F in the above table will be more appropriate from the clinical patient point of view. At RAL-UK, an activity per laser pulse of 10^5 Bq of ^{18}F was produced, while for PET application activities a few times 10^8 Bq is required. One possible solution (Ross et al., 1999; Chekhlov et al., 2005) is to build a high power high repetition rate (~kHz) laser that produces about 10^{20} W/cm^2 per pulse.

Another medical application for proton beam is radiotherapy. This device is presently developed with conventional accelerators. The proton beam requirements are proton energy 230–250 MeV, with an intensity of few times 10^{10} protons per second. The advantage of this radiotherapy is that the characteristics of the proton beam stopping power, namely the energy losses of the protons, are localized and thus avoid undesirable irradiation of healthy tissues. Present lasers are not suitable for the production of monoenergetic, high-energy high intensity proton beams.

5.6   Two Counterstreaming Laser Pulses

Before turning our discussion to the enhanced nuclear processes in a dense plasma, let us describe some simulations of the two laser beams configuration. We assume a uniform plasma slap with a width of $25c/\omega_{pe}$. Two counterpropagating laser pulses interact synchronously with the two opposite plasma boundaries. We assume normalized electric field amplitudes of $a_0 = 30$ and $a_0 = 100$, nearly corresponding to laser intensities of $I_0 = 10^{21}$ and 10^{22} W/cm^2, with a pulse duration of 30 fs. The laser frequency is taken as $\omega = 10\omega_{pe}$, which means that we are considering a low-density plasma. This is appropriate for the case of a gas jet. The behavior of thin solid targets can be assumed as similar, as long as the two ion shock fronts that can be created inside the overdense plasma are able to merge, and the two plasma regions are made to interact within the duration of the laser pulses. Two plane waves with equal amplitudes, linearly polarized along the axis x_3, propagate toward the target in the positive and negative x_1 directions. The simulation box has 630×630 cells, and the maximum dimensions of the target are $x_{1\,max} = x_{2\,max} = 50.4$ (c/ω_{pe}). The electron and ion densities are normalized to $n_{e0} = n_{i0} = 1(\omega_{pe}^2 m_e / 4\pi e^2)$.

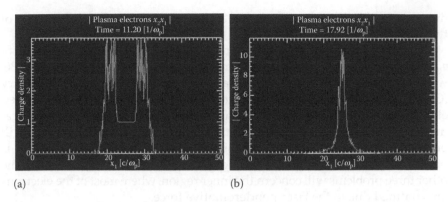

(a) (b)

FIGURE 5.2
Electron density profile in the plasma target, for (a) $t = 11.20\omega_{pe}$ and (b) $t = 17.92\omega_{pe}$.

The simulation results are illustrated in Figures 5.2 and 5.3. The first figure represents the electron density profiles $n_e(x_1)$, produced by the two laser pulses at two different instants of time, one $t_1 = 11.2/\omega_{pe}$ at the early stages of their interaction with the opposite boundaries, and the other $t_2 = 17.92/\omega_{pe}$ when the two pulses merge at the center of the target. Figure 5.3 shows the corresponding longitudinal electric field amplitude $E_1(x_1)$, for the same two instants. We clearly see that the plasma electrons are pushed by the radiation pressure of the two pulses and accumulate at the center, with a maximum density that is an order of magnitude larger than the initial uniform density. At the same time, very high longitudinal electrostatic fields of the order of 10% of the transverse field amplitude are generated as a result of charge separation. This dramatic compression of the plasma electrons at the center of the target creates two distinct target zones, a positively charged and a negatively charged one, where different nuclear processes can be enhanced, as explained below. Furthermore, the nuclear processes can be studied in this configuration with an improved space and time resolution. This is a direct result of the creation of these two distinct zones, with typical width and

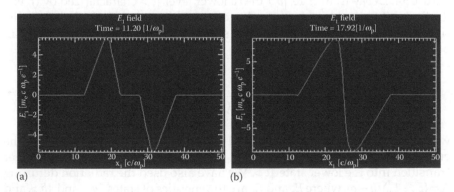

(a) (b)

FIGURE 5.3
Longitudinal electric field in the plasma target, for (a) $t = 11.20\omega_{pe}$ and (b) $t = 17.92\omega_{pe}$.

duration for the charge accumulation at the target center of the same order of the laser pulse width and duration.

The above results correspond to a rather extreme laser field amplitude of $a_0 = 100$, but code runs done using a more conservative amplitude of $a_0 = 30$, well in the range of the Astra-Gemini laser system at the Rutherford laboratory (Collier et al., 2004), show very similar results, with a maximum normalized electron density at the target center only reduced from 10.5 to 8. The practical interest of this laser target configuration for enhanced nuclear physics will be illustrated next, by considering four different problems. The first one is pertinent to the outer region of the target, almost devoid of electrons. The other three problems will concern the inner region, where most of the electrons are confined due to the laser ponderomotive force.

5.7   Enhanced Muon Fusion

The absence of electrons in the outer region makes it very useful for the study of stimulated muon catalyzed processes, because the stripped ions of deuterium (or tritium) will easily capture an incoming flux of muons. The process of molecular formation can be enhanced by the incoming laser field. The nuclear processes will occur in the following way. First, the background deuterium (or tritium) ions in the positively charged regions will capture a muon and create a $(d\mu)$ atomic state. The muonic atoms or ions then collide with a second deuterium ion to create a muon molecule. The intense photon background can stimulate the formation of muonic molecules and therefore accelerate the well-known spontaneous fusion process (Bracci and Fiorentini, 1982; Eliezer and Henis, 1994). This is due to stimulated radiative transitions from a dissociated state in the continuum of the interacting $(d\mu) + d$ to a bound state of the molecule $dd\mu$, characterized by the quantum numbers $(J, v) = (1, 1)$. These states are denoted by $|a>$ and $|b>$ in the scheme of Figure 5.4. Decay from state $|b>$ into a lower level $|c>$ characterized by $(J, v) = (0, 1)$, due to spontaneous or to Auger transitions, is also included in the figure. The transition probability for stimulated free-bound transitions can be determined by (Eliezer and Henis, 1988)

$$P_{ab} = \frac{2\pi e^2}{3c\hbar^2} I_0 \frac{\gamma_b \tau}{\Delta^2 + (\gamma_b / 2)^2} |d_{ab}|^2 \qquad (5.7)$$

where I_0 and τ are the intensity and duration of the laser pulse, γ_b is the energy lifetime of state $|b>$ associated to its spontaneous radiation and Auger transition into the lower state $|c>$. We have also used the radiation detuning $\Delta = (E_a - E_b)/\hbar - \omega$, where E_a and E_b are the energies of states $|a>$ and $|b>$, and d_{ab} are the matrix elements for dipolar transitions between these two states.

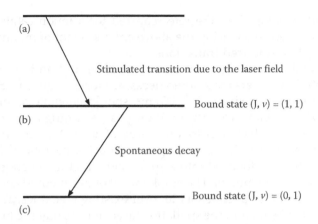

(a)

Stimulated transition due to the laser field

Bound state $(J, v) = (1, 1)$

(b)

Spontaneous decay

Bound state $(J, v) = (0, 1)$

(c)

FIGURE 5.4
Scheme for laser-induced muon molecule formation.

Using the above transition probability, we can determine the enhancement factor associated with the induced radiation

$$\eta \frac{1}{2} = \frac{P_{ab}}{P_0} = \frac{P_{ab}}{\tau \gamma_a} \tag{5.8}$$

where γ_a is the energy lifetime of the state $|a>$ of the continuum due to the spontaneous molecular formation. Notice that the energy distance between the states $|a>$ and $|b>$ for the $dd\mu$ molecule is equal to $E_a - E_b = 1.975\,eV$, which is very close to the photon energy of the second harmonic of the intense laser system (basic frequency $\hbar\omega_0 \approx 1\,eV$, therefore second harmonic photons with energy $\hbar\omega_0 \approx 2\,eV$).

Using the tabulated value of the matrix element d_{ab} for the molecule $dd\mu$, and considering the resonance condition $\Delta = 0$, we obtain the impressive enhancement factor of $\eta \approx 10^{21}$, for a laser intensity of $I_0 = 10^{20}\,W/cm^2$. For a 10% detuning, $\Delta = 0.1\,(E_a - E_b)/\hbar$, the enhancement factor reduces to $\eta = 10^{10}$. However, notice that for very short pulses, with a large spectrum of frequencies, we will have exact resonant conditions for a significant energy range of unbounded states $|a>$. This shows that in future experiments of laser-induced muon catalyzed fusion, we expect to observe an intense burst of fusion processes within the duration of the laser pulse. The muon fusion catalyzed processes in the double laser beam configuration will be enhanced further by the larger probability for muon capture in the positive region of the target.

It should be noticed that the existing muon sources only produce a small number of particles, of about 10^5 muons per second (Matsuzaki et al., 2001). These are generated in a volume of $25\,cm^3$, which is clearly much larger than the focal volume of the counterpropagating PW laser beams considered here. The present day spatially integrated production rate is therefore many orders of magnitude smaller than the rate required for a muon catalyzed fusion reactor, even when considering the above enhancement of muonic molecule formation by laser-induced transitions.

One way that the muon generation rate could be substantially enhanced is the use of 70 kJ PW lasers considered necessary for fast ignition. Such a laser pulse, focused to intensities $>10^{23}$ W/cm^2, should generate as many as 10^{14} protons of about 1 GeV from thin foil targets (assuming a reasonable conversion efficiency of PW laser to ion energy of 20%). A proton beam of this charge, generated over 10 ps, corresponds to a kA current of ions. These beams could then be focused into a spallation target to generate pions that eventually decay into muons. The employment of this method could increase the production rate by as much as 10^7 over existing machines. The ability to focus the proton beam into the spallation target means that the luminosity of the muon source should be increased, although this clearly needs verification by Monte Carlo simulation methods in the future.

The enhancement described here can be very useful for an understanding of the temporal scales of the molecular formation and subsequent fusion reactions associated with the muon catalysis chain. A different type of laser-induced muon processes has previously been considered (Chelkowski et al., 2004) based on the tunnel dissociation of the muonic molecules. This process has, in principle, a much better time resolution, of the order of the laser period, but its transition probabilities are quite low and will be submerged by the more efficient process considered here. However, the above simulations suggest an interesting extension of the tunneling process, by considering the additional tunneling due to the longitudinal electric fields that are created in the central region of the target. These fields are weaker than the laser oscillating field, but are quasistatic and could lead to a nonnegligible increase of tunnel ionization.

5.8   Enhanced Fission Reactions

Let us now turn our attention to the central region of the target where most of plasma electrons have been accumulated due to the counterpropagating laser ponderomotive forces. In this region, in addition to the increasing number of electrons, we have an additional enhancement factor for electron collisions with the nuclei. This is due to the focusing effect of the Coulomb field on the electron wave function (Milosevic et al., 2004). It should be noticed that this reference erroneously attributes to the Lorentz force the reduction of close collisions that is due to the laser ponderomotive force. The Coulomb field focusing effect has previously been considered for lower energy atomic processes. This Coulomb focusing field leads to an increase in the electron–nucleus collision cross section in the central region of the target. The associated enhancement factor can be estimated as $f_{enh} \sim 10^{3-4}$ (Milosevic et al., 2004). Notice however that this enhancement factor can only be applied to the electrons that were not affected by the ponderomotive force compression and are therefore free to move back to the proximity of their nuclei. As an important application of Coulomb focusing, we consider fission reactions of the type

$$e^- + A \rightarrow f_1 + f_2 + \bar{v}n + e^- \tag{5.9}$$

where
    A is the nucleus
    f_1 and f_2 are the resulting fission fragments

In the central region of the target, the electron current driven by the laser field can be written as

$$J = -ec[n_0 f_{enh} + (n_e - n_0)] \tag{5.10}$$

where
    n_0 is the initial plasma density before laser compression
    n_e is the actual electron pick density observed in the center

This expression is valid for the inner plasma region, where $n_e > n_0$.

As an example, let us consider the case of uranium fission, $A = {}^{238}U$. The electrofission cross section is $\sigma_{ef} \sim 1\,mb$, and the energy threshold is of the order of 10 MeV (Boyer et al., 1988). The transition probability for fission events can be estimated by $P_{ef} = |J|\ \sigma_{ef}\tau$, where τ is the duration of the pulse. The resulting fission yield can then be determined by $Y_{ef} = \sigma_0 \delta P_{ef}\varepsilon$, where σ_0 is the density of the uranium atoms in the target, δ the width of the target, and ε the energy release per fission reaction. In a thin solid target, for a laser intensity of 10^{21} W/cm², we obtain $Y_{ef} \sim 20\,\mu J$, which would correspond to the impressive number of 10^6 events per laser shot. This confirms the results of Boyer et al. (1988), enhanced here by Coulomb focusing. This shows that the double PW laser beam is a very promising configuration for laser-enhanced nuclear fission reactions. As a final remark, we should remind here that the fission reactions for uranium have a threshold at 10 MeV. We have therefore to assume that the maximum kinetic energy of the plasma electrons in the PW laser field, determined by $\varepsilon = m_e c^2 (1 + a_0^2)^{1/2}$, is higher than the fission reaction threshold, which is compatible with moderately high values of the laser parameter a_0.

5.9 Enhanced Pair Production

As a third example, we consider the case of electron–positron pair creation by laser–plasma interaction at the center of a laser-compressed target. This was first proposed by Liang et al. (1998), but we can consider different plasma conditions and study the influence of two enhancement factors: electron density increase at the center of the target and Coulomb focusing of the uncompressed electron population, when the electrons oscillate around the nearly immobile high Z ions. It has been shown that, for high intensity laser–plasma interactions, the main pair production mechanism is that due to close electron–ion collisions (Nakashima and Takabe, 2002). This is the so-called Trident process

$$e^- + Z \rightarrow e^+ + 2e^- + Z \tag{5.11}$$

which dominates over all the other pair creation processes. The production rate of electron–positron pairs by the Trident process, in the central plasma region, can be defined by

$$\frac{\mathrm{d}}{\mathrm{d}t} n_\pm = [n_p + n_0 f_{\text{enh}} + (n_e - n_0)] < n_0 f(v) v \sigma_{\text{ei}} > \tag{5.12}$$

where
   $n_\pm$ is the density of pairs
   n_p and n_0 are the densities of positrons and ions
   $f(v)$ is the electron distribution function
   v the electron velocity

The electron–ion cross section for pair creation is given by (Heitler, 1954) $\sigma_{\text{ei}} = 1.4 \times 10^{-30} Z^2 (\ln \beta \gamma)^3$, where Z is the number of electrons in the neutral atom of the target, $\beta = v/c$, and $\gamma = (1 - \beta^2)^{-1/2}$. For a laser intensity of 10^{21} W/cm^2, we get a pair creation rate of nearly 10^{11} s^{-1}, which means that, for a pulse duration of 50 fs we could obtain a positron density of $n_\pm = n_p \sim 10^{-3} n_e$. This estimate is similar to that given by Liang et al. (1998), but for a much shorter laser pulse. The enhancement factor valid in our configuration can compensate much shorter pulse durations. The double PW laser configuration of the Astra-Gemini system is therefore a plausible candidate for the intense production of electron–positron pairs, with densities approaching 0.1% of the electron density at the center of a compressed target.

5.10   Stimulated Gamma-Ray Emission

Let us now consider, as a last but important example of nuclear physics in the PW laser regime, the possible occurrence of stimulated emission of gamma ray photons from isomeric nuclei. The reduction of the nuclear isomer lifetimes by intense laser was proposed some years ago (Arad et al., 1979; Becker et al., 1984; Eliezer et al., 1995). It is based on the principle of induced nuclear photoemission that has been established experimentally (Collins et al., 1988) by the depopulation of the isomeric state ^{180}Ta^m in the reaction ^{180}Ta^m (γ, γ')^{180}Ta. In this experiment, hard x-rays were obtained from accelerator-induced transitions. However, we can replace the accelerator by the intense laser pulse, which accelerates the plasma electrons existing at the center of the target to several MeV.

The number of excited nuclei N_γ capable of providing the appropriate gamma-ray emission is given by the well-known formula

$$N_\gamma = N_A F \tau \sigma \tag{5.13}$$

where

N_A is the number of atoms that are irradiated

F is the flux of incident particles (e.g., photons or electrons) crossing the unit cross-sectional area per unit time

τ is the time duration of the flux

σ is the cross section of the process under consideration

Let us consider a three-level nuclear process. The cross section for nuclear anti-Stokes scattering (see Figure 5.5) is given by

$$\sigma = \frac{\lambda^2}{8\pi} \Gamma_{ba} \Gamma_{ac} \left(\frac{\hbar\omega}{E_b - E_a} \right)^{2L_1+1} \left(\frac{E_c - E_a + \hbar\omega}{E_b - E_a} \right)^{2L_2+1} \left[(\hbar\Delta)^2 + \left(\frac{\Gamma_b}{2} \right)^2 \right]^{-1} \tag{5.14}$$

Here we have considered an isomeric state "a," defined by the energy level E_a, line-width $\Gamma_a = \hbar/\tau_a$ (thus, τ_a is the lifetime of level "a"), spin J_a, and parity π_a (which can be denoted by $J_a^{\pi a}$ in the diagram of Figure 5.5). This nucleus absorbs a low-energy photon and changes into an upper-energy state "c," with energy E_c, line-width Γ_c, and spin–parity $J_c^{\pi c}$. Following this transition, level "c" will decay radiatively into the final state "b," with energy E_b, line-width Γ_b, and spin–parity $J_b^{\pi b}$, emitting a high-energy photon.

The a–c and c–b transitions have appropriate line-widths of $\Gamma_{ac} = \Gamma_a + \Gamma_c$ and $\Gamma_{cb} = \Gamma_b + \Gamma_c$. The cross section (Equation 5.14) describes the absorption of a photon with an energy $\hbar\omega = hc/\lambda$, where λ is the photon wavelength, ω is its angular frequency, and h is Planck's constant. The integers L_1 and L_2 are the multipolarities of the electromagnetic transitions between the levels a–c and

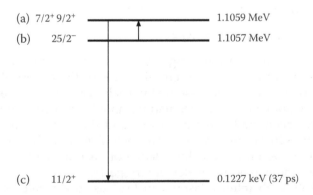

(a) $7/2^+ 9/2^+$ ——— 1.1059 MeV

(b) $25/2^-$ ——— 1.1057 MeV

(c) $11/2^+$ ——— 0.1227 keV (37 ps)

FIGURE 5.5
Scheme for stimulated gamma-ray emission.

c–b, respectively. The detuning factor $\hbar\Delta = E_c - E_a - \hbar\omega$, which is the energy difference from the resonance, plays a crucial role in the numerical evaluation of the anti-Stokes resonance. In resonant conditions, defined by

$$\hbar\Delta = \gamma_c, \quad \hbar\omega = E_c - E_a \tag{5.15}$$

Equation 5.14 yields, for our example, the following value for the cross section:

$$\sigma = \left(\frac{\lambda^2}{10\pi}\right)\frac{\tau_c}{\tau_a} \simeq (1.23 \times 10^{-14}\,\text{cm}^2)4.62 \times 10^{-16} \simeq 5.68 \times 10^{-30}\,\text{cm}^2 \tag{5.16}$$

In Figure 5.5, the anti-Stokes transition of the isomer ^{179}Hf is considered. This isomer has an excited state with energy 1.1057 MeV and a half-life of 25.1 days. Very close to this nuclear state, at a distance of 200 eV above this state "b," there is another energy level, which is unstable and decays by emitting a gamma ray with energies of 0.8195, 0.9832, and 1.1059 MeV, with a ratio of 6%, 24%, and 70% respectively. In the figure only the fastest radiative transition is represented. The multipolarity of the excitation is $L_1 = 8$ (or 9), and from the excited state to the ground state the polarity is $L_2 = 1$.

The ^{179}Hf isotope does not occur in nature and therefore it as to be produced by an accelerator. The isomers have been created by the nuclear reactions (Hubel et al., 1970; Hubel et al., 1971) ^{181}Ta(d, α)^{179}Hf and ^{176}Yb(α, n)^{179}Hf. A number as large as $N_a = 5 \times 10^{15}$ of isotope atoms can be obtained in a sample, which can be used as a thin target for the two counterpropagating PW laser beams. This would create a strong burst of gamma-ray emission, which could be highly relevant to gamma ray laser research. Experimental work on isomer nuclear physics in ultraintense laser fields has already been accessed (Andreev et al., 2000).

5.11 QED Vacuum Processes

Vacuum is one of the most exciting concepts in contemporary physics. Not only because it conveys strange reminiscences of the historically outdated concept of ether, but also because real particles can materialize in vacuum under intense applied fields. Speculations about the energy contents of the vacuum have been made over the years, and the Casimir force, resulting from differences in vacuum pressure at a material interface, has been predicted and measured. Definitely, our knowledge of vacuum will be an essential aspect of our understanding of elementary particles.

Experiments with intense lasers can lead us to a better exploration of quantum vacuum, specially the photon and electron aspects of quantum vacuum, which are described by the theory of QED. We can discern three kinds of QED phenomena that can be studied with the next generation of intense laser systems. For simplicity, we will call them (1) the low-energy range of photon QED vacuum, (2) the intermediate-energy range of virtual pair creation, and (3) the high-energy range of vacuum disruption. After a short presentation of these three ranges of phenomena, we discuss several different vacuum processes associated with them.

The low-energy QED covers the domain of the so-called quantum optics. It only involves photon emission phenomena and the virtual electron–positron pairs can be neglected. In such a domain, three distinct processes can be considered: (1) dynamical Casimir effect (Moore, 1970; Dodonov et al., 1989, 1993; Uhlmann et al., 2004), (2) Unruh–Davies radiation (Davies, 1975; Unruh, 1976; Birrel and Davies, 1982), and (3) time refraction (Mendonça et al., 2000, 2003). Dynamical Casimir results from an extension of the double plate geometry of the famous Casimir effect (Casimir, 1948; Milton, 2004), which revealed the energetic contents of vacuum. Unruh–Davies radiation (usually called Unruh radiation) demonstrates the existence of thermal radiation seen in an accelerated reference frame in vacuum. It explores the equivalence between gravitation and acceleration and is intimately related to the Hawking radiation (Bekenstein, 1973; Hawking, 1974,1975). Finally, time refraction is the temporal version of refraction. It can be seen as the basic mechanism leading to photon acceleration (Mendonça, 2001). As a result of time refraction, superluminal frames with constant velocity can also observe a radiation spectrum resembling the Unruh radiation (Guerreiro et al., 2005).

For intermediate energies, corresponding to laser intensities well below the Schwinger limit, but where the influence of virtual electron–positron pairs is retained, photon–photon interactions in vacuum can be studied (Heisenberg and Euler, 1936; Schwinger, 1951; Dittrich and Gier, 2000). Due to the excitation of virtual pairs, vacuum behaves as a nonlinear optical medium. Such nonlinearities are extremely weak, but can eventually be observed with the existing or with the proposed ultraintense laser systems. These nonlinear effects include photon splitting (Bialynicka-Birula and Bialynicka-Birula, 1970; Adler, 1971), second harmonic generation (Kaplan and Ding, 2000), self-focusing (Soljacic and Segev, 2000), nonlinear wave mixing (Brodin et al., 2001), and sideband generation by rotating magnetic fields (Adler, 2007; Mendonça et al., 2006b; Mendonça, 2007b). Finally, for laser energies approaching the Schwinger limit, production of real electron–positron pairs can start, and vacuum disruption can eventually be attained. Some of these processes will be described next.

5.12 Unruh Radiation

According to Unruh (1976) and Davies (1975), an observer moving in vacuum with uniform acceleration $\dot{v}$ will see thermal radiation at a temperature proportional to the acceleration, as given by this surprisingly simple formula:

$$k_B T = \frac{\hbar \dot{v}}{2\pi c} \tag{5.17}$$

This is a kinematic effect associated with the accelerated reference frame, and its real existence has been under debate (Ford and O'Connell, 2005). However, there is a large consensus that any detector moving in the accelerated frame could be able to observe such thermal radiation. The detector could be an electron, an atom, or an accelerated ionization front.

One of the main difficulties related to the possible observation of this effect is the small value of the constant multiplying the acceleration $\dot{v}$ in the above equation. An observable thermal effect would need an extremely high acceleration, but several experimental schemes have been proposed. In what concerns the use of intense lasers, two different schemes should be retained. One uses the accelerated ionization fronts produced by intense lasers in gaseous or solid targets (Yablonovich, 1989). In this case, the particles of the medium could stay at rest, while the boundary between the neutral and the ionized part of the target would be accelerated up to relativistic speeds.

The other proposed experimental scheme concerns the use of two identical ultraintense laser beams in counterpropagation (Chen and Tajima, 1999), in a similar configuration to that considered above for the enhanced nuclear phenomena, but now with vacuum replacing the solid target. At the antinodes of the resulting standing wave, an electron could be accelerated to relativistic speeds at every laser cycle $2\pi/\omega_0$. For PW lasers this would lead to a maximum acceleration of 10^{25} g. An estimate of the number of radiated photons can be made by using a quasiclassical approach, where the electrons are assumed to interact with classical vacuum fluctuations (Boyer, 1980). The expected number of photons with frequency ω radiated per unit of time is (Chen and Tajima, 1999)

$$\frac{dN(\omega)}{dt} = \frac{3e^2}{2m_e c^3} \frac{\omega_a^2}{[\exp(2\pi\omega/\omega_a)-1]} \left[2 + \left(\frac{\omega}{\omega_a} \right)^2 \right] \tag{5.18}$$

where
$\omega_a = 2a_0\omega_0$
$a_0 = eE_0/m_e c\omega_0$ is the normalized laser electric field amplitude

This Unruh radiation process is only relevant over a short time interval inside of each laser period, which is of the order of $1/a_0$. On the other hand, this radiation mechanism competes with the usual Larmor radiation, associated with the accelerated charged particle. Because of the well-known angular dependence of the Larmor radiation, Unruh process would still dominate inside a very small solid angle of the order of $10^{-3}/a_0$. In a typical PW experiment with $a_0 \sim 10^2$, this would make the experiment very difficult to perform. Progress in our theoretical understanding of the Unruh process has still to be made. For instance, a recent quantum model has shown that, in contrast with Larmor radiation, photons produced by the Unruh effect are emitted in pairs (Schutzhold et al., 2006), which would help to discriminate between the two processes. This property is actually shared with the photons emitted from vacuum by time refraction (Mendonça and Guerreiro, 2005), which are also emitted in pairs.

5.13 Time Refraction

Let us now consider time refraction due to the temporal change in the refractive index of an arbitrary optical medium. This can be discussed both at classical and quantum levels. The classical properties imply the occurrence of beam splitting and photon reflection, while the quantum description predicts radiation of pairs of photons from vacuum. This new radiation process can be related with Unruh, but with a fundamentally different nature, because it is not related to the existence of any accelerated frame.

Let us also assume that the refractive index of an optical medium starts changing at time $t = 0$, due to some external agent. The medium is considered unbounded and uniform, but conversion to a bounded medium is straightforward, and the case of an empty cavity will be discussed later. We can describe the variation of the refractive index by a generic function of time, $n(t)$. For a given electromagnetic field mode propagating in the Ox-direction in such a medium, the electric field is described by

$$\vec{E}(x,t) = \left[\vec{E}(t)e^{-i\phi(t)} + \vec{E}'(t)e^{i\phi(t)} \right] e^{ikx} + cc \qquad (5.19)$$

with the phase function

$$\phi(t) = \int_0^t \omega(t')dt' \qquad (5.20)$$

Here, $\vec{E}$ and $\vec{E}'$ are the field amplitudes for waves propagating in the positive and negative Ox-directions, respectively. The time-dependent value for the mode frequency ω satisfies the instantaneous dispersion relation

$$\omega(t) = kc/n(t) \qquad (5.21)$$

This expression can be interpreted as a temporal Snell's law (Mendonça, 2001), because it relates the wave frequencies at two different times $t_1 \neq t_2$, as $\omega(t_1)n(t_1) = \omega(t_2)n(t_2)$, whereas the usual Snell's law for (space) refraction relates the wavevectors in two different media. According to our work (Mendonça and Guerreiro, 2005), the temporal evolution of the electric field is determined by

$$\frac{dE}{dt} = -\frac{1}{2n}\frac{dn}{dt}\left[3E + E' \exp\left(+2i\int \omega(t')dt'\right)\right] \qquad (5.22)$$

and

$$\frac{dE'}{dt} = -\frac{1}{2n}\frac{dn}{dt}\left[3E' + E \exp\left(-2i\int \omega(t')dt'\right)\right] \qquad (5.23)$$

The physical meaning of these equations can be understood by considering the initial conditions with $E' = 0$, where the waves initially propagate in the forward direction. For arbitrary times, we assume that the wave propagating in the forward direction still dominates over the wave propagating in the backward direction, $|E| \gg |E'|$, which is valid for a slowly varying medium. In this case we get

$$E(t) \simeq \left[1 - \frac{3}{2}\int_0^t \frac{1}{n(t')}\frac{dn}{dt'}dt'\right]E(0) \equiv T(t)E(0) \qquad (5.24)$$

where we can define a temporal transmission coefficient, $T(t) \sim 1$. Considering now Equation 5.23, and assuming that $E(t) \sim E(0) = $ const., we can also obtain the reflected field resulting from the nonstationarity of the medium, of the form $E'(t) = R(t)E(0)$, where the temporal reflection coefficient is determined by

$$R(t) \simeq -\frac{E(0)}{2}\int_0^t \frac{1}{n(t')}\frac{dn}{dt'}\exp\left(-2i\int^{t'}\omega(t'')dt''\right)dt' \qquad (5.25)$$

These expressions for the temporal transmission and reflection coefficients, $T(t)$ and $R(t)$, are formally analogous to the well-known coefficients for stationarity by nonhomogenous media $T(x)$ and $R(x)$ (Ginzburg, 1961), but with the space coordinate x replaced by the time coordinate t (Figure 5.6).

We can now consider the quantum regime of time refraction. A temporal change in the refractive index will lead to the creation of photon pairs. For an arbitrarily time varying refractive index, $n(t)$, the average number of photon pairs created at a given wavelength $2\pi/k$, per unit volume, is given by (Mendonça and Guerreiro, 2005)

$$N_k(t) = \sinh^2[r(t)] \qquad (5.26)$$

where the argument $r(t)$ is determined by

$$r(t) = \int_0^t \left(\frac{d}{dt'}\ln n\right)^2 \frac{dt'}{\omega(t')} \qquad (5.27)$$

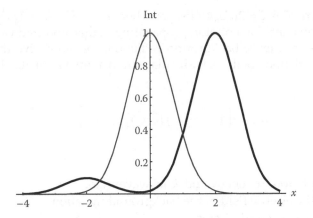

FIGURE 5.6
Time refraction: an initial Gaussian pulse (at $t = 0$) suffers a temporal split that can be observed (at a time $t > 0$), as shown by the curve in bold.

This result is valid for an unbounded medium. However if we consider an optical cavity, it can be related to the dynamic Casimir effect.

On the other hand, time refraction can be seen as a limiting case of a medium with an infinite moving boundary. By using a Lorentz transformation, time refraction can be transformed into a medium with a superluminal boundary, with velocity $u > c$. The velocity of the Lorentz frame is simply $v_\infty = -c^2/u$. It can then be shown that a superluminal boundary moving with constant velocity is capable of emitting light from vacuum (Guerreiro et al., 2005). This type of radiation is similar, but in some sense complementary to the Unruh radiation. Acceleration is not required and is replaced by superluminality. It is known that ionization fronts with arbitrary and eventually superluminal velocities can be produced by intense lasers in a target, and they could be used to study these quantum vacuum properties in the low-energy regime. The high- and intermediate-energy regimes will involve the existence of real or virtual pair electron–positron pairs, as discussed next.

5.14 Nonlinear Vacuum

The nonlinear QED effects associated with the excitation of virtual electron–positron pairs in vacuum can be described by the Heisenberg–Euler Lagrangian density. This can be written as a nonlinear quantum correction $\delta\mathcal{L}$ to the usual classical electromagnetic Lagrangian density $\mathcal{L}_0$, as (Itzykson and Zuber, 1980) $\mathcal{L} = \mathcal{L}_0 + \delta\mathcal{L} = \varepsilon_0\mathcal{F} + \zeta(4\mathcal{F}^2 + 7\mathcal{G}^2)$, where the two relativistic invariants $\mathcal{F}$ and $\mathcal{G}$ are determined by $\mathcal{F} \equiv \mathcal{L}_0/\varepsilon_0 = \frac{1}{2}(E^2 - c^2B^2)$ and $G = c(\vec{E}\cdot\vec{B})$. Here $\vec{E}$ and $\vec{B}$ are the electric and magnetic fields. The nonlinear parameter appearing in $\delta\mathcal{L}$ is $\zeta = 2\alpha^2\varepsilon_0^2\hbar^3/45m_e^4c^5$, where $\alpha = e^2/2\varepsilon_0\,hc \approx 1/137$ is the fine structure constant. For a photon propagating in this modified vacuum in the presence of an intense background radiation, we can derive the following dispersion relation (Bialynicka-Birula and Bialynicka-Birula, 1970; Brodin et al., 2001)

$$\omega = kc\left[1 - \frac{1}{2}\varepsilon_0\lambda_\pm f(\vec{k},\vec{k}')\,|\,E(\vec{k}')\,|^2\right], \qquad (5.28)$$

where

ω is the photon frequency and $\vec{k}$ its wavevector
$\vec{E}(\vec{k}')$ is the electric field of the background radiation
$f(\vec{k},\vec{k}')$ is a geometric factor
$\lambda_+ = 14\zeta$ or $\lambda_- = 8\zeta$, according to the photon polarization state

The resulting vacuum refractive index $n = kc/\omega$ is then given by

$$n(\omega,\vec{k},t) = 1 + \frac{2\hbar\lambda_\pm}{c}\int f(\vec{k},\vec{k}')k'N(\vec{r},\vec{k}',t)\frac{d\vec{k}'}{(2\pi)^3} \qquad (5.29)$$

where we have introduced the photon occupation number, $N(\vec{k}') = (\varepsilon_0 c/4\hbar k')$ $|\,E(\vec{k}')\,|^2$.

5.15 Birefringence

Let us now consider a static magnetic field $\vec{B}_e \equiv \vec{B}_0$ in vacuum. In this case, the dispersion relation for a photon state with frequency ω and wave vector $\vec{k}$ also depends on the direction of its polarization vector $\hat{e} = \vec{E}/|\,E$. For polarizations parallel and perpendicular to the static magnetic field, the photon refractive index will be

$$n_\| = 1 + 7\mu_0 c^4\zeta B_0^2, \quad n_\perp = 1 + 4\mu_0 c^4\zeta B_0^2\sin^2\alpha \qquad (5.30)$$

where α is the angle between the static field $\vec{B}_0$ and the direction of propagation, defined by $\vec{k}$. This is known as vacuum birefringence (Bialynicka-Birula and Bialynicka-Birula, 1970; Adler, 1971). For propagation along the static field we reduce to $n_\| = n_\perp = 1$, which means that Faraday rotation is forbidden.

The situation suffers an important qualitative change if we consider a rotation magnetic field. Let us then assume the case of a static but rotating magnetic field $\vec{B}_e(t) = B_0\hat{b}_0(t)$ with a constant amplitude B_0, where the unit vector $\hat{b}_0(t)$ rotates with a very small angular frequency $\omega_0 \ll \omega$ in the plane perpendicular to the photon wavevector $\vec{k}$. In other words, it rotates in the plane of wave polarization. In this case the photon dispersion relation in the rotating magnetized vacuum is given by

$$\left(k^2 - \frac{\omega^2}{c^2}\right) = 11\frac{\omega^2}{\varepsilon_0}\zeta\,|B_0\,|^2 \tag{5.31}$$

which corresponds to a modified refractive index of

$$n = 1 + \frac{11}{2}\mu_0 c^4 \zeta \left|B_0^2\right| \tag{5.32}$$

We can see that rotation of the magnetic field inhibits birefringence. This results from the fact that, in a rotating field, the parallel and perpendicular polarization states change with time and are mixed together. Therefore, if we replace B_0 by $B_e(t)$ in Equation 5.30, and average over a rotation period $2\pi/\omega_0$, we get Equation 5.32.

5.16 Sidebands

However, another effect also occurs in the rotating magnetic field $\vec{B}_e(t)$. The nonlinear terms in the Heisenberg–Euler Lagrangian will couple photon modes with frequencies $\omega_n = \omega + 2n\omega_0$ and generate a cascade of sidebands. The different mode amplitudes are determined by (Mendonça, 2007b)

$$\frac{d}{d\tau}E_n = iw_n(E_{n-1}e^{i2\phi} + E_{n+1}e^{-i2\phi}) \tag{5.33}$$

where the phase mismatch is determined by $\varphi = -\omega_0 z/c$, and the coupling constants are

$$w_n = \frac{3}{4}\frac{c^2}{\varepsilon_0}\zeta\omega_n\,|B_0\,|^2 \tag{5.34}$$

For small interacting distances, $z \ll \lambda(2\omega_0/\omega)$, we can neglect the phase mismatch and obtain simple analytical solutions for the electric field amplitudes E_n. Solutions compatible with initial conditions such that $E(\omega, \tau = 0) \equiv E_0$ and no sidebands initially exist, $E_n(\tau = 0) = 0$ are then given by

$$E_n(\tau) = E_0 J_n(w\tau)e^{in\pi/2} \tag{5.35}$$

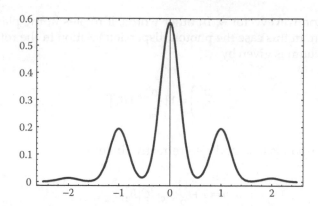

FIGURE 5.7
Laser sidebands generated in vacuum: relative intensity versus twice the sideband frequency $2\omega_0$, for the extreme case $w\tau = 1$.

where J_n are the Bessel functions of the first kind. Using the asymptotic expressions for the Bessel functions with small arguments we can estimate for the first sideband $n = 1$ the following amplitude

$$\left| \frac{E_1}{E_0} \right| = \frac{3}{8} \frac{\alpha}{45} \left(\frac{B_0}{B_{\text{crit}}} \right)^2 \frac{\Delta z}{\lambda} \qquad (5.36)$$

where
  Δz is the optical path
  α is the fine structure constant
  $B_{\text{crit}} = m^2 c^2 / e\hbar$ is the critical field

This could be applied to the PVLAS (Polarizzatione del Vuoto con LASer, in Italian) configuration (Zavattini et al., 2006). But a more convenient configuration for intense laser experiments could be that using high harmonics (produced with a solid target), in order to minimize the wavelength λ, and a very intense magnetic field over a distance of a few meters. With improved accuracy, the observation of the second sideband E_2 would eventually help to clarify the observations (see Figure 5.7).

5.17 Photon Acceleration in Vacuum

In the geometric optics approximation, the dynamics of photons in vacuum can be described by the ray equations, which can be stated in the canonical form as

$$\frac{d\vec{r}}{dt} = \frac{\partial \omega}{\partial \vec{k}}, \quad \frac{d\vec{k}}{dt} = -\frac{\partial \omega}{\partial \vec{r}} \qquad (5.37)$$

where the Hamiltonian ω is determined by Equation 5.28. Let us consider the important particular case where the background radiation is dominated by a single photon beam propagating in direction Oz. This means that we can make $\vec{k}' = k'\vec{e}_z$. If the probe photon described by these equations of motion propagates at a given angle θ with respect to the z-axis, we have $(\vec{n} \cdot \vec{n}') = \cos\theta$ and $(\vec{n} \cdot \vec{e}') = \sin\theta \cos\psi$, where ψ is the second angle necessary to define the direction of $\vec{n}$ with respect to the background electric field. In this case, the geometric factor is $f(\vec{k},\vec{k}') \equiv f(\theta) = [2(1-\cos\theta) - \sin^2\theta]$. The photon dispersion relation becomes

$$\omega(\vec{r},\vec{k},t) = kc\left[1 - \frac{1}{2}\lambda^* f(\theta)I(\vec{r},t)\right] \qquad (5.38)$$

where $\lambda^* = 4\hbar\lambda_+/c$, the angle θ can vary in space and time according to the photon dynamics described by Equation 5.37, and the background field intensity $I(\vec{r};t)$ is determined by the integral.

$$I(\vec{r},t) = \int k'N(\vec{r},\vec{k}',t)\frac{d\vec{k}'}{(2\pi)^3} \qquad (5.39)$$

Let us assume that the beam is modulated in intensity along the propagation direction, and that we can write $I(z - ut, \vec{r}_\perp)$. For a very large beam waist, we can neglect the dependence over the transverse direction and assume that $u = c$. From the equations of motion (Equation 5.37), we can establish a relation between the initial value of the frequency $\omega_0 \equiv \omega(t = 0)$ and a subsequent value $\omega(t)$, such that (Mendonça et al., 2006)

$$\omega(t) = \omega_0 \frac{1 - \delta n(t)}{(1 - \delta n_0)} \frac{(1 - \cos\theta_0 - \delta n_0)}{1 - \cos\theta(t) - \delta n(t)}. \qquad (5.40)$$

This expression can be used to calculate the frequency shift of a test photon interacting with the radiation background. Such a frequency shift can be associated with a process of photon acceleration. Notice that the group velocity of the photon as determined by the nonlinear dispersion relation, and given by $\vec{v} = \vec{n}c[1 - \delta n(t)]$, is independent of the photon frequency, but changes with time due to the change in the angle between the photon and the background beam.

The very weak nonlinear QED force acting on the test photon and due to the gradient of the beam intensity (represented by the derivative of background intensity profile) acts on the probe photon over very large distances, because it travels with a parallel velocity nearly equal to the light speed c, eventually leading to a nonnegligible frequency upshift. This could eventually be observed in future experiments using ultraintense laser pulses.

A closely related problem is that of time refraction discussed above, but in this case it is for pure vacuum conditions. For waves propagating in vacuum, the changes in refractive index can only be due to the nonlinear QED effects. Assuming that the probe photons propagate perpendicularly to a strong laser pulse, two different effects can take place: frequency shift and photon reflection. These effects are extremely small, but if observed they would give us direct information of the nonlinear vacuum. The frequency shift is determined by the temporal Snell's law (Equation 5.21), or

$$\frac{\delta\omega(t)}{\omega(0)} = \delta n(t) = -\frac{\lambda^*}{2} f(\theta)I(\vec{r}, t) \tag{5.41}$$

This is a very tiny frequency shift, which can however be amplified with interferometric techniques. If the frequency shifted signal propagates along one of the arms of an interferometer, with a total length L, and assuming that the reference signal propagating in the other arm was not frequency shifted, we will obtain a total amount of fringe displacement given by

$$\delta\phi(t) = \frac{L}{c}\delta\omega(t) = \frac{2\pi L}{\lambda(0)}\delta n(t) \tag{5.42}$$

Notice that we have assumed that the frequency shift only occurs locally, at the intersection of the probe and the intense laser beam, and not along the whole propagation length. The amplification factor resulting from the interferometric technique is of order $L/\lambda(0)$ and can be very large. For L of the order of 1 m, and for near-infrared photons, this factor could be as large as 10^6.

5.18　Search for Pseudoscalar Particles

The axions were originally proposed to explain charge–parity in strong interactions, which is known as the strong CP problem (Peccei and Quinn, 1977; Weinberg, 1978; Wilczek, 1978). These hypothetical elementary particles would have a very small mass (in the meV range) and would couple very weakly with quarks, leptons, and photons. Due to these rather extreme properties, they could also be used to explain dark matter in the universe (Raffelt, 1996; Khlopov and Rubin, 2004). Different axion experiments in operation have not yet been able to prove their existence. (Zioutas et al., 2005; Fairbairn et al., 2006; Lamoreaux, 2006; van Bibber and Rosenberg, 2006; Zavattini et al., 2006; Melissinos, 2007; Mohapatra and Nasri, 2007). New experimental configurations are therefore needed. Schemes based on x-ray lasers were recently proposed (Rabadan et al., 2006). We have also recently considered the direct axion–photon coupling, which could lead to a new experimental concept using intense lasers (Mendonça, 2007a).

The interaction of axions with photons in the presence of a static magnetic field is theoretically well understood (Maiani et al., 1986; Gasperini, 1987; Raffelt and Stodolski, 1988). This corresponds to an indirect interaction, mediated by quarks, which is known as the Primakov effect. However, we can also consider a more general Primakov process, where the static field is absent. Axions are elementary excitations of a pseudoscalar field a. Photons couple to the axion field through the Lagrangian density (Raffelt, 1996).

$$L_{int} = g_{a\gamma}(\vec{E} \cdot \vec{B})a \qquad (5.43)$$

where $g_{a\gamma}$ is the coupling constant, with $\vec{E}$ and $\vec{B}$ the electric and magnetic fields. It is obvious that, under a CP transformation, this interaction Lagrangian will remain invariant because $\vec{E}$ is a polar vector and $\vec{B}(\vec{E} \cdot \vec{B})$ is an axial one, and therefore $(\vec{E} \cdot \vec{B})$ is a pseudoscalar. Equation 5.43 shows that an axion can couple to two photons, or in alternative to one photon in the presence of a static electric or magnetic field. The first option is considered here. The coupled equations for the electromagnetic and the axion fields can be written (using $c = 1$) as

$$(\partial_t^2 - \nabla^2)\vec{E} = -g_{a\gamma}\vec{B}\partial_t^2 a, \quad (\partial_t^2 - \nabla^2 + m_a^2)a = g_{a\gamma}(\vec{E} \cdot \vec{B}) \qquad (5.44)$$

where
$\partial_t \equiv \partial/\partial t$
m_a is the axion mass

We can see from these equations that two electromagnetic waves with frequencies ω and ω' can only interact with the axion field if their polarizations are nearly orthogonal to each other, $\vec{E}_0 \cdot \vec{E}' \approx 0$. In this case they can exchange energy with the pseudoscalar field a with frequency ω_a. Let us examine the difference in frequency case $\omega_a = \omega' - \omega$. It is possible to show that the field amplitudes will evolve according to (Mendonça, 2007a)

$$E(z) \simeq E(0)\sum_{n=0} J_{2n}(\tau_{2n}(z))e^{i\Delta_{2n}z/2 + 2in(\delta - \pi/2)} \qquad (5.45)$$

and

$$a(z) \simeq \frac{2}{\omega_n}E(0)\sum_{n=0} J_{2n+1}(\tau_{2n+1}(z))e^{i\Delta_{2n}z/2 + i(2l+1)(\delta - \pi/2)} \qquad (5.46)$$

This means that we will generate odd axion field components a_{2n+1} and even sidebands of the photon field with amplitudes E_{2n}. Due to the smallness of $g_{a\gamma}^2 |E'|^2$, in a plausible experimental condition only the first two sidebands will eventually be seen. However, this sideband spectrum would be a signature of an axion field.

An estimate of the signal associated with the first sideband $\omega_2 = \omega + 2\omega'$ can be obtained from the above solutions, for $\Omega_r\tau \ll 1$, as $E_2(z) \sim (g_{a\gamma}z)^2 B_r^2 E_0/4$, in agreement with previous estimates (Raffelt and Stodolski, 1988). Comparing the relative amplitude variation $\delta_{static} = E_2/E_0$ as in current experiments using static magnetic field and low intensity lasers (Zavattini et al., 2006) with that due to the parametric axion excitation by strong laser fields. The result is

$$\delta_{\text{static}} / \delta_{\text{laser}} \simeq \left(\frac{B_{\text{r}}}{B'}\right)^2 \left(\frac{z_{\text{static}}}{z_{\text{laser}}}\right)^2 \qquad (5.47)$$

where B' is the magnetic field associated with the intense pump laser. For the experiment reported by Zavattini et al., 2006, the static magnetic field B_{r} is of the order of 5 T. In contrast, the magnetic field of a PW laser can attain 10^5 T. However, the interaction length z_{static} is about 10^4 m for the static magnetic field experiments, whereas the interaction length for PW laser experiments is smaller than 1 m. The combined balance between field strength and interaction distance makes these two experimental concepts of nearly equal efficiency. However, new EW laser systems (Mourou et al., 2006) will significantly improve the axion experiments and will eventually give a convincing answer to the existence of such a hypothetical particle. This could become a new area of application for the future ultraintense laser systems.

5.19   Vacuum Disruption

It was first shown by Schwinger (1951) that electron–positron pairs can be created in pure vacuum in the presence of an intense electric field, E. If this amplitude is much smaller than the critical field E_c, the rate of production of real pairs per unit volume is approximately determined by

$$\frac{dN_{\text{pairs}}}{dt} \simeq \frac{\alpha^2 \omega_0^4}{\pi^3} \left(\frac{E}{E_c}\right)^4 \qquad (5.48)$$

This expression is valid for electric fields with a duration larger than the inverse of the Compton frequency. This expression gives a very small creation rate for the present laser systems. For two counterpropagating laser beams, the threshold intensity leading to the emission of a single electron–positron pair can be estimated as

$$I_{\text{thr}} \sim \frac{m_{\text{e}} c^3}{2\pi w \lambda_{\text{C}}^2} \qquad (5.49)$$

where
   w is the laser spot size
   $\lambda_{\text{C}} = \hbar/m_{\text{e}}c$ is the Compton wavelength

If the laser beams are focused near the diffraction limit we get for a laser wavelength of $\lambda = 800$ nm a threshold intensity of nearly 10^{21} W/cm². This is still quite marginal for the existing laser systems, which makes such experimental proposals very risky, even considering that their scientific purpose is ambitious and exciting.

These laser experiments are aimed at observing direct pair creation by using low-energy photons. However, indirect pair creation, which is not a pure vacuum effect, was already observed at SLAC (Stanford Linear Accelerator Center) (Burke et al., 1997), where gamma-ray photons with an energy of 30 GeV, created from laser interaction with high-energy electrons, were used to produce real electron–positron pairs. Another indirect and more remote evidence of vacuum pair creation and annihilation was the observation of 0.51 MeV gamma-ray emission from the center of our galaxy (Riegler et al., 1981).

Various schemes have been proposed to increase the electric field amplitude, by using plasma effects, in order to approach the Schwinger limit. One is to compress the laser pulse duration by using the self-reaction on the laser of the plasma wakefield. This is a consequence of the photon acceleration processes that take place inside the wakefield, leading to an increase in the laser bandwidth (Murphy et al., 2006), while maintaining the spectral phase coherence. Some evidence of laser pulse compression can be found in recent experiments with moderately intense laser pulses in the 100 TW regime (Faure et al., 2005). Strong wakefields excited by intense laser pulses in magnetized plasmas could also approach the critical field and excite direct pair production from pure vacuum (Rios et al., 2006). Another process is related to the high harmonics emitted by laser solid target interactions. After focusing, the high harmonic radiation could eventually become closer to the critical limit than the original laser pulse field itself (Zepf et al., 2007).

More speculative scenarios, anticipating the eventual vacuum disruption by laser fields close to the Schwinger limit, have already been discussed theoretically. In this case, the electron–positron plasma is described by Vlasov equations with a source term. Simulations show that pair creation will excite electron plasma waves with increasing frequency (because the electron and positron densities are increasing), accompanied by a decrease in the electric field due to particle screening (Kluger et al., 1991; Blaschke et al., 2006) and laser pulse damping (Bulanov et al., 2005).

5.20 Conclusion

We have discussed recent and possible future nuclear and quantum vacuum experiments using ultraintense lasers. First, we have reviewed recent work on nuclear and particle physics. We have also discussed a plausible double-beam configuration for laser-enhanced nuclear reactions. In such a configuration we expect the formation of a strongly compressed electron population at the center of the target. Two distinct plasma regions can then be identified. One, near the plasma boundary, almost devoid of electrons, and therefore suited for enhanced muon catalyzed fusion; the other at the center, where an enhancement for nuclear cross sections due to Coulomb focusing (Milosevic et al., 2004) can eventually be present. The implications of this double laser beam configuration were discussed for several different nuclear processes, and estimates for realistic laser and plasma parameters were given. This can eventually stimulate future research in the area of nuclear physics using ultraintense laser facilities. Enhanced muon catalyzed fusion, enhanced fission reactions, electron–positron pair creation processes, and stimulated gamma-ray emission were described. Other interesting nuclear processes could equally be considered, such as the electron capture or inverse beta decay of the nuclei. These other processes could lead to the creation of a short pulse neutrino source.

Quantum vacuum effects were also considered. We have described several new processes of moderately high-energy physics, associated with the nonlinear optical properties of the electromagnetic vacuum. In most of them we have assumed that light propagates in pure vacuum, with no material support of the interaction between different photon modes, except for the QED nonlinearities of vacuum. These processes include optical birefringence in a static magnetic field, optical sideband cascades in a rotating field, photon acceleration of probe photons propagating in the presence of an intense radiation background, and time refraction in the perturbed vacuum. QED vacuum is a virtual plasma medium, where the nonlinear effects are due to the existence of virtual electron–positron pairs, excited by intense fields. This is a difficult but promising new area, which can help in understanding the physical background of particles and fields. Possible experimental configurations using ultraintense laser systems, in the PW and the EW regimes, can be seen as promising candidates for the observation of nonlinear effects in the QED vacuum. Other areas of particle physics were also covered, such as the Unruh effect and the search for axions.

Acknowledgments

We would like to thank R. Bingham, G. Brodin, L. Ferreira, M. Marklund, P. Norreys, E. Ribeiro, and P.K. Shukla for stimulating discussions and J. Martins for help with the Osiris code.

References

Adler S (1971), Photon splitting and photon dispersion in a strong magnetic field, *Ann. Phys.*, **67**, 599.
Adler S L (2007), Vacuum birefringence in a rotating magnetic field, *J. Phys. A*, **40**, F143.

Andreev A et al. (2000), Excitation and decay of low-lying nuclear states in a dense plasma produced by a subpicosecond laser, *J. Exp. Theor. Phys.*, **91**, 1163.

Arad B, Eliezer S, and Paiss Y (1979), Nuclear anti-stokes transitions induced by laser radiation, *Phys. Lett. A*, **74**, 395.

Becker W, Schlicher R, and Scully M O (1984), Laser-induced nuclear anti-Stokes transitions revisited, *Phys. Lett. A*, **106**, 441.

Bekenstein J D (1973), Extraction of energy and charge from a black hole, *Phys. Rev. D*, **7**, 949.

Bertrand P, Ghizzo A, Karttunen S J, Pttikangas T J H, Salomaa R R E, and Shoucri M (1994), Generation of ultrafast electrons by simultaneous stimulated Raman backward and forward scattering, *Phys. Rev. E*, **49**, 5656.

Bialynicka-Birula Z and Bialynicka-Birula I (1970), Nonlinear effects in quantum electrodynamics. Photon propagation and photon splitting in an external field, *Phys. Rev. D*, **2**, 2341.

Birrel N D and Davies P C W (1982), *Quantum Fields in Curved Space*, Cambridge University Press, Cambridge.

Blaschke D B et al. (2006), Pair production and optical lasers, *Phys. Rev. Lett.*, **96**, 140402.

Borghesi M et al. (1997), Relativistic channeling of a picosecond laser pulse in a near-critical preformed plasma, *Phys. Rev. Lett.*, **78**, 879.

Boyer T H (1980), Thermal effects of acceleration through random classical radiation, *Phys. Rev. D*, **21**, 2137.

Boyer K, Luk T Y S, and Rhodes C K (1988), Possibility of optically induced nuclear fission, *Phys. Rev. Lett.*, **60**, 557.

Bracci L and Fiorentini G (1982), Mesic molecules and muon catalyses fusion, *Phys. Rep.*, **86**, 169.

Brodin G, Marklund M, and Stenflo L (2001), Proposal for detection of QED vacuum nonlinearities in Maxwell's equations by the use of waveguides, *Phys. Rev. Lett.*, **87**, 171801.

Bulanov S S, Fedotov A M, and Pegoraro F (2005), Damping of electromagnetic waves due to electron-positron pair production, *Phys. Rev. E*, **71**, 016404.

Burke et al. (1997), Positron production in multiphoton light-by-light scattering, *Phys. Rev. Lett.*, **79**, 1626.

Casimir H B K (1948), On the attraction between two perfectly conducting plates, *Proc. K. Ned. Akad. Wet. B*, **51**, 795.

Chekhlov O et al. (2005) High energy broadband ultra-short pulse OPCPA (optical parametric chirped pulse amplification) system, Central Laser Facility, UK, Annual Report 2004/2005, p. 253.

Chelkowski S, Bandrauk A D, and Corkum P B (2004), Muonic molecules in superintense laser fields, *Phys. Rev. Lett.*, **93**, 083602.

Chen P and Tajima T (1999), Testing Unruh radiation with ultraintense lasers, *Phys. Rev. Lett.*, **83**, 256.

Clark E L, Krushelnick K, Davies J R, Zepf M, Tatarakis M, Beg F N, Machacek A, Norreys P A, Santala M I K, Watts I, and Dangor A E (2000), Measurements of energetic proton transport through magnetized plasma from intense laser interactions with solids, *Phys. Rev. Lett.*, **84**, 670.

Collier J et al. (2004), The Astra Gemini project, in Central Laser Facility RAL Annual Report, RAL-TR-2004-025.

Collins C B, Eberhard C D, Glesener J W, and Anderson J A (1988), Depopulation of the isomeric state ^{180}Ta^m by the reaction ^{180}Ta$^m(y, y)^{180}$Ta, *Phys. Rev. C*, **37**, 2267.

Cowan et al. (2000), Photonuclear fission from high energy electrons from ultra-intense laser-solid interactions, *Phys. Rev. Lett.*, **84**, 903.

Davies P C W (1975), Scalar production in Schwarzschild and Rindler metrics, *J Phys A: Math. Gen.*, **8**, 609.

Davies J R (2002), How wrong is collisional Monte Carlo modeling of fast electrons transport in high intensity laser-solid interactions?, *Phys. Rev. E*, **65**, 026407.

Dias J M et al. (1997), Experimental evidence of photon acceleration of ultra-short laser pulses in relativistic ionization fronts, *Phys. Rev. Lett.*, **78**, 4773.

Disdier L, Garonnet J-P, Malka G, and Miquel J-L (1999), Fast neutron emission from a high-energy ion beam produced by a high-intensity sub-picosecond laser pulse, *Phys. Rev. Lett*, **82**, 1454.

Dittrich W and Gies H (2000), *Probing the Quantum Vacuum*, Springer Tracts in Physics, Springer, Berlin.

Dodonov V V, Klimov A B, and Man'ko V I (1989), Generation of squeezed states in a resonator with a moving wall, *Phys. Lett. A*, **142**, 511.

Dodonov V V, Klimov A B, and Nikonov D E (1993), Quantum phenomena in nonstationary media, *Phys. Rev. A*, **47**, 4422.

Eliezer S (2002), *The Interaction of High-Power Lasers with Plasmas*, Institute of Physics, Bristol, UK.

Eliezer S and Henis Z (1988), Enhancement of μ-molecular resonant formation by laser induced stokes transitions, *Phys. Lett. A*, **131**, 361.

Eliezer S and Henis H (1994), Muon-catalyzed fusion and energy production perspective, *Fusion Technol.*, **26**, 46.

Eliezer S and Hora H (1989), Double layers in laser-produced plasmas, *Phys. Rep.*, **172**, 339.

Eliezer S, Martinez-Val J M, Paiss Y, and Velarde G (1995), Induced Stokes and anti-Stokes nuclear transitions, *Quantum Electron.*, **25**, 1106.

Fairbairn M, Rashbar T, and Troitsky S (2006), Shining light through the Sun, *arXiv:astro-ph/0610844v1*.

Faure J et al. (2005), Observation of laser-pulse shortening in nonlinear plasma waves, *Phys. Rev. Lett.*, **95**, 205003.

Ford G W and O'Connell R F (2005), Is there Unruh radiation, *Phys. Lett. A*, **350**, 17.

Gasperini M (1987), Axion production by electromagnetic fields, *Phys. Rev. Lett.*, **59**, 396.

Ginzburg V L (1961), *Propagation of Electromagnetic Waves in Plasma*, Gordon and Breach, New York, Chapter IV.

Guerreiro A, Mendonça J T, and Martins A M (2005), New mechanism of vacuum radiation from non-accelerated moving boundaries, *J. Opt. B: Quantum Semiclassical Opt.*, **7**, S69.

Hawking S W (1974), Black hole explosions? *Nature*, **248**, 30.

Hawking S W (1975), Particle creation by black holes, *Commun. Math. Phys.*, **43**, 199.

Heisenberg W and Euler H (1936), Consequences of Dirac's theory of the positron, *Z. Phys.*, **98**, 714.

Heitler W (1954), Quantum Theory of Radiation, Oxford, London.

Hemker R G (2000), PhD thesis, UCLA; R.A. Fonseca et al., in *Lecture Notes in Computer Science*, vol. 2329, (Springer-Verlag, Heidelberg, 2002), p. III–342.

Hilscher D, Berndt O, Enke M, Jahnke U, Nickles P V, Ruhl H, and Sandner W (2001), Neutron energy spectra from the laser-induced $D(d, n)^3 He$ reaction, *Phys. Rev. E*, **64**, 016414.

Hubel H, Naumann R A, and Hopke P K (1970), Isomeric transitions in 29-day $^{179m}_2 Hf$, *Phys. Rev. C*, **2**, 1447.

Hubel H et al. (1971), Magnetic moment of the $I^\pi = 25/2^-$ isomer in the ^{179}Hf and the anomalous orbital proton g factor, *Phys. Rev. C*, **12**, 2013.

Itzykson C and Zuber J-B (1980), *Quantum Field Theory*, McGraw-Hill, New York.

Kaplan A E and Ding Y J (2000), Field-gradient-induced second-harmonic generation in magnetized vacuum, *Phys. Rev. A*, **62**, 043805.

Khlopov M Yu and Rubin S G (2004), *Cosmological Pattern of Microphysics in Inflationary Universe*, Kluwer Academic Publishers, Dordrecht.

Kluger Y et al. (1991), Pair production in a strong electric field, *Phys. Rev. Lett.*, **67**, 2427.

Krushelnick K, et al. (1999), Multi-MeV ion production from high-intensity laser interactions with under dense plasmas, *Phys. Rev. Lett.*, **83**, 737.

Lamoreaux S (2006), Particle physics: the first axion? *Nature*, **441**, 31.

Ledingham K W D et al. (2000), Photonuclear physics when a multi-terawatt laser pulse interacts with solid targets, *Phys. Rev. Lett.*, **84**, 899.

Ledingham K W D, Singhal R P, McKenna P, and Spencer I (2002), Laser induced nuclear physics and applications, *Europhys. News*, July/August issue, 120.

Ledingham K W D, McKenna P, and Singhal R P (2003), Applications for nuclear phenomena generated by ultra-intense lasers, *Science*, **300**, 1107.

Liang E P, Wilks S C, and Tabak M (1998), Pair production by ultraintense lasers, *Phys. Rev. Lett.*, **81**, 4887.

Maiani L, Petronzio R, and Zavattini E (1986), Effects of nearly massless, spin-zero particles on light propagation in a magnetic field, *Phys. Lett. B*, **175**, 359.

Mackinnon A J, Borghesi M, Hatchett S, Key M H, Patel P K, Campbell H, Schiavi A, Snavely A, Wilks S C, and Willi O (2001), Effect of plasma scale length on multi-MeV proton production by intense laser pulses, *Phys. Rev. Lett.*, **86**, 1769.

Maksimchuk A, Gu S, Flippo K, Umstadter D, and Bychenkov V Yu (2000), Forward ion acceleration in thin films driven by a high-intensity laser, *Phys. Rev. Lett.*, **84**, 4108.

Malka G, Fuchs J, Amiranoff F, Baton S D, Gaillard R, Miquel J L, Ppin H, Rousseaux C, Bonnaud G, Busquet M, and Lours L (1997), Suprathermal electron generation and channel formation by an ultra relativistic laser pulse in an under dense preformed plasma, *Phys. Rev. Lett.*, **79**, 2053.

Marklund M and Shukla P K (2006), Nonlinear collective effects in photon-photon and photon-plasma interactions, *Rev. Mod. Phys.*, **78**, 591.

Marklund M, Eliasson B, and Shukla P K (2004), Numerical investigation of logarithmic corrections in two-dimensional spin models, *JETP Lett.*, **79**, 208.

Marklund M, Shukla P K, and Eliasson B (2005), The intense radiation gas, *Europhys. Lett.*, **70**, 327.

Matsuzaki et al. (2001), The RIKEN-RAL pulsed muon factory, *Nucl. Instr. Meth. A*, **465**, 365.

Melissinos A C (2007), Experimental observation of optical rotation generated in vacuum by a magnetic field, *arXiv:hep-ph/0702135v1* (comment).

Mendonça J T (2001), *Theory of Photon Acceleration*, Institute of Physics Publishing, Bristol.

Mendonça J T (2007a), Axion excitation by intense laser fields, *Europhys. Lett.*, **79**, 21001.

Mendonça J T (2007b), QED vacuum in a rotating magnetic field, unpublished.

Mendonça J T and Guerreiro A (2005), Time refraction and the quantum properties of vacuum, *Phys. Rev. A*, **72**, 044512.

Mendonça J T, Guerreiro A, and Martins A M (2000), Quantum theory of time refraction, *Phys. Rev. A*, **62**, 033805.

Mendonça J T, Davies J R, and Eloy M (2001), Proton and neutron sources using Tera-Watt lasers, *Meas. Sci. Technol.*, **12**, 1801.

Mendonça J T, Martins A M, and Guerreiro A (2003), Temporal beam splitter and temporal interference, *Phys. Rev. A*, **68**, 043801.

Mendonça J T, Marklund M, Shukla P K, and Brodin G (2006a), Photon acceleration in vacuum, *Phys. Lett. A*, **359**, 700.

Mendonça J T, Dias de Deus J, and Ferreira P C (2006b), Higher harmonics in vacuum from nonlinear QED effects without low-mass intermediate particles, *Phys. Rev. Lett.*, **96**, 100403; Erratum, **97**, 269901.

Milosevic N, Corkum P B, and Brabec T (2004), How to use lasers for imaging attosecond dynamics of nuclear processes, *Phys. Rev. Lett.*, **92**, 013002.

Milton K A (2004), The Casimir effect: Recent controversies and progress, *J. Phys. A: Math. Gen.*, **37**, R209.

Mohapatra R N and Nasri S (2007), Reconciling the CAST and the PVLAS results, *Phys. Rev. Lett.*, **98**, 050402.

Moore G T (1970), Quantum theory of the electromagnetic field in a variable-length one-dimensional cavity, *J. Math. Phys.*, **11**, 2679.

Mourou G A, Tajima T, and Bulanov S V (2006), Optics in the relativistic regime, *Rev. Mod. Phys.*, **78**, 309.

Murphy C et al. (2006), Photon acceleration by laser wakefields, *Phys. Plasmas*, **13**, 033108.

Nakashima K and Takabe H (2002), Numerical study of pair creation by ultraintense lasers, *Phys. Plasmas*, **9**, 1505.

Norreys P A (2002), Physics with peta-watt lasers, *Phys. World*, September issue, 39.

Norreys P A, Fews A P, Beg F N, Bell A R, Dangor A E, Lee P, Nelson M B, Schmidt H, Tatarakis M, and Cable M D (1998), Neutron production from picosecond laser irradiation of deuterated targets at intensities of 10^{19}W/cm^2, *Plasma Phys. Control Fusion*, **40**, 175.

Peccei R D and Quinn H R (1977), CP conservation in the presence of pseudoparticles, *Phys. Rev. Lett.*, **38**, 1440.

Rabadan R, Ringwald A, and Sigurdson K (2006), Photon regeneration from pseudo-scalars at X-ray laser facilities, *Phys. Rev. Lett.*, **96**, 110407.

Raffelt G G (1996), *Stars as Laboratories for Fundamental Physics*, University of Chicago Press, Chicago.

Raffelt G and Stodolski L (1988), Mixing of the photon with low-mass particles, *Phys. Rev. D*, **37**, 1237.

RAL (2003), Central laser facility RAL annual report, RAL-TR-2003-018.

Riegler G R et al. (1981), Variable positron annihilation radiation from the galactic center region, *Astrophys. J.*, **248**, L13.

Rios L A et al. (2006), Pair production by strong wakefield excited by lasers in a magnetized plasma *J. Exp. Theor. Phys.*, **103**, 47.

Ross I N, Matousek P, Towrie M, Langley A J, Collier J L, Danson C N, Gomez C H, Neely D, and Osvay K (1999), Prospects for multi-PW source using optical parametric chirped pulse amplifiers, *Laser Part. Beams*, **17**, 331.

Sakabe S, Shimizu S, Hashida M, Sato F, Tsuyukushi T, Nishihara K, Okihara S, Kagawa T, Izawa Y, Imasaki K, and Iida T (2004), Generation of high-energy protons from the Coulomb explosion of hydrogen clusters by intense femto-second laser pulses, *Phys Rev A*, **69**, 023203.

Santala M I K et al. (2001), Production of radioactive nuclides by energetic protons generated from intense laser-plasma interactions, *Appl. Phys. Lett.*, **78**, 19.

Sarkisov G S, Bychenkov V Yu, Novikov V N, Tikhonchuk V T, Maksimchuk A, Chen S-Y, Wagner R, Mourou G, and Umstadter D (1999), Self-focusing, channel formation, and high-energy ion generation in interaction of an intense short laser pulse with a He jet, *Phys. Rev. E*, **59**, 7042.

Schutzhold R, Schaller G, and Habs D (2006), Signatures of the Unruh effect from electrons accelerated by ultrastrong laser fields, *Phys. Rev. Lett.*, **97**, 121302.

Schwinger J (1951), On gauge invariance and vacuum polarization, *Phys. Rev.*, **82**, 664.

Shukla P K and Eliasson B (2004), Modulational and filamentational instabilities of intense photon pulses and their dynamics in a photon gas, *Phys. Rev. Lett.*, **92** 073601.

Shukla P K, Marklund M, Tskhakaya D D, and Eliasson B (2004), Nonlinear effects associated with interactions of intense photons with a photon gas, *Phys. Plasmas*, **11**, 3767.

Snavely R A et al. (2000), Intense high-energy proton beams from petawatt-laser irradiation of solids, *Phys. Rev. Lett.*, **85**, 2945.

Spencer I et al. (2002), A nearly real-time high temperature laser-plasma diagnostic using photonuclear reactions in tantalum, *Rev. Sci. Instrum.*, **73**, 3801.

Soljacic M and Segev M (2000), Self-trapping of electromagnetic beams in vacuum supported by QED nonlinear effects, *Phys. Rev. A*, **62**, 043817.

Uhlmann M et al. (2004), Resonant cavity photon creation via dynamical Casimir effect, *Phys. Rev. Lett.*, **93**, 193601.

Umstadter D (2001), Review of physics and applications of relativistic plasmas driven by ultra-intense lasers, *Phys. Plasmas*, **8**, 1774.

Unruh W G (1976), Notes on black-hole evaporation, *Phys. Rev. D*, **14**, 870.

van Bibber K and Rosenberg L J (2006), Ultrasensitive searches for the axion, *Phys. Today*, **59**, 30 August.

Weinberg S (1978), A new light boson?, *Phys. Rev. Lett.*, **40**, 223.

Wilczek F (1978), Problem of strong P and T invariance in the presence of instantons, *Phys. Rev. Lett.*, **40**, 279.

Yablonovich E (1989), Accelerating reference frame for electromagnetic waves in a rapidly growing plasma: Unruh-Davies-Fulling-DeWitt radiation and the non-adiabatic Casimir effect, *Phys. Rev. Lett.*, **62**, 1742.

Zavattini E et al. (2006). Experimental observation of optical rotation generated in vacuum by a magnetic field, *Phys. Rev. Lett.*, **96**, 110406.

Zepf M et al. (2007), Bright quasi-phase-matched soft-X-ray harmonic radiation from Argon ions, *Phys. Rev. Lett.*, **99**, 143901

Zioutas K et al. (2005), First results from the CERN Axion Solar Telescope, *Phys. Rev. Lett.*, **94**, 121301.

6

Equations of State

Shalom Eliezer and Zohar Henis

CONTENTS

6.1 Introduction

The equation of state (EOS) describes a physical system by the relation between its thermodynamic quantities, such as pressure, energy, density, entropy, specific heat, etc., and is related to both fundamental physics and applied sciences. The knowledge of EOS is necessary to understand the science of extreme concentration of energy and matter, phase transitions, strongly coupled plasma systems, etc. It is also required for many applications such as inertial fusion energy, astrophysics, geophysics and planetary science, new materials including nanoparticles, etc. For example, Figure 6.1 schematically describes various systems in the temperature–electron density domain. A route to inertial confinement fusion (ICF) leading to a solution of the energy problem is shown in this figure. As can be seen from this schematic figure, the well-known low-density ideal gas EOS as well as the extremely high-density Thomas–Fermi (TF) EOS cannot correctly describe the ICF problem and therefore more complicated (i.e., realistic) EOS are needed.

The EOS describes nature over all possible values of pressure, density, and temperatures where local thermodynamic equilibrium can be sustained. Since it is not yet known how to quantitatively describe material at every available thermodynamic state, including all phases of matter, it is necessary to introduce simplified methods whose range of applicability is limited.

6.1.1 Thermodynamic Potentials and EOS

We assume that X describes the state of a system defined by a potential $F(X)$. The conjugate variable of X is $P = dF/dX$. If X is replaced by P as independent variable by the Legendre transformation, then $\Psi(P)$ defined by

FIGURE 6.1
Schematic description of electron density–temperature domains for various systems and a route for ICF energy solution. The ideal gas EOS and TF EOS domains are shown.

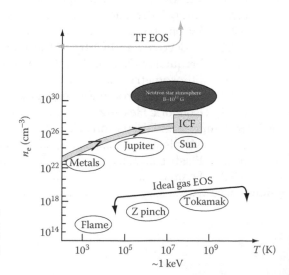

$$\Psi(P) = F - PX$$

which is also a potential. The Legendre transformation for several variables is defined by

$$\Psi(P_1, P_2, \dots, P_n) = F(X_1, X_2, \dots, X_n) - \sum_{i=1}^{n} P_i X_i; \quad P_i = \frac{dF}{dX_i}$$

$$d\Psi(P_1, P_2, \dots, P_n) = dF - \sum_{i=1}^{n} (P_i dX_i + dP_i X_i)$$

For example, the conjugate variables of entropy and specific volume (S,V) are the temperature and pressure (T,P), respectively. Assuming a system with a constant number of particles, $N = \text{const.}$, the Gibbs potential G is derived from the internal energy E by the following Legendre transformation:

$$G(T,P) = E(S,V) - \left[\left(\frac{\partial E}{\partial S} \right)_V S + \left(\frac{\partial E}{\partial V} \right)_S V \right]; \Rightarrow G = E - TS + PV$$

A summary of the thermodynamic potentials, derived from each other by a Legendre transformation, is given in Table 6.1. The thermodynamic potentials are E, internal energy; H, enthalpy; F, Helmholtz' free energy; G, Gibbs' free energy; and Φ, the grand potential. The appropriate variables of the potentials are denoted by V, specific volume $= 1/\rho$, where ρ, density; T, temperature; P, pressure; S, entropy; μ, chemical potential; and N, the number of particles.

TABLE 6.1

Thermodynamic Potentials

Quantity	Variables	Relations
E (internal energy)	S,V,N	$E = TS - PV + \mu N$
H (enthalpy)	S,P,N	$H = E + PV$
F (Helmholtz' free energy)	T,V,N	$F = E - TS$
		$F = PV + \mu N$
G (Gibb's free energy)	T,P,N	$G = \mu N$
Φ (grand potential)	T,V,μ	$\Phi = -PV$
		$\Phi = F - \mu N$

Note: E, internal energy; H, enthalpy; F, Helmholtz' free energy; G, Gibbs' free energy; and Φ, the grand potential as a function of their appropriate variables: V, specific volume $= 1/\rho$, where ρ is the density; T, temperature; P, pressure; S, entropy; μ, chemical potential; and N, the number of particles.

The various EOS derived from these potentials are summarized in Table 6.2. As one can see from this table there are many possible presentations of EOS. Some specific examples are used in this chapter. In particular, for ideal gas EOS the following relations are given: the Helmholtz' free energy F, the pressure P, the internal energy E, the heat capacity at constant volume C_V, the entropy S, and the Gibbs' free energy G:

$$F(T,V,N) = -Nk_B T \ln\left[\left(\frac{mk_B T}{2\pi\hbar^2}\right)^{3/2} V\right]$$

$$P = -\left(\frac{\partial F}{\partial V}\right)_T = \left(\frac{N}{V}\right)k_B T \tag{6.1}$$

$$E = -T^2\left[\frac{\partial(F/T)}{\partial T}\right]_V = \frac{3}{2}Nk_B T; \quad C_v = \left(\frac{\partial E}{\partial T}\right)_V = \frac{3}{2}Nk_B$$

$$S = -\left(\frac{\partial F}{\partial T}\right)_V = Nk_B \ln\left[\left(\frac{mk_B T}{2\pi\hbar^2}\right)^{3/2} V\right] + \frac{3}{2}Nk_B$$

$$G(T,P,N) = F + PV = -Nk_B T \ln\left[\left(\frac{mk_B T}{2\pi\hbar^2}\right)^{3/2}\left(\frac{k_B T}{P}\right)\right] \tag{6.2}$$

TABLE 6.2

Relations between the Potentials and EOS

E	$dE = TdS - PdV + \mu dN$	EOS
		$\mu = \left(\dfrac{\partial E}{\partial N}\right)_{S,V}; \quad T = \left(\dfrac{\partial E}{\partial S}\right)_{V,N}; \quad P = -\left(\dfrac{\partial E}{\partial V}\right)_{S,N}$
H	$dH = TdS + VdP + \mu dN$	EOS
		$\mu = \left(\dfrac{\partial H}{\partial N}\right)_{S,P}; \quad T = \left(\dfrac{\partial H}{\partial S}\right)_{P,N}; \quad V = \left(\dfrac{\partial H}{\partial P}\right)_{S,N}$
F	$dF = -SdT - PdV + \mu dN$	EOS
		$\mu = \left(\dfrac{\partial F}{\partial N}\right)_{T,V}; \quad S = -\left(\dfrac{\partial F}{\partial T}\right)_{V,N}; \quad P = -\left(\dfrac{\partial F}{\partial V}\right)_{T,N}$
G	$dG = -SdT + VdP + \mu dN$	EOS
		$\mu = \left(\dfrac{\partial G}{\partial N}\right)_{T,P}; \quad S = -\left(\dfrac{\partial G}{\partial T}\right)_{P,N}; \quad V = \left(\dfrac{\partial G}{\partial P}\right)_{T,N}$
Φ	$d\Phi = -SdT - PdV - Nd\mu$	EOS
		$N = -\left(\dfrac{\partial \Phi}{\partial \mu}\right)_{T,V}; \quad S = -\left(\dfrac{\partial \Phi}{\partial T}\right)_{V,\mu}; \quad P = -\left(\dfrac{\partial \Phi}{\partial V}\right)_{T,\mu}$

Note: The potentials and their appropriate variables are defined in Table 6.1.

where

k_B and $2\pi\hbar = h$ are Boltzmann and Planck constants, respectively
m is the mass of the ideal gas particles
all other variables are defined above

6.1.2 Hydrodynamic Equations

The science of EOS (Eliezer et al., 1986; Eliezer and Ricci, 1991; Bushman et al., 1993) is studied experimentally in the laboratory by using static and dynamic techniques. In static experiments the sample is squeezed between pistons or anvils. The conditions in these static experiments are limited by the strength of the construction materials. In dynamic experiments shock waves (SWs) are created (Zeldovich and Raizer, 1966). Since the passage time of the shock is short in comparison with the disassembly time of the shocked sample, one can do SW research for any pressure that can be supplied by a driver, assuming that a proper diagnostic is available. In the scientific literature, the following shock-wave generators are discussed: a variety of guns (such as rail guns and two-stage light-gas gun) that accelerate a foil to collide with a target, exploding foils, magnetic compression, chemical explosives, nuclear explosions, and high-power lasers (Eliezer, 2002). The conventional high explosive and gun drivers can achieve pressures up to a few megabars, while with high-power lasers up to 1 Gbar pressure has been achieved.

The dimension of pressure is given by the scale defined by the pressure of 1 atm at standard conditions ≈ 1 bar $= 10^6$ dyn/cm^2 (in CGS units) $= 10^5$ Pa (in MKS units, Pa = N/m^2).

In 1974 the first direct observation of a laser-driven SW was reported (van Kessel and Sigel, 1974). A planar solid hydrogen target (taken from a liquid-helium cooled cryostat and inserted into an evacuated interaction chamber) was irradiated with a 10 J, 5 ns, Nd laser (1.06 μm wavelength) and the spatial development of the laser-driven SW was measured using high-speed photography. The estimated pressure in this pioneer experiment was 2 Mbar.

Twenty years after the first published experiment, The Nova laser from Livermore laboratories in United States created a pressure of 750 ± 200 Mbar (Cauble et al., 1993, 1994). This was achieved in a collision between two gold foils, where the flyer (Au foil) was accelerated by a high intensity x-ray flux created by the laser–plasma interaction.

The science of high pressure is usually analyzed in a medium that has been compressed by a one-dimensional (1D) SW. For a 1D SW traversing a known medium (density and temperature are known before the SW passes through) the density, pressure, and energy of the shocked material are uniquely determined from the conservation of mass, momentum, and energy, and the measurement of the shock and particle flow velocities.

The starting points in analyzing the 1D SWs are the fluid equations described by the conservation laws of mass, momentum, and energy

$$\frac{\partial \rho}{\partial t} = -\frac{\partial}{\partial x}(\rho u)$$

$$\frac{\partial}{\partial t}(\rho u) = -\frac{\partial}{\partial x}(P + \rho u^2) \tag{6.3}$$

$$\frac{\partial}{\partial t}\left(\rho E + \frac{1}{2}\rho u^2\right) = -\frac{\partial}{\partial x}\left(\rho E u + P u + \frac{1}{2}\rho u^3\right)$$

The usual notation is used, where x–t is space–time, ρ is the density, u is the flow velocity, P is the pressure, and E is the internal energy per unit mass. These three equations have four unknowns ρ–P–E–u and therefore another equation, namely the EOS, is needed to solve the problem.

6.1.3 Heat Wave

We analyze a medium dominated by its flow and the disturbances that propagate with a speed of the order of the sound velocity. The research of SWs requires that the heat wave velocity u_{hw} is much smaller than the shock velocity u_s; this condition is satisfied if u_{hw} is smaller than the sound velocity c_s (Figure 6.2), which in turn is smaller than u_s.

Since the laser pulse duration is not negligible, to a good approximation one may assume that during the laser pulse duration a constant flux (the laser absorbed flux I) is supplied by the laser to the fluid at the boundary $x = 0$:

$$I = -\kappa\left(\frac{\partial T}{\partial x}\right)_{x=0} = -c_V \rho a T''\left(\frac{\partial T}{\partial x}\right)_{x=0} = \text{const.}$$

$$I \sim \frac{c_V \rho a T_0^{n+1}}{x_f} \tag{6.4}$$

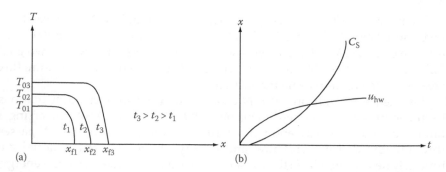

FIGURE 6.2
(a) Propagation of a heat wave during the laser pulse duration, where the laser absorbed flux is deposited at $x = 0$. (b) The positions of the heat wave front and the acoustic wave disturbance.

where

a is a constant

c_V is the heat capacity at constant volume

ρ is the density of the medium

x_f is the front position of the heat wave

T_0 is the temperature at the origin ($x = 0$), i.e., how far the heat wave front is from the origin $x = 0$ and how much the temperature increases there during the laser pulse duration

The heat wave equation (energy conservation) can be approximated by

$$\rho c_V \frac{\partial T}{\partial t} \sim \rho c_V \frac{T_0}{t} \sim \frac{I}{x_f}$$

Solving these equations yields the order of magnitude solutions:

$$T_0 \sim \left(\frac{I}{\rho c_V}\right)^{2/2(n+2)} a^{-1/(n+2)} t^{1/(n+2)}; \quad x_f \sim \left(\frac{I}{\rho c_V}\right)^{n/(n+2)} a^{1/(n+2)} t^{(n+1)/(n+2)} \tag{6.5}$$

These scaling laws were derived by dimensional analysis. The problem under consideration is defined by two parameters: $[I/(\rho c_V)]$ with dimension $[\text{cm} \cdot \text{K/s}]$ and the diffusivity of the medium $\chi = a T^n$, defined by the coefficient a with dimension $[\text{cm}^2/(\text{K}^n \cdot \text{s})]$. The exact solutions are given by these equations up to a dimensionless constant, usually of the order 1.

$$u_{sw} = \frac{dx_f}{dt} \propto t^{-1/(n+2)}; \quad c_s \propto T_0^{1/2} \propto t^{1/2(n+2)} \tag{6.6}$$

From these equations one can see that the heat wave speed decreases with time while the acoustic wave increases with time, so that the acoustic disturbance overtakes the thermal wave. An exact calculation of this problem (Babuel-Peyrissac et al., 1969) shows that for solid deuterium, with a density $0.2\,\text{g/cm}^3$, and for an absorbed laser flux of $10^{15}\,\text{W/cm}^2$, the sound wave overtakes the heat wave at 10 ps. From the above scaling laws it is evident that this time decreases for higher densities and lower absorbed laser intensity. Therefore, in laser interaction with a solid target the hydrodynamic effects predominate over the heat wave for a laser pulse duration larger than 100 ps.

6.1.4 Sound Waves

The motion of the fluid and the changes in the density of the medium caused by a small pressure change describe the physics of sound waves (Courant and Friedrichs, 1948). For equilibrium pressure P_0 and density ρ_0, the changes in pressure ΔP and density $\Delta \rho$ due to the existence of a sound wave are extremely small.

The motion in a sound wave is isentropic, $S(x) = $ const.; therefore, the change in the pressure is given by

$$\Delta P = \left(\frac{\partial P}{\partial \rho}\right)_{S} \Delta \rho \equiv c_{s}^{2} \Delta \rho \tag{6.7}$$

The mass and momentum conservations together with the last equation yield the wave equation

$$\frac{\partial^{2}(\Delta \rho)}{\partial t^{2}} - c_{s}^{2} \frac{\partial^{2}(\Delta \rho)}{\partial x^{2}} = 0 \tag{6.8}$$

From this equation it is evident that c_{s} is the velocity of the $\Delta \rho$ disturbance, which is defined as the velocity of the sound wave. The pressure change ΔP and the flow velocity u satisfy a similar wave equation. Equation 6.8 has two families of solutions f and g:

$$\left.\begin{array}{r}\Delta \rho \\ \Delta P \\ \Delta u\end{array}\right\} \sim f(x - c_{s}t) \quad \text{and} \quad g(x + c_{s}t) \tag{6.9}$$

where the speed of sound is

$$c_{s} = +\sqrt{\left(\frac{\partial P}{\partial \rho}\right)_{S}} \tag{6.10}$$

The disturbances $f(x - c_{s}t)$ are moving in the positive x-direction while $g(x + c_{s}t)$ propagates in the negative x-direction.

For an ideal gas the pressure is

$$P = n(Z+1)k_{B}T = \frac{\rho}{m}(Z+1)k_{B}T \tag{6.11}$$

where m and Z are the ion mass and ionization, respectively. The ideal gas EOS for an adiabatic process is

$$P = P_{0}\left(\frac{\rho}{\rho_{0}}\right)^{\gamma} ; \quad \gamma = \frac{C_{P}}{C_{V}} \tag{6.12}$$

where C_{P} and C_{V} are the heat capacities at constant pressure and constant volume, respectively. The speed of sound for the ideal gas is

$$c_{s} = \left(\frac{\gamma P}{\rho}\right)^{1/2} = \left[\frac{\gamma(Z+1)k_{B}T}{m}\right]^{1/2} \tag{6.13}$$

If the undisturbed gas is not stationary, then the flow stream carries the waves. A transformation from the coordinates moving with the flow (velocity u in $+x$ direction) to the laboratory coordinates means that the sound wave is traveling with a velocity $u + c_s$ in the $+x$ direction and $u - c_s$ in the $-x$ direction. The curves in the x–t plane in the direction of propagation of small disturbances are called characteristic curves. For isentropic flow there are two families of characteristics described by

$$C_+ : \frac{dx}{dt} = u + c_s; \quad C_- : \frac{dx}{dt} = u - c_s \tag{6.14}$$

so that the physical quantities P and ρ do not change along these stream lines. There is also a characteristic curve C_0, where $dx/dt = u$ (not discussed here). The $C_\pm$ characteristics are curved lines in the x–t plane because u and c_s are, in general, functions of x and t.

Using the mass conservation and the momentum conservation given in Equation 6.3 for an isotropic process ($S = $ const.), we get

$$\left[\frac{\partial u}{\partial t} + (u + c_s)\frac{\partial u}{\partial x} \right] + \frac{1}{\rho c_s}\left[\frac{\partial P}{\partial t} + (u + c_s)\frac{\partial P}{\partial x} \right] = 0$$
$$\left[\frac{\partial u}{\partial t} + (u - c_s)\frac{\partial u}{\partial x} \right] - \frac{1}{\rho c_s}\left[\frac{\partial P}{\partial t} + (u - c_s)\frac{\partial P}{\partial x} \right] = 0 \tag{6.15}$$

The brackets [] in these equations are derivatives along the characteristics $C_\pm$. In order to comprehend this statement, we define a curve $x = y(t)$. The derivative of a function $F(x,t)$ along the curve $y(t)$ is given by

$$\left(\frac{dF}{dt} \right)_y = \frac{\partial F}{\partial t} + \frac{\partial F}{\partial x}\frac{dx}{dt} = \frac{\partial F}{\partial t} + y'\frac{\partial F}{\partial x}$$
$$dF = \int \left(\frac{dF}{dt} \right)_y dt; \quad x = y(t); \quad y' \equiv \frac{dx}{dt} \tag{6.16}$$

For example, for the derivative of F along $x = $ const. one has $y' = 0$, along a stream line $y' = u$, along the characteristic C_+ we have $y' = u + c_s$, etc. Therefore, Equations 6.15 and 6.16 can be rewritten as

$$dJ_+ \equiv du + \frac{1}{\rho c_s}dP = 0 \quad \text{along } C_+ \left(\frac{dx}{dt} = u + c_s \right)$$
$$dJ_- \equiv du - \frac{1}{\rho c_s}dP = 0 \quad \text{along } C_- \left(\frac{dx}{dt} = u - c_s \right) \tag{6.17}$$

These equations describe an isentropic flow in 1D with planar symmetry. $dJ_\pm$ are total differentials and can be integrated:

$$J_+ = u + \int \frac{dP}{\rho c_s} = u + \int \frac{c_s d\rho}{\rho}; \quad J_- = u - \int \frac{dP}{\rho c_s} = u - \int \frac{c_s d\rho}{\rho} \tag{6.18}$$

For the second relation on the right-hand side of these, Equation 6.7 has been used. J_+ and J_- are called Riemann invariants and are occasionally used to numerically solve the flow equations for (and only for) an isentropic process since J_+ and J_- are constants along the characteristics C_+ and C_-, respectively. For a nonisentropic flow the density and the speed of sound are functions of two variables, dJ_+ and dJ_- are not total differentials, and therefore J_+ and J_- have no physical (or mathematical) meaning.

6.1.5 Rarefaction Waves

We now analyze the case where the pressure is suddenly dropped in an isentropic process, for example, after the high-power laser is switched off and the ablation pressure drops. Another interesting case is after the laser-induced high-pressure wave has reached the backside of a target and at the interface with the vacuum there is a sudden drop in pressure. In these cases, if one follows the variation in time for a given fluid element one gets

$$\frac{D\rho}{Dt} < 0, \quad \frac{DP}{Dt} < 0; \quad \frac{D}{Dt} \equiv \frac{\partial}{\partial t} + u\frac{\partial}{\partial x} \tag{6.19}$$

We consider the behavior of a gas caused by a receding piston, to visualize the phenomenon of a rarefaction wave (RW). The piston is moving in the $-x$ direction so that the gas is continually rarefied as it flows (in the $-x$ direction). The disturbance, called an RW, is moving forward, in the $+x$ direction. One can consider the RW to be represented by a sequence of jumps $d\rho$, dP, du, etc., so that we can use the Riemann invariant to solve the problem. The forward RW moves into an undisturbed material defined by pressure P_0, density ρ_0, flow u_0, and the speed of sound c_{s0}. Using Equation 6.18 we get

$$u - u_0 = \int_{P_0}^{P} \frac{dP}{\rho c_s} = \int_{\rho_0}^{\rho} \frac{c_s d\rho}{\rho} \quad \text{rarefaction moving in } +x \text{ direction}$$

$$\tag{6.20}$$

$$u - u_0 = -\int_{P_0}^{P} \frac{dP}{\rho c_s} = \int_{\rho_0}^{\rho} \frac{c_s d\rho}{\rho} \quad \text{rarefaction moving in } -x \text{ direction}$$

As an example, we calculate some physical quantities for an RW in an ideal gas. Since in an RW the entropy is constant, one has the following relation between the pressure, the density, and the speed of sound:

$$\frac{P}{P_0} = \left(\frac{\rho}{\rho_0}\right)^{\gamma}; \quad \frac{c_s}{c_{s0}} = \left(\frac{\rho}{\rho_0}\right)^{(\gamma-1)/2} \tag{6.21}$$

Substituting Equation 6.21 into Equation 6.20 and doing the integral, one gets

$$u - u_0 = \int_{\rho_0}^{\rho} \frac{c_s d\rho}{\rho} = \int_{c_{s0}}^{c_s} \frac{2 dc_s}{(\gamma - 1)} = \frac{2}{(\gamma - 1)}(c_s - c_{s0})$$

$$\Rightarrow c_s = c_{s0} + \frac{1}{2}(\gamma - 1)(u - u_0)$$

(6.22)

In this example the piston is moving in the $-x$ direction and at the gas–piston interface, the gas velocity is the same as the piston velocity, which has a negative value. The absolute value of the negative gas flow velocity is limited since the speed of sound is positive:

$$c_{s0} + \frac{1}{2}(\gamma - 1)(u - u_0) \geq 0; \quad \Rightarrow \frac{2c_{s0}}{(\gamma - 1)} \geq -u + u_0 \geq -u = |u|$$

(6.23)

From Equation 6.22 it is evident that the speed of sound is decreased since u is negative. This implies, according to Equation 6.21, that the density and the pressure are decreasing as expressed mathematically in Equation 6.19.

6.1.6 Shock Waves

The development of singularities, in the form of SWs, in a wave profile due to the nonlinear nature of the conservation Equation 6.3 has been already discussed by B. Riemann, W. J. M. Rankine, and H. Hugoniot in the second half of the nineteenth century (1860–1890).

An SW is created in a medium that suffers a sudden impact (e.g., a collision between an accelerated foil and a target) or in a medium that releases large amounts of energy in a short period of time (e.g., high explosives). When a pulsed high-power laser interacts with matter, very hot plasma is created. This plasma exerts a high pressure on the surrounding material, leading to the formation of an intense SW, moving into the interior of the target. The momentum of the out flowing plasma balances the momentum imparted to the compressed medium behind the shock front. For very high laser intensities ($I > 10^{15}$ W/cm^2) also the laser momentum I/c (where c is the speed of light) has to be taken into account (Eliezer, 2002). The thermal pressure together with the laser momentum and the momentum of the ablated material drives the SW.

It is convenient to analyze an SW by inspecting a gas compressed by a piston moving into it with a constant velocity u. The medium has initially (the undisturbed medium) a density ρ_0, a pressure P_0, and when it is at rest, $u_0 = 0$. An SW starts moving into the material with a velocity denoted by u_s. Behind the shock front the medium is compressed to a density ρ_1 and a pressure P_1. The gas flow velocity, denoted by u_p and usually called the particle velocity, in the compressed region is equal to the piston velocity, $u = u_p$. The initial mass before it is compressed, $\rho_0 A u_s t$ (A is the cross-sectional area of the tube), equals the mass after compression, $\rho_1 A(u_s - u_p)t$, implying the mass conservation law:

$$\rho_0 u_s = \rho_1 (u_s - u_p) \tag{6.24}$$

The momentum of the gas put into motion $(\rho_0 A u_s t) u_p$ equals the impulse due to the pressure forces $(P_1 - P_0)At$ yielding the momentum conservation law (or Newton's second law)

$$\rho_0 u_s u_p = P_1 - P_0 \tag{6.25}$$

The increase in internal energy E(energy/mass) and of kinetic energy per unit mass due to the piston-induced motion is $(\rho_0 A u_s t)(E_1 - E_2 + u_p^2/2)$. This increase in energy is supplied by the piston work, thus the energy conservation implies

$$\rho_0 u_s \left(E_1 - E_0 + \frac{1}{2} u_p^2 \right) = P_1 u_p \tag{6.26}$$

In the SW frame of reference, the undisturbed gas flows into the shock discontinuity with a velocity $v_0 = -u_s$ and leaves this discontinuity with a velocity $v_1 = -(u_s - u_p)$

$$v_0 = -u_s; \quad v_1 = -(u_s - u_p) \tag{6.27}$$

Substituting Equation 6.27 into Equations 6.24 through 6.26, one gets the conservation laws of mass, momentum, and energy as seen from the shock-wave front frame of reference:

$$\text{shock wave frame} \begin{cases} \rho_1 v_1 = \rho_0 v_0 \\ P_1 + \rho_1 v_1^2 = P_0 + \rho_0 v_0^2 \\ E_1 + \dfrac{P_1}{\rho_1} + \dfrac{v_1^2}{2} = E_0 + \dfrac{P_0}{\rho_0} + \dfrac{v_0^2}{2} \end{cases} \tag{6.28}$$

These equations can be obtained more rigorously from the fluid equation (Equation 6.3) by integrating them over the SW layer. Assuming that the thickness of the SW front tends to zero, then for two points x_0 and x_1 lying on both sides of the shock discontinuity, one can write

$$\lim_{x_1 \to x_0} \int_{x_0}^{x_1} \frac{\partial}{\partial t} [\cdots] \, dx = 0; \quad \lim_{x_1 \to x_0} \int_{x_0}^{x_1} \frac{\partial}{\partial t} [\cdots] \, dx = [\cdots]_{x_1} - [\cdots]_{x_0} \tag{6.29}$$

where [···] represents the different terms of equations (Equation 6.3). The procedure described in Equation 6.29 yields the jump conditions in the SW frame of reference as given by Equation 6.28.

The jump conditions in the laboratory frame of reference are given in Equations 6.24 through 6.26 for a fluid initially at rest. In a more general case, the material is set into motion before the arrival of the SW (e.g., by another SW). If the initial flow velocity is $u_0 \neq 0$, then the conservation laws (mass, momentum, and energy) in the laboratory frame of reference can be written

$$\text{laboratory frame } \begin{cases} \rho_0(u_s - u_0) = \rho_1(u_s - u_p) \\ \rho_0(u_s - u_0)(u_p - u_0) = P_1 - P_0 \\ \rho_0(u_s - u_0)\left(E_1 - E_0 + \dfrac{1}{2}u_p^2 - \dfrac{1}{2}u_0^2\right) = P_1 u_p - P_0 u_0 \end{cases} \quad (6.30)$$

Equations 6.29 or 6.30 are occasionally called the Rankine–Hugoniot relations. These relations are used to determine the state of the compressed solid behind the shock front. Assuming that the initial state is well defined and the quantities E_0, u_0, P_0, and ρ_0 are known, one has five unknowns E_1, u_p, P_1, ρ_1, and u_s with three equations (Equation 6.30). Usually the SW velocity is measured experimentally, and if the EOS is known (in this case one has four equations) $E = E(P,\rho)$, then the quantities of the compressed state can be calculated. If the EOS is not known, then one has to measure experimentally two quantities of the shocked material, for example u_s and u_p, in order to solve the problem.

From the first two terms of Equation 6.28 and using Equation 6.27, the following general relations can be written

$$\frac{v_0}{v_1} = \frac{\rho_1}{\rho_0} \equiv \frac{V_0}{V_1}; \quad V \equiv \frac{1}{\rho}$$

$$v_0 = V_0 \left(\frac{P_1 - P_0}{V_0 - V_1}\right)^{1/2} = |u_s|; \quad v_1 = V_1 \left(\frac{P_1 - P_0}{V_0 - V_1}\right)^{1/2} \quad (6.31)$$

$$v_0 - v_1 = \left[(P_1 - P_0)(V_0 - V_1)\right]^{1/2} = |u_p|$$

Substituting v_0 and v_1 from these relations into the third relation of Equation 6.28, one gets

$$E_1(V_1, P_1) - E_0(V_0, P_0) = \frac{1}{2}(P_1 + P_0)(V_0 - V_1) \quad (6.32)$$

The thermodynamic relation $E(V,P)$ is the EOS of the material under consideration, and assuming the knowledge of this function, then Equation 6.32 yields a graph (the notation of P_1 is changed to P_H and V_1 is V)

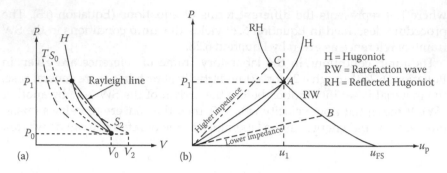

FIGURE 6.3
(a) The Hugoniot P–V curve, denoted by H, an isentrope S_0, and an isotherm denoted by T, both starting like the Hugoniot at V_0, and an isentrope S_1 starting at the final shock pressure P_1. (b) The Hugoniot P–u_p curve. RW is the release wave, while RH is the reflected Hugoniot.

$$P_H = P_H(V; V_0, P_0) \qquad (6.33)$$

This curve is known in the literature as the Hugoniot curve. The Hugoniot curve, shown schematically in Figure 6.3a, is a two-parameter (V_0, P_0) family of curves, so that for each initial condition (V_0, P_0) there is a different curve. The Hugoniot curve is not a thermodynamic function, it does not show the pressure–volume (or density) trajectory of an SW development, but it is a plot of all possible final shocked (compressed) states for a given initial state (V_0, P_0). For example, the Hugoniot curve is different from the isentropic curves of the pressure $P_S(V)$, which describes the thermodynamic trajectory of pressure–volume (or density) for any given entropy S. In Figure 6.3a, two isentropes and one isotherm are plotted schematically. It is interesting to note that for a given final pressure the compression $(\rho/\rho_0 = V_0/V)$ is higher for an isentrope relative to the Hugoniot and the isothermal compression is the highest.

It is useful to consider the SW relations for an ideal gas with constant specific heats. In this case the EOS are

$$E = C_V T = \frac{PV}{\gamma - 1}; \quad S = C_V \ln\left(PV^\gamma\right) \qquad (6.34)$$

where γ is defined in Equation 6.12. Substituting Equation 6.34 into Equation 6.32, the Hugoniot curve for an ideal gas EOS is obtained:

$$\frac{P_1}{P_0} = \frac{(\gamma + 1)V_0 - (\gamma - 1)V_1}{(\gamma + 1)V_1 - (\gamma - 1)V_0} \qquad (6.35)$$

From this equation it follows that the compression ratio $\rho_1/\rho_0 = V_0/V_1$ cannot increase above a certain value determined by γ. The pressure in the shocked ideal gas tends to infinity for the maximum possible compression given by

$$\frac{P_1}{P_0} \to \infty \Rightarrow \left(\frac{P_1}{\rho_0}\right)_{\max} = \left(\frac{V_0}{V_1}\right)_{\max} = \frac{\gamma+1}{\gamma-1} \qquad (6.36)$$

For example, the maximum compression caused by a planar SW in a medium with $\gamma = 5/3$ is $4\rho_0$.

Using Equation 6.13, for the speed of sound c_s, and Equation 6.35, both for the ideal EOS fluid, one obtains from Equation 6.31

$$M_0^2 \equiv \left(\frac{u_s}{c_{s0}}\right)^2 = \left(\frac{v_0}{c_{s0}}\right)^2 = \frac{1}{2\gamma}\left[(\gamma-1)+(\gamma+1)\frac{P_1}{P_0}\right] > 1$$

$$M_1^2 \equiv \left(\frac{v_1}{c_{s1}}\right)^2 = \frac{1}{2\gamma}\left[(\gamma-1)+(\gamma+1)\frac{P_0}{P_1}\right] < 1 \qquad (6.37)$$

The ratio M of the flow velocity to the sound velocity is known as the Mach number. In the limit of a weak SW, defined by $P_1 \approx P_0$, one has $M_1 \approx M_0 \approx 1$. However, the equation (Equation 6.37) yields in general $M_1 < 1$ and $M_0 > 1$. The meaning of these relations is that in the shock frame of reference, the fluid flows into the shock front at a supersonic velocity ($M_0 > 1$) and flows out at a subsonic velocity ($M_1 < 1$). In the laboratory frame of reference, one has the well-known result that the SW propagates at a supersonic speed (with respect to the undisturbed medium) and at a subsonic speed with respect to the compressed material behind the shock. Although this phenomenon has been proven here for an ideal gas EOS, this result is true for any medium, independent of the EOS (Landau and Lifshitz, 1987).

In an SW the entropy always increases. For example, an ideal gas equation (Equations 6.34 and 6.35) gives the increase in entropy during an SW process

$$S_1 - S_0 = C_v \ln\left(\frac{P_1 V_1^\gamma}{P_0 V_0^\gamma}\right)$$

$$= \left[\frac{P_0 V_0}{(\gamma-1)T_0}\right] \ln\left\{\left(\frac{P_1}{P_0}\right)\left[\frac{(\gamma-1)\frac{P_1}{P_0}+(\gamma+1)}{(\gamma+1)\frac{P_1}{P_0}+(\gamma-1)}\right]^\gamma\right\} > 0 \qquad (6.38)$$

The increase in entropy indicates that an SW is not a reversible process, but a dissipative phenomenon. The entropy jump of a medium (compressed by SW) increases with the strength of the SW (defined by the ratio P_1/P_0). The larger the P_1/P_0 the larger is $S_1 - S_0 = \Delta S$. The value of ΔS is determined by the conservation laws (mass, momentum, and energy) and by the EOS; however, the mechanism of this change is described by viscosity and thermal conductivity (Zeldovich and Raizer, 1966).

So far we have discussed a 1D, steady-state SW passing through a gas or a fluid medium, with a thermal equilibrium ahead and behind the SW front. When an SW is created in a solid material, one has to take into account the elastic–plastic properties of the target (Asay and Shahinpoor, 1993). For SWs with a pressure larger than 20 kbar, all the material strength properties are negligible and the solid target may be considered a liquid. Since we are interested in high pressures, larger than 20 kbar, the fluid jump equations (Equations 6.28 and 6.30) describe the experiments correctly.

6.1.7 $u_s - u_p$ EOS

It was found experimentally (McQueen, 1991) that for many solid materials, initially at rest, the following linear relation between the shock velocity u_s and the particle velocity u_p is valid to a very good approximation:

$$u_s = c_0 + \alpha u_p \qquad (6.39)$$

where c_0 and α are constants. c_0 is the bulk sound speed at standard conditions, where c_0 is related to the longitudinal (c_L) and transversal (c_t) velocities of the sound:

$$c_0 = \sqrt{c_L^2 - \frac{4}{3}c_t^2}.$$

In Table 6.3, the experimental values of c_0 and α are given (Steinberg, 1996) for some elements that fit Equation 6.39. ρ_0 is the initial density of the elements (with an atomic number Z). Equation 6.39 is occasionally called the (Hugoniot) EOS.

It is convenient to describe the Hugoniot curve in the pressure-particle speed space, P-u_p. In particular, for the Equation 6.39 EOS, the Hugoniot is a parabola in this space (Figure 6.3b). When the SW reaches the back surface of the solid target, the free surface starts moving (into the vacuum or the surrounding atmosphere) with a velocity u_{FS} and a release wave, in the form of an RW, is backscattered into the medium. Note that if the target is positioned in vacuum, then the pressure of the back surface (denoted in the literature as the free surface) is zero, a boundary value fixed by the vacuum. If an atmosphere surrounds the target, then an SW will run into this atmosphere. In our analysis we do not consider this effect and take $P = 0$ at the free surface. This approximation is justified for analyzing the high-pressure shocked target that is considered here.

If the target A is bounded by another solid target B (Figure 6.4a), then an SW passes from A into B and a wave is backscattered (into A). The impedances

TABLE 6.3

Experimental Fit to $u_s = c_0 + \alpha u_p$ on the Hugoniot Curve

Element		Z	ρ_0 (g/cm³)	c_0 (cm/µs)	α
Li	Lithium	3	0.534	0.477	1.066
Be	Beryllium (S200)	4	1.85	0.800	1.124
Mg	Magnesium	12	1.78	0.452	1.242
Al	Aluminum (6061-T6)	13	2.703	0.524	1.40
V	Vanadium	23	6.10	0.5077	1.201
Ni	Nickel	28	8.90	0.465	1.445
Cu	Copper	29	8.93	0.394	1.489
Zn	Zinc	30	7.139	0.303	1.55
Nb	Niobium	41	8.59	0.444	1.207
Mo	Molybdenum	42	10.2	0.5143	1.255
Ag	Silver	47	10.49	0.327	1.55
Cd	Cadmium	48	8.639	0.248	1.64
Sn	Tin	50	7.287	0.259	1.49
Ta	Tantalum	73	16.69	0.341	1.2
W	Tungsten	74	19.3	0.403	1.237
Pt	Platinum	78	21.44	0.364	1.54
Au	Gold	79	19.3	0.308	1.56
Th	Thorium	90	11.7	0.213	1.278
U	Uranium	92	19.05	0.248	1.53

Note: u_s is the shock wave velocity, u_p is the particle flow velocity, and ρ_0 is the initial density of the element with an atomic number Z.

$Z = \rho_0 u_s$ of A and B are responsible for the character of this reflected wave. If $Z_A > Z_B$ then an RW is backscattered (into A) while in the $Z_A < Z_B$ case an SW is backscattered at the interface between A and B. Note that in both cases an SW goes through (into medium B). These possibilities are shown schematically in Figure 6.3b. The main laser beam creates an SW. H denotes the Hugoniot of A, and point A describes the pressure and particle flow velocity of the SW (just) before reaching the interface between the targets. If $Z_A > Z_B$ then the lower impedance line (the line $P = Zu_p$) meets the RW curve at point B, while point C describes the case $Z_A < Z_B$ (a higher impedance), where at the interface an SW is backscattered. The final pressure and final flow velocity (just) after the wave passes the interface are determined by point C for the higher impedance ($Z_B > Z_A$) and by point B for the lower impedance ($Z_B < Z_A$). The latter case is shown in detail in Figure 6.4b.

If the impedances of A and B are not very different, impedance matching, then to a very good approximation the RW curve in Figure 6.4b and RH–RW curve in Figure 6.3b are the mirror reflection (with respect to the vertical line at $u_1 = $ const.) of the Hugoniot H curve. In this case, from Figure 6.4b, one has

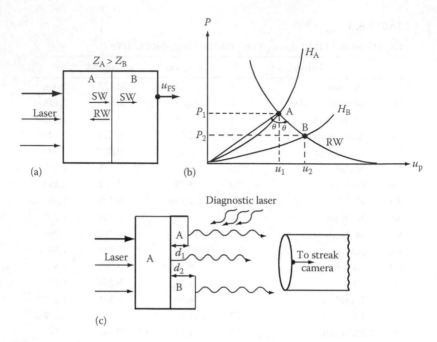

FIGURE 6.4
(a) A laser induces a shock wave into target A with impedance smaller than target B. An RW is reflected at the interface A–B. (b) The Hugoniot curves H_A and H_B for shock waves in A and B. The reflected RW is also shown. (c) A schematic setup for an impedance matching experiment.

$$Z \equiv \rho_0 u_s; \quad u_{sA} = c_{0A} + \alpha_A u_1$$
$$P_1 = Z_A u_1 = \rho_{0A} u_{sA} u_1; \quad P_2 = Z_B u_2$$
$$\tan\theta = \frac{u_2 - u_1}{P_2 - P_1} = \frac{u_1}{P_1} \Rightarrow \frac{P_2}{P_1} = \frac{2 Z_B}{Z_A + Z_B} \approx \frac{2 \rho_{0B} c_{0B}}{\rho_{0A} c_{0A} + \rho_{0B} c_{0B}} \tag{6.40}$$

where the last approximate equality is for weak SWs. A similar result is obtained in the case with higher impedance ($Z_B > Z_A$).

In Figure 6.4c a schematic setup of an impedance matching experiment is given. When an SW reaches the interface with the vacuum (or the surrounding atmosphere) it irradiates according to the temperature of the SW heated medium. If the SW temperature is high (~few thousands degrees K) then the self-illumination may be large enough to be detected by a streak camera (or other appropriate optical collecting device with a fast information recording). If the detecting devices are not sensitive to the self-illumination, then the measurement of a reflected (diagnostic) laser may be more useful, since the reflection changes significantly with the arrival of the SW. The SW velocities in A and B are directly measured in this way by recording the signal of shock breakthrough from the base of A and from the external surfaces of the stepped targets. The time t_1 that the SW travels through a distance d_1 in A and the time t_2 that the SW travels through a distance d_2 in B (Figure 6.4c) yield the appropriate shock velocities in both targets. Since the initial densities are known, the impedances of A and B are directly measured. The EOS of A is known and the Hugoniot of A is plotted. Using Equation 6.40 P_1 is known (from the measurement of u_{sA}), and P_2 is directly calculated from the measurements of both impedances. In this way, the difficult task of measuring two parameters in the unknown (EOS) material B is avoided.

It seems quite straightforward to measure directly the SW velocity if the SW is steady, 1D, and the measurement device is accurate. It is also possible to measure indirectly the particle flow velocity by measuring the free surface velocity. Accurate optical devices (Moshe et al., 1996), called velocity interferometer system for any reflector (VISAR) and optically recording velocity interferometer system (ORVIS), practically, very fast recording "radar" devices in the optical spectrum, have been developed to accurately measure the free surface velocity. After the SW reaches the back surface of the target, a release wave with the characteristics of an RW is back scattered into the target. Since this isentrope is almost a mirror image of the Hugoniot (Figure 6.4b), one gets

$$u_1 = \frac{1}{2}u_{FS} \tag{6.41}$$

Therefore, the measurement of u_s and u_{FS} determines all the parameters in the compressed medium (assuming the initial state is accurately known).

6.1.8 Impact by an Accelerating Foil

The highest laser-induced pressures, ~10^8–10^9 atm have been obtained during the collision of a target with an accelerating foil. This acceleration was achieved by laser-produced plasma or by x-rays from a cavity produced by laser–plasma interactions. It is therefore now shown how the pressure is calculated in a planar collision between a flyer and a target.

The flyer has a known (i.e., measured experimentally) initial velocity before impact, u_f. The initial state before collision is for the target (B) $u_p = 0$ and $P = 0$, while for the flyer (A) $u_p = u_f$ and $P = 0$. Upon impact, an SW moves forward into B, and another SW goes into the flyer in the opposite direction. The pressure and the particle velocity are continuous at the interface of target–flyer. Therefore, the particle velocity of the target changes from 0 to u, while the particle velocity in the flyer changes from u_f to u. Moreover, the pressure in the flyer plate A equals the pressure in the target plate B, and if the EOSs are known and given by Equation 6.39, and the second term of (Equation 6.30) is used to calculate the pressure, one gets

$$P_H = \rho_{0B}u\left(c_{0B} + \alpha_B u\right) = \rho_{0A}\left(u_f - u\right)\left[c_{0A} + \alpha_A\left(u_f - u\right)\right] \tag{6.42}$$

This is a quadratic equation in u, with the following solution:

$$u = \frac{-b - \sqrt{b^2 - 4ac}}{2a}$$

$$a \equiv \rho_{0A}\alpha_A - \rho_{0B}\alpha_B; \quad c \equiv (c_{0A}\rho_{0A} + \rho_{0A}\alpha_A u_f)u_f$$

$$b \equiv -(c_{0A}\rho_{0A} + c_{0B}\rho_{0B} + 2\rho_{0A}\alpha_A u_f)$$

(6.43)

Once the flow velocity u is known, it is substituted into Equation 6.42 to derive the pressure.

If the EOS of the target is not known, then it is necessary to measure the SW velocity u_{sB} as explained here. In this case, the pressure equality in the flyer and the target yields

$$P_H = \rho_{0B}u_{sB}u = \rho_{0A}\left[c_{0A} + \alpha_A (u_f - u)\right](u_f - u)$$

(6.44)

Note that in this equation u_{sB} is known. The solution of this equation is

$$u = u_f + w - \left[w^2 + \left(\frac{\rho_{0B}}{\rho_{0A}\alpha_A}\right)u_{sB}u_f\right]^{1/2}$$

$$w \equiv \frac{1}{2\alpha_A}\left(c_{0A} + \frac{\rho_{0B}u_{sB}}{\rho_{0A}}\right)$$

(6.45)

In these types of experiments it is occasionally convenient to measure the free surface velocity of the target to calculate the particle velocity.

Foils are used to impact stationary targets and reach pressures of hundreds of Mbar. In this method, the flyer stores kinetic energy from the driver, which may be laser or x-rays, and delivers it much rapidly as thermal energy in collision. In addition, the flyer serves as a preheat shield so that the target remains on a lower adiabat if it were exposed to the driver. The flyer technique was demonstrated using laser as a driver (Obenschain et al., 1983) and later applied to achieve pressures of more than 100 Mbar (Fabbro et al., 1986).

The flyer technique was used with x-ray created by Nova and produced over 700 Mbar in a planar shock (Cauble et al., 1993), enabling first experimental measurements of EOS in the gigabar regime.

6.1.9   SW Stability

We end this section with a short comment on SW stability. One can see from Equation 6.21 that different disturbances of density travel with different velocities, so that the larger the density ρ the faster the wave travels. Therefore, an initial profile $\rho(x,0)$ becomes distorted with time. This is true not only for the density but also for the pressure $P(x,0)$, for the flow velocity $u(x,0)$, etc. In this way a smooth function of these parameters will steepen in time due to the nonlinear effect of the wave propagation (higher amplitudes move faster). Therefore, a compression wave is steepened into an SW because in most solids the sound velocity is an increasing function of the pressure. In the laboratory frame of reference, the speed of a disturbance is the sum of the flow velocity and the sound velocity ($c_s + u$). Therefore, a higher-pressure disturbance will catch up with the lower-pressure disturbance causing a sharpening profile of the wave. In reality there are also dissipative mechanisms such as viscosity and thermal transport. Therefore, the sharpening profile mechanism can only increase until the dissipative forces become significant, and they begin to cancel out the effect of increasing sound speed with pressure. When the sum of these opposing mechanisms cancels out, the wave profile does not change in time anymore and it becomes a steady SW.

The dissipative phenomena are nonlinear functions of the strain rate (strain is the relative distortion of a solid ~dl/l and the strain rate is the time derivative of the strain). Therefore, for very fast laser-induced phenomena the rise time of an SW is very small, and it can be significantly less than 1 ns.

As already stated above, a disturbance moves at the speed $c_s + u$ in a compression wave. Therefore, a disturbance behind the shock front cannot be slower than the shock velocity; because in this case it will not be able to catch the wave front, and the shock would decay (that is, the shock is unstable to small disturbances behind it). Similarly, a small compressive disturbance ahead of the shock must move slower than the shock front in order not to create another SW. Thus, the conditions for a stable SW can be summarized in the following way:

$$\frac{dc_s}{dP} > 0; \quad c_s + u_p \geq u_s; \quad u_s > c_{s0} \tag{6.46}$$

The first of these equations states that the speed of sound increases with increasing pressure. The second equation describes the fact that the SW is subsonic (Mach number smaller than 1) with respect to the shocked medium. The last equation of Equation 46 is the well-known phenomenon that an SW is supersonic (Mach number larger than 1) with respect to the unshocked medium. Using Equations 6.10 and 6.31

$$c_s = \sqrt{\left(\frac{\partial P}{\partial \rho}\right)_s} = V\sqrt{-\left(\frac{\partial P}{\partial V}\right)_s}; \quad u_s = V_0\sqrt{\frac{P_1 - P_0}{V_0 - V_1}} \tag{6.47}$$

one gets from the last part of Equation 6.46 the following shock stability criterion:

$$\frac{P_1 - P_0}{V_0 - V_1} > -\left(\frac{\partial P}{\partial V}\right)_{s,0} \tag{6.48}$$

This inequality states that the slope of the Rayleigh line (Figure 6.3a) is larger than the slope of the isentrope at the initial state. Similarly, one can show that the second part of Equation 6.46 yields the criterion that the Hugoniot at the final state must be steeper than the Rayleigh line:

$$\frac{P_1 - P_0}{V_0 - V_1} \leq -\left(\frac{dP}{dV}\right)_{H,1} \tag{6.49}$$

It is important to point out that these criteria are not always satisfied. For example, the speed of sound of fused silica decreases with pressure in a domain of low pressures, so that in this case the SW is not stable. Moreover, in the regime of phase transitions (solid–solid due to change in symmetry or solid–liquid) the SW can split into two or more SWs. However, in these cases the stability criteria can be satisfied for each individual SW.

6.1.10 Critical Problems for Laser-Induced SWs

The problems with laser-induced SWs are the small size of the targets (~100 μm), the short laser pulse duration (~1 ns), the poor spatial uniformity of a coherent electromagnetic pulse (the laser), and therefore the nonuniformity of the created pressure. All these difficulties have been addressed in the literature and the subject of laser-induced SW research has been developed significantly. The critical problems are summarized below and their possible solutions are discussed.

1. Planarity (1D) of the SW regardless of the laser irradiance non-uniformity.
2. Steady SW during the diagnostic measurements in spite of the laser short pulse duration.
3. Well-known initial conditions of the shocked medium. This requires the control (namely, to avoid) of the fast electron and x-ray pre-heating.
4. Good accuracy (~1%) of the measurements.

The planarity of the SW is achieved by using optical smoothing techniques (Lehmberg and Obenschain, 1983; Kato et al., 1984; Skupsky et al., 1989; Koenig et al., 1994; Batani et al., 1996). With these devices the laser is deposited into the target uniformly, within ~2% of energy deposition. For example (Lehmberg and Obenschain, 1983), one technique denoted as induced spatial incoherence consists of breaking each laser beam into a large number of beamlets by reflecting the beam off a large number of echelons. The size of each beamlet is chosen in such a way that its diffraction-limited spot size is about the target diameter. All of the beamlets are independently focused and overlapped on the target. Another technique (Kato et al., 1984) divides the beam into many elements that have a random phase shift. This is achieved

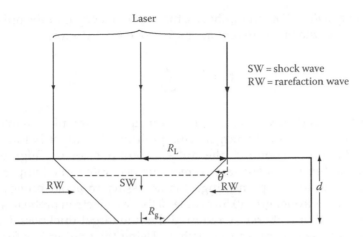

FIGURE 6.5
A planar target profile irradiated by a laser (the target thickness is d). The 1D spot area = πR_g^2 (g for good) in comparison with the laser irradiation area = πR_L^2. RW and SW stand for rarefaction wave and shock wave, respectively.

by passing the laser beam through a phase plate with a randomly phase shifted mask.

The focal spot of the laser beam on the target has to be much larger than the target thickness in order to achieve a 1D steady-state SW. In Figure 6.5 one can see a schematic profile of a planar target with thickness d, irradiated by a laser with a focal spot area = πR_L^2. A lateral RW enters the shocked area and reduces the pressure and density of the shocked area. This effect distorts the 1D character of the wave, since the shock front is bent in such a way that for very large distances (>>d) the SW front becomes spherical. The RW propagates toward the symmetry axis with the speed of sound c_s (in the shock-compressed area), which is larger than the SW velocity u_s. Therefore, the good (i.e., undisturbed by the RW) 1D shock area = πR_g^2 is limited by

$$R_g = R_L - \left(\frac{d}{u_s}\right) c_s \leq R_L - d; \quad A \equiv \pi R_L^2 \geq 10d^2 \qquad (6.50)$$

where d/u_s is the time that the SW reaches the back surface. Therefore, in order to have a 1D SW one requires that $R_g \sim d$, implying an $R_L \sim 2d$ at least, so that the laser focal spot area $A \sim 10d^2$. This constraint implies very large laser focal spots for thick targets.

The second constraint requires a steady SW, particularly the shock velocity has to be constant as it traverses the target. A rarefaction (RW), initiated at a time $\Delta\tau \sim \tau_L$ (the laser pulse duration) after the end of the laser pulse, follows an SW into the target. It is necessary that the RW does not overtake the SW at position $x = d$ (the back surface) during the measurement of the SW velocity, implying $\tau_L > d/u_s$. For strong shocks, the shock velocity is of the order of the square root of the pressure (Equations 6.31 or 6.47); therefore,

$$\tau_L > \frac{d}{u_s} \propto \frac{d}{\sqrt{P}} \tag{6.51}$$

Hot electrons and/or x-rays can appear during the laser–plasma interaction causing preheating of the target. This preheats the target before the SW arrives, therefore "spoiling" the initial conditions for the high-pressure experiment. Since it is not easy to measure accurately the temperature of the target due to this preheating, it is necessary to avoid preheating. By using shorter wavelengths (0.5 μm or less) the fast electron preheat is significantly reduced. It is therefore required that the target thickness d is larger than the hot electron mean free path λ_e. Using the scaling law for the hot electron temperature T_h one has

$$d \gg \lambda_e \propto T_h^2 \propto \left(I_L \lambda_L\right)^{0.6} \tag{6.52}$$

Taking into account all of the constraints given by the relations in Equations 6.50 through 6.52 and using the experimental scaling law

$$P \propto I_L^{0.8} \tag{6.53}$$

one gets the scaling of the laser energy W_L

$$W_L = I_L A \tau_L \propto I_L^{2.4} \propto P^3 \tag{6.54}$$

Therefore, to increase the 1D SW pressure by a factor of two it is necessary to increase the laser energy by an order of magnitude.

A more elegant and efficient way to overcome this problem is to accelerate a thin foil. The foil absorbs the laser, plasma is created (ablation), and the foil is accelerated like in a rocket. In this way, the flyer stores kinetic energy from the laser during the laser pulse duration (the acceleration time) and delivers it, in a shorter time during the collision with a target, in the form of thermal energy. The flyer is effectively shielding the target so that the target initial conditions are not changed by fast electrons or by laser-produced x-rays. For these reasons the laser-driven flyer can achieve much higher pressures on impact than the directly laser-induced SW (Fabbro et al., 1986; Cauble et al., 1993).

The accuracy of measurements in the study of laser-induced high-pressure physics requires diagnostics with a time resolution better than 100 ps, and occasionally better than 10 ps, and a spatial resolution of the order of a few microns. The accurate measurements of SW speed and particle flow velocity are usually obtained with optical devices, including a streak camera (Veeser et al., 1979; Cottet et al., 1985; Ng et al., 1987; Evans et al., 1996b) and velocity interferometers (Moshe et al., 1996; Celliers et al., 1998; Swift et al., 2004).

6.1.11 SWs and EOS Studies

The phenomena of SW propagation through condensed matter are strongly related to the physics of the EOS (Altshuler, 1965; Eliezer et al., 1986; Eliezer and Ricci, 1991; Bushman et al., 1993). Atoms in a solid state are attracted to each other at large distances and repel each other at short distances. At zero temperature and zero pressure the equilibrium is obtained at a specific volume V_{0c}, reached at the minimum of the (potential) energy. The atoms can be separated to a large distance by supplying the binding energy (known also as the cohesion energy) of the order of a few eV/atom, while compressing a condensed matter requires energy to overcome the repulsive forces.

Small pressures, of the order of 10^2 atmospheres, can compress a gas to a few times its initial density. However, the compression of a metal by only 10% requires an external pressure (static or SW) of the order of 10^5 atm. This is evident from the values of the compressibility at standard conditions $\sim 10^{-6}$ atm^{-1}, where κ is defined by

$$\kappa_S = -\frac{1}{V}\left(\frac{\partial V}{\partial P}\right)_S \tag{6.55}$$

Note that the bulk modulus B_S is the inverse of the compressibility, $B_S = 1/\kappa_S$.

The P–V in Figure 6.6a describes the Hugoniot curve $P_H(V)$ for the initial conditions V_0, T_0, and $P_0 = 0$. Note that the atmospheric pressure is negligibly

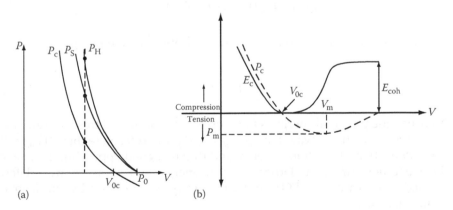

FIGURE 6.6
(a) The cold pressure P_c, the Hugoniot pressure P_H, and the isentrope P_s. The initial conditions for P_H and P_s are $V_0 = 1/\rho_0$, $P_0 = 0$, and T_0 (room temperature). (b) A schematic representation of the cold energy E_c and pressure P_c. The cohesion energy E_{coh} and the minimum pressure P_m, both at $T = 0$, are also shown. For $P_c > 0$ the solid is compressed while for $P_c < 0$ a tension is applied on the medium.

small in comparison with the SW pressures under consideration, so that $P_0 = 0$ or 1 atm is an equivalent initial condition for SW experiments in solids. The difference in Figure 6.5a at any particular value of V between the cold pressure P_c and the Hugoniot pressure P_H is equal to the thermal pressure P_T that is caused by the SW. Since P_T is positive, the cold pressure at room temperature is negative, $P_c(V_0) < 0$. In this figure also the isentrope P_S that passes through the beginning of the Hugoniot is shown, so that the mutual positions of P_H, P_S, and P_c are illustrated.

The pressure and the internal energy of a material can be divided into two contributions.

1. The cold term (temperature independent) caused by the interaction forces between the atoms of the material. The appropriate pressure P_c and energy E_c are defined as "cold pressure" and "cold energy," respectively.
2. The thermal term is usually divided into two parts: one part describes the motion of the atoms in the lattice while the second part is due to the electron thermal motion.

P_a and E_a denote the thermal lattice pressure and energy, respectively, and the appropriate electron contributions are P_e and E_e. Therefore, the EOS is usually written in the following form:

$$
\begin{aligned}
P(V,T) &= P_c(V) + P_T(V,T) \\
P_T(V,T) &= P_a(V,T) + P_e(V,T) \\
E(V,T) &= E_c(V) + E_T(V,T) \\
E_T(V,T) &= E_a(V,T) + E_e(V,T)
\end{aligned}
\tag{6.56}
$$

The thermodynamic relation relates the cold pressure and energy

$$
E_c(V) = -\int_{V_{0c}}^{V} P_c(V)\,dV
\tag{6.57}
$$

As usual, the potential energy is fixed up to a constant, and this constant has been chosen in such a way that $E_c(V_{0c}) = 0$ (Figure 6.6b). Note that $P_c(V_{0c}) = 0$ is the definition of V_{0c}. It has been found that to a good approximation, near the Hugoniot curve, the lattice thermal energy and pressure are related through the Gruneisen EOS, and the energy is described by the Debye solid-state theory.

$$
E_a = C_V(T - T_0) + E_0
$$

$$
P_a(V,T) = \frac{\gamma_a(V)}{V} E_a
\tag{6.58}
$$

where

γ_a is the Gruneisen coefficient (assumed to be a function of V)

C_V is the lattice-specific heat at constant volume

T_0 is the initial temperature in the SW experiment, usually the room temperature

E_0 is the internal energy at T_0, defined by

$$E_0 = \int_0^{T_0} c_V(T)\,dT \qquad (6.59)$$

At room temperature and for slightly lower and higher temperatures, the specific heat C_V has the Dulong–Petit value:

$$C_V = \frac{3R}{<A>} \left[\frac{erg}{g\ deg} \right]; \quad R = 8.314 \times 10^7 \left[\frac{erg}{mol\ deg} \right]$$

$$<A> = \text{average atomic weight} \left[\frac{g}{mol} \right] \qquad (6.60)$$

For example, $<A>$ for aluminum and copper is 27 and 63.55, respectively.

Assuming for the electron thermal part an EOS similar to the Gruneisen EOS for the lattice, one has

$$P_e = \frac{\gamma_e}{V} E_e \qquad (6.61)$$

To the first approximation, the thermal electron energy is taken from the free electron model

$$E_e = \frac{1}{2}\beta T^2 \qquad (6.62)$$

and β is related to the specific volume V by the thermodynamic relation

$$\left(\frac{\partial E}{\partial V} \right)_T = T\left(\frac{\partial P}{\partial T} \right)_V - P \qquad (6.63)$$

Substituting Equations 6.61 and 6.62 into Equation 6.63 yields

$$\frac{\partial \beta}{\partial V} = \frac{\gamma_e \beta}{V} \Rightarrow \beta = \gamma_e \exp\left(\int_{V_0}^{V} \frac{\gamma_e dV}{V} \right)$$

$$\beta = \beta_0 \left(\frac{V}{V_0} \right)^{1/2}, \quad \text{for } \gamma_e = 0.5 \qquad (6.64)$$

The value of $\gamma_e = 0.5$ is a good approximation for many metals up to SW data of few megabars pressure.

Combining Equations 6.56 through 6.58, 6.61, and 6.64, one gets

$$P = P_c + \frac{\gamma_a}{V}\left[c_V\left(T - T_0\right) + E_0\right] + \frac{1}{4}\left(\frac{\beta_0}{V_0}\right)\left(\frac{V_0}{V}\right)^{1/2} T^2$$

$$E = -\int_{}^{V} P_c dV + c_V\left(T - T_0\right) + E_0 + \frac{\beta_0}{2}\left(\frac{V}{V_0}\right)^{1/2} T^2$$

(6.65)

The cold pressure can be described by a polynomial in $(V_{0c}/V)^{1/3}$ in such a way that it fits the experimental values of the bulk modulus, the derivative of the bulk modulus at zero pressure, the cohesion energy, and the TF model (or modified models) at very high compressions.

Although the EOS described in Equation 6.65 has 10 parameters, it does not describe phase transitions, such as solid to solid with different symmetry or melting and evaporation. In order to describe these phenomena many more parameters based on theoretical models and experimental data are required.

In an SW experiment one can measure the SW velocity (u_s) and the particle velocity (u_p). It has been attempted to also measure the temperature associated with an SW; however, the experimental data so far are not very accurate.

The SW velocity is directly measured in a stepped target (velocity = measured step length/measured time difference) using an optical streak camera. The particle velocity is more difficult to measure, and u_p is deduced, for example, from impedance matching experiments or by the optical measuring of the free surface (i.e., back surface) velocity. The measured EOS is given in the form of the linear $u_s - u_p$ relation, Equation 6.39. The velocity measurements should have an accuracy of about 1% in order to receive good accuracy (~10%) in other variables such as pressure, energy, etc. It is important to point out that the linear $u_s - u_p$ relation, Equation 6.39, is not always correct for all solid elements. Sometimes a higher polynomial in u_p is necessary to fit the experimental data. For example (Steinberg, 1996),

$$\text{Hafnium: } u_s\,(\text{cm}/\mu\text{s}) = 0.298 + 1.448u_p - 3.852\left(\frac{u_p}{u_s}\right)u_p + 7.0\left(\frac{u_p}{u_s}\right)^2 u_p$$

$$\text{Lead: } u_s\,(\text{cm}/\mu\text{s}) = 0.2006 + 1.429u_p + 0.8506\left(\frac{u_p}{u_s}\right)u_p - 1.64\left(\frac{u_p}{u_s}\right)^2 u_p \qquad (6.66)$$

$$\text{Titanium: } u_s\,(\text{cm}/\mu\text{s}) = 0.502 + 1.536u_p - 5.138\left(\frac{u_p}{u_s}\right)u_p + 10.82\left(\frac{u_p}{u_s}\right)^2 u_p$$

In the following part of this section it is shown how one can study the thermodynamic EOS in the neighborhood of the Hugoniot from the measurements of the shock velocity and the particle velocity. For simplicity the further discussion assumes the linear $u_s - u_p$ relation (Equation 6.39), valid for many metals.

It is convenient to define the parameter η related to the compression ρ/ρ_0 in such a way that $\eta = 0$ for no compression and $\eta = 1$ for an infinite ρ/ρ_0:

$$\eta \equiv 1 - \frac{V}{V_0} = 1 - \frac{\rho_0}{\rho} \tag{6.67}$$

Using Equations 6.31 and 6.39 the particle velocity and the shock velocity can be presented in the following way:

$$\left.\begin{array}{l} \dfrac{u_p}{u_s} = \dfrac{V_0 - V}{V_0} = \eta \\[2mm] u_s = c_0 + \alpha u_p \end{array}\right\} \Rightarrow \quad \begin{array}{l} u_p = \dfrac{\eta c_0}{1 - \alpha\eta} \\[3mm] u_s = \dfrac{c_0}{1 - \alpha\eta} \end{array} \tag{6.68}$$

Substituting these results into the Hugoniot pressure, one gets P_H as a function of V (or η), the experimentally measured quantities c_0 and α, and the initial condition V_0:

$$P_H = \rho_0 u_s u_p = \frac{\eta c_0^2}{V_0 (1 - \alpha\eta)^2} = \frac{c_0^2 (V_0 - V)}{\left[V_0 - \alpha(V_0 - V)\right]^2} \tag{6.69}$$

It is interesting to point out that from this equation the maximum possible compression is obtained for an infinite P_H, implying

$$\frac{\rho_{max}}{\rho_0} = \frac{\alpha}{\alpha - 1} \tag{6.70}$$

For example (Table 6.3), α of aluminum equals 1.4 implying that in an SW the compression ρ/ρ_0 is less than 3.5.

The derivative of this pressure with respect to the specific volume V is

$$P_H' \equiv \left(\frac{dP}{dV}\right)_H = \frac{dP_H}{dV} = -\left(\frac{c_0}{V_0}\right)^2 \frac{1 + \alpha\eta}{(1 - \alpha\eta)^3} = -c_0^2 \left\{\frac{V_0 + \alpha(V_0 - V)}{\left[V_0 - \alpha(V_0 - V)\right]^3}\right\} \tag{6.71}$$

Using this relation the bulk modulus on the Hugoniot B_H, is obtained

$$B_H = -V\left(\frac{dP}{dV}\right)_H \tag{6.72}$$

and its derivatives are derived:

$$B_H\left(P_0,V_0\right)=B_S\left(P_0,V_0\right)=\frac{c_0^2}{V_0}=\rho_0 c_0^2$$

$$B_H'\left(P_0,V_0\right)=B_S'\left(P_0,V_0\right)=4\alpha-1 \tag{6.73}$$

From SW data (EOS) one knows c_0 and α and therefore the bulk modulus and its pressure derivative. The last two physical quantities, B_S and $dB_S/dP = B_S'$, are very important in calculating the cold pressure P_c and energy E_c. B_S, of the order of 1 Mbar, can also be measured in static experiments thanks to the technological advances in reaching high pressures (up to a few megabars). However, it is difficult to measure B_S' in static experiments to a good accuracy since this quantity measures the curvature of the P–V curve. On the other hand, the SW experiments in this domain of pressure are easily and accurately performed.

We now calculate the speed of sound:

$$c_s = \sqrt{\left(\frac{\partial P}{\partial \rho}\right)_S} = V\sqrt{-\left(\frac{\partial P}{\partial V}\right)_S} \equiv V\sqrt{-\frac{dP_S(V)}{dV}} \tag{6.74}$$

For this purpose one needs the isentrope pressure curve $P_S(V)$, which can be calculated from the knowledge of the Hugoniot pressure $P_H(V)$ using the Gruneisen EOS. In order to do so, the electron thermal contribution P_e should be neglected. For most of the solid materials, $P_e \ll P_a$ for pressures less than a few hundreds of kilobars. For example, we demonstrate this for aluminum at an SW pressure of 580 kbar. At this Hugoniot point, the compression is 1.40 and the temperature is 1480 K. Using the following experimental data at standard conditions,

$$\text{Aluminum:}\quad V_0^{-1}=\rho_0=2.70\ \ [\text{g/cm}^3]$$
$$c_V=8.96\times10^6\ [\text{erg/(g deg)}];\quad B_S=1.36\times10^{12}\ [\text{dyn/cm}^2 \tag{6.75}$$
$$E_0=10^7\ [\text{erg/g}];\quad \gamma_a=2.09;\quad \beta_0=500\ [\text{erg/(g deg}^2)]$$

we get from the EOS Equation 6.65

$$\text{Aluminum:}\quad \frac{V_0}{V}=\frac{\rho}{\rho_0}=1.40,\quad T_H=1480\ \text{K},\quad P_H=580\ \text{kbars}$$
$$\Rightarrow P_c=495.6\ \text{kbars},\quad P_a=83.5\ \text{kbars},\quad P_e=0.9\ \text{kbars} \tag{6.76}$$

From this example one can see that even in an SW with 580 kbar of pressure in aluminum, the electronic thermal part contributes about 0.15% to the total pressure. Therefore, in the following analysis the electronic thermal part is neglected.

The Gruneisen EOS (Equation 6.58) can be written in the form

$$P(V,E) = P_c(V) + \frac{\gamma_a(V)}{V}[E - E_c(V)]$$

(6.77)

Substituting $P = P_H$ in this equation and subtracting it from P, one gets

$$P - P_H = \frac{\gamma_a}{V}(E - E_H)$$

(6.78)

For $P = P_S$, the derivative of Equation 6.78 yields (after using $dE_S/dV = -P_S$)

$$P_S = -\frac{dE_H}{dV} + (P_H - P_S)\frac{d}{dV}\left(\frac{V}{\gamma_a}\right) + \frac{V}{\gamma_a}\left(\frac{dP_H}{dV} - \frac{dP_S}{dV}\right)$$

(6.79)

Taking the derivative of the Hugoniot energy relation, Equation 6.32, gives

$$\frac{dE_H}{dV} = \left(\frac{V_0 - V}{2}\right)\frac{dP_H}{dV} - \left(\frac{P_0 + P_H}{2}\right)$$

(6.80)

Substituting Equation 6.80 into Equation 6.79 yields

$$\frac{dP_S}{dV} = \left[1 - \frac{\gamma_a}{2V}(V_0 - V)\right]\frac{dP_H}{dV} + \frac{\gamma_a}{2V}(P_0 + P_H)$$
$$+ \frac{\gamma_a}{V}(P_H - P_S)\frac{d}{dV}\left(\frac{V}{\gamma_a}\right) - P_S\left(\frac{\gamma_a}{V}\right)$$

(6.81)

Using Equations 6.74 and 6.81 and defining the speed of sound on the Hugoniot points, i.e., c_H is equal to c_S for $P_S = P_H$, one gets

$$c_H^2 = \left(\frac{\gamma_a V}{2}\right)P_H + \left[\left(\frac{\gamma_a V}{2}\right)(V_0 - V) - V^2\right]\frac{dP_H}{dV}$$
$$c_S^2 = c_H^2 + \gamma_a V(P_S - P_H)\left[1 + \frac{d}{dV}\left(\frac{V}{\gamma_a}\right)\right]$$

(6.82)

In the second equation of Equation 6.82 the speed of sound c_S is along the release isentrope off the Hugoniot (i.e., an isentrope starting after the SW point with pressure P_H). It was found experimentally that the Gruneisen coefficient satisfies the relation

$$\frac{\gamma_a(V)}{V} = \frac{\gamma_0}{V_0} = \text{const.}$$

(6.83)

Using this relation in Equation 6.82 yields the approximate relation

$$c_S^2 \approx c_H^2 + \gamma_0 \left(\frac{V^2}{V_0} \right) (P_S - P_H) \tag{6.84}$$

From Equation 6.84, or more accurately from Equation 6.82, the Gruneisen parameter may be determined if the Hugoniot pressure and the speed of sound are measured experimentally. By placing a transparent material coupled to the target under consideration, the rarefaction overtakes the SW and reduces significantly the shock pressure. By measuring this phenomenon, the change in light radiance caused by the decrease in the shock pressure and the speed of sound in the shocked material can be measured.

6.1.12 Phase Transitions

Any chemical composition can exist in the four states of matter: solid, liquid, gas, and plasma. Furthermore every solid can have few different states; in particular 11 different crystalline (solids) water structures were observed. The different states differ in density, heat capacity, and other thermodynamic quantities as well as in mechanical and optical properties. There are also phase transitions to superconductivity, superfluidity, ferromagnetism (i.e., paramagnetic to ferromagnetic), liquid crystal, etc.

Some phase transitions are induced by modification of atomic or molecular arrangements (e.g., solid–solid, solid–liquid), while others are induced by modification of electronic properties (e.g., ferromagnetism, superconductivity).

There are transitions with latent heat and transitions without latent heat ($L = 0$).

L is the heat that the system absorbs (or emits) without increasing its temperature. P. Ehrenfest proposed first and second transitions according to the discontinuity of the derivatives (first with nonzero L and second with $L = 0$) of the thermodynamic potentials.

For example, in a first-order phase transition, the Gibbs potential $G(P,T)$ is continuous at the phase transition while its derivatives, entropy (minus derivative of G with respect to T while P is constant) and specific volume V (derivative of G with respect to P while T is constant), are not continuous. In a second-order phase transition, the first derivatives of G, namely S and V, are also continuous while the second derivatives, the heat capacity at constant pressure C_p and compressibility at constant temperature κ_T), are not continuous.

L. D. Landau suggested in 1937 that a phase transition with $L = 0$ is accompanied by a change in symmetry (with the exception of liquid–gas transition). Landau introduced the order parameter, a physical quantity of extensive character, which is zero in the most symmetric (i.e., most disordered) phase and nonzero in the least symmetric (i.e., ordered) phase. Transitions with no order parameter (the symmetry group of the phases are not included one into another) are of the first kind with L nonzero. In a transition with order parameter, the symmetry group of the least symmetric phase is a subgroup of the symmetry group of the most symmetric group. In this case $L = 0$ and is the second kind.

Liquid–solid (solidification) and solid–liquid (melting) phase transitions are the most widespread in nature with very important implications. Although melting and solidification have been well-known phenomena for a very long time, the theory of this phase transition is not yet understood in a satisfactory way. Therefore, a combination of theory, models, and phenomenology is necessary.

A coexistence line in the P–T plane corresponding to the equation $\mu_S = \mu_L$, where μ_S and μ_L are the chemical potential of the solid and liquid, respectively, represents the equilibrium between a liquid and a solid. This relation yields the famous Clausius–Clapeyron equation for the phase boundary in P–T plane, for a first-order phase transition

$$\frac{dP}{dT} = \frac{S_2 - S_1}{V_2 - V_1} = \frac{l_f}{T_m \left(V_2 - V_1 \right)}$$

where

T_m is the temperature of phase transition
L_f is the latent heat and during this phase transition the specific volume and entropy change are $V_2 - V_1$ and $S_2 - S_1$, respectively

Most of the materials satisfy $V_L > V_S$, implying $dP/dT > 0$. For anomalous matter, $V_L < V_S$ and $dP/dT < 0$, e.g., the density of ice is less than the density of liquid water by 8% at 0°C. The melting temperature is a function of pressure, $T_m(P)$, and for some materials dP/dT can change sign with the change in pressure; some materials are normal in some pressure domains and anomalous in other pressure domains.

The Lindemann model for melting is based on Einstein model of a solid. Einstein developed in 1907 a simple model to explain the experimental observations that the solid heat capacity decreases at low temperatures below the Dulong–Petit value of $3R/\text{mol}$. Einstein assumed that a lattice of N atoms vibrates as a set of $3N$ independent harmonic oscillators in 1D, where each oscillator has a frequency v. The energy of the oscillator is hv, where h is Planck constant, and the Einstein temperature T_E is defined by $k_B T_E = hv$. The Lindemann model is further based on the classical harmonic oscillator model of atoms with mass M, amplitude A (of vibration around the equilibrium), spring constant K, and energy ε

$$\varepsilon = \frac{1}{2} K A^2 = \frac{1}{2} \left(M 4\pi^2 v^2 \right) A^2 = \frac{1}{2} \left(\frac{M k_B^2 T_E^2}{\hbar^2} \right) A^2$$

The average thermal energy is $\varepsilon = k_B T$ and at melting $\varepsilon = k_B T_m$. Lindemann assumed that all solids melt (at a temperature T_m) when the amplitude of oscillation is a constant f (~0.07 for solids with fcc symmetry) times the inter-atomic distance a, namely $A = fa$. Therefore, comparing the classical energy with $A = fa$ and the thermal energy, one gets

$$T_m = \left(\frac{Mk_B T_E^2 f^2}{2\hbar^2}\right)a^2$$

For a solid with a number density of atoms $n_a = 1/a^3$ and density $\rho = Mn_a$, one gets

$$T_m = \left(\frac{Mk_B T_E^2 f^2}{2\hbar^2}\right)n_a^{-2/3} = \left(\frac{M^{5/3}k_B T_E^2 f^2}{2\hbar^2}\right)\rho^{-2/3} \qquad (6.85)$$

The Lindemann model has the advantage of simplicity and is a good approximation for monoatomic crystalline solids. However, this model does not provide any fundamental explanation for a solid structure suddenly losing its stability at temperature T_m. On the other hand this model is useful to test modern theories of melting, none of which is satisfactory.

Van der Waals was the first to propose in 1881 an EOS that can describe the transition from a gas to a liquid. Van der Waals EOS for 1 mol is

$$\left(P+\frac{N^2 a}{V^2}\right)(V-Nb)= Nk_B T \qquad (6.86)$$

$N_A k_B = R = 8.31$ [J/mol·K]; $N = N_A$ is Avogadro's number ~6×10^{23} and k_B is Boltzmann's constant. Nb accounts for the volume of the molecules (or atoms) and $N^2 a/V^2$ accounts for the long-range attraction of other molecules (or atoms). This force is zero for a molecule inside the volume but has a positive contribution on the surface where the resultant force is directed toward the inside of the liquid.

This EOS does not represent in general experiments; however, it represents qualitatively the isotherms. Mathematically, any third-degree equation allows the liquid–gas transition (van der Waals EOS is a cubic equation in V).

Many materials have phase transitions of the first kind, like solid–solid or solid–liquid. These transitions have a change in volume and release (or absorb) latent heat.

An example: iron at $P = 1$ bar, $T = 1183\,K$, is transformed from α-phase to the γ-phase; this change is accompanied by a 2.5% decrease in volume and an absorption of 853 J/mol. Furthermore, at $P = 130$ kbar the α-iron is transformed to ε-iron.

During a phase transition V changes and $P = $ const., $T = $ const.; however, for the $S = $ const. line in a P–V diagram, the P is not constant, unless $T = 0$.

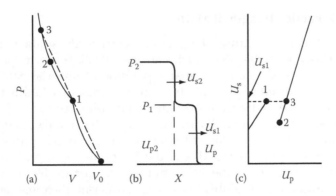

FIGURE 6.7
(a) P–V diagram in the domain of phase transition, (b) the shock wave splitting, and (c) the u_s – u_p in the domain of phase transition. u_s is the shock velocity and u_p is the particle flow velocity of the shocked material.

In Figure 6.7 we describe schematically the Hugoniot curves with a first-order phase transition in (Figure 6.7a) $P - V$ and (Figure 6.7c) $u_s - u_p$ spaces, where u_s and u_p are the shock velocity and the particle velocity, respectively. In Figure 6.7b the stable case is shown where $u_{s1} > u_{s2}$ with an SW splitting, i.e., the double shock structure is shown.
Stability of the SWs requires

$$u_{s1} \geq u_{s2}$$

proof (in the laboratory frame):

$u_{s1} \geq u_{s2}$
proof (in the laboratory frame):
$$u_{s1} = V_0 \left(\frac{P_1 - P_0}{V_0 - V_1} \right)^{1/2}$$

$$u_{s2}* (\text{frame where } u_{p1}=0) = V_1 \left(\frac{P_2 - P_1}{V_1 - V_2} \right)^{1/2} \tag{6.87}$$

$$u_{s2} = u_{s2}* + u_{p1} = V_1 \left(\frac{P_2 - P_1}{V_1 - V_2} \right)^{1/2} + \left[(P_1 - P_0)(V_0 - V_1) \right]^{1/2}$$

$$u_{s2} \leq u_{s1} \Rightarrow V_1 \left(\frac{P_2 - P_1}{V_1 - V_2} \right)^{1/2} \leq V_1 \left(\frac{P_1 - P_0}{V_0 - V_1} \right)^{1/2} \tag{6.88}$$

The last inequality is correct since in Figure 6.7a the slope of line 0–1 is greater than the slope of line 1–2.

6.1.13 Dynamic Strength of Matter

Spall is a dynamic fracture of materials, extensively studied in ballistic research (Rosenberg et al., 1983; Bushman et al., 1993). The term spall, as used in SW research, is defined as planar separation of material parallel to the wave front as a result of dynamic tension perpendicular to this plane. The reflection of an SW pulse from the rear surface (the free surface) of a target causes the appearance of an RW into the target. Tension (i.e., negative pressure) is induced within the target by the crossing of two opposite RWs, one coming from the front surface due to the fall of the input pressure and the second due to reflection of the SW from the back surface (Figure 6.8a). If the magnitude and duration of this tension are sufficient, then internal rupture, called spall, occurs. In this section we would like to point out that this type of research can also be done with laser beams (Gilath et al., 1988; Eliezer et al.,

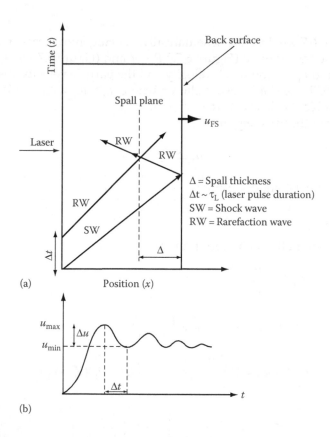

FIGURE 6.8
(a) x–t (space–time) diagram of a laser-induced shock wave. A rarefaction wave (RW), initiated after the end of the laser pulse, follows an SW into the target. After the shock wave reaches the back surface, a rarefaction is reflected. A spall may be created at the intersection of the two RWs. (b) Typical free surface velocity when a spall is created.

1990; Boustie and Cottet, 1991; Fortov et al., 1991; De Resseguier and Cottet, 1995; Eliaz et al., 2000; Moshe et al., 2000).

Spall in ductile materials is controlled by localized plastic deformation around small voids that grow and coalesce to form the spall plane. In very fast phenomena and high tension, like in high-power pulsed laser interaction with a target, new voids are created due to fluctuations and also contribute to the spall formation. Spall in brittle materials takes place by dynamic crack propagation without large-scale plastic deformation. In this section we do not describe the physics of dynamic failure (Grady, 1988; Dekel et al., 1998).

In Figure 6.8a the x–t schematic diagram for spallation is given, while in Figure 6.9 the metallurgical cross section for a spall in an aluminum (6061) target, 100 μm thick, is shown. A laser-created SW in the aluminum target induced the spall. This typical metallurgical cross section was taken after the experiment was finished. The strain ε that has been formed at the spall area is defined by

$$\varepsilon(1D) = \frac{\Delta l}{l}; \quad \varepsilon(3D) = \frac{\Delta V}{V} = -\frac{\Delta \rho}{\rho} \approx 3\varepsilon(1D) \tag{6.89}$$

where Δl is the difference between the final and original lengths of the target in 1D and l is the original length, while in 3D the strain is defined by the relative change in the volume. From the cross section of Figure 6.9 one can measure the dynamic strain directly. One of the important parameters, for the different models describing the spall creation, is the strain rate:

$$\dot{\varepsilon} = \frac{d\varepsilon}{dt} \tag{6.90}$$

High-power short pulse lasers have been used to create strain rates as high as 5×10^8 s^{-1}.

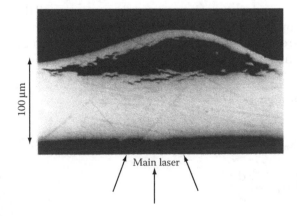

FIGURE 6.9
Cross section in an Al 6061 target after the spall creation.

When the SW reaches the back surface of the solid target bounded by the vacuum (or the atmosphere) the free surface develops a velocity $u_{FS}(t)$. This velocity is given by the sum of the particle flow velocity u_p and the RW velocity U_r. The material velocity increase U_r is given by the Riemann integral along an isentrope from some point on the Hugoniot (pressure P_H) to zero pressure (Equations 6.20 and 6.74)

$$u_{FS} = u_p + U_r$$

$$U_r = \int_0^{P_H} \frac{dP}{\rho c_s} = \int_{\rho_0}^{\rho} \frac{c_s d\rho}{\rho} = \int_{V(P_H)}^{V(P=0)} \left(-\frac{dP_S}{dV} \right)^{1/2} dV \qquad (6.91)$$

The derivative dP_S/dV is given in Equation 6.81, so that the knowledge of P_S gives the value of U_r. Layers of the target adjacent to the free surface go into motion under the influence of the SW transition from V_0, $P_0 = 0$ to V, P_H, and subsequent isentropic expansion in the reflected RW from P_H, V to $P_0 = 0$, V_2, where $V_2 > V_0$. Although these two processes are not the same, it turns out that for pressures less than 1Mbar one has a very good approximation,

$$u_p \approx U_r \Rightarrow u_{FS} \approx 2u_p \qquad (6.92)$$

It was found experimentally, for many materials, that this relation is very good (within 1%) up to SW pressures of about 1 Mbar. Therefore, from the free surface velocity measurements, one can calculate the particle flow velocity of the SW-compressed material. This free surface velocity together with the experimental measurement of the SW velocity might serve as the two necessary quantities, out of five (P_H, $V = 1/\rho$, E_H, u_s, u_p), to fix a point on the Hugoniot.

A typical free surface velocity measurement, in the case of the creation of a spall, is given in Figure 6.8b. u_{max} (related to u_p in the above discussion) in this figure is the maximum free surface velocity. Later, the free surface velocity decreases to u_{min} until a second shock arrives from the spall at the new free surface. When an RW reaches the internal rupture of the target (the spall), an SW is reflected toward the free surface, causing an increase in the free surface velocity. These reverberation phenomena are repeated until the free surface reaches an asymptotic constant velocity. The spall strength Σ, namely the negative pressure (tension) with which a spall occurs, can be calculated using the Riemann invariants

$$u(P = 0, u = u_{min}) \equiv u_{min} = u_0' - \int_\Sigma^0 \frac{dP}{\rho c_s} \Bigg\}$$

$$u(P = 0, u = u_{max}) \equiv u_{max} = u_0' + \int_\Sigma^0 \frac{dP}{\rho c_s} \Bigg\} \Rightarrow \qquad (6.93)$$

$$\Delta u \equiv u_{max} - u_{min} = 2\int_\Sigma^0 \frac{dP}{\rho c_s}$$

Assuming that the negative pressure is not too large, then to a good approximation $\rho = \rho_0$ and $c_s = c_0$, implying that the spall strength is

$$\Sigma = -\frac{\rho_0 c_0 \left(u_{max} - u_{min}\right)}{2} \equiv -\sigma_{spall} \tag{6.94}$$

The minimum (i.e., the maximum absolute value) possible Σ is given by the EOS, as shown in Figure 6.6b, by the value P_m.

The strain ε and the strain rate $d\varepsilon/dt$ can be approximated by

$$\varepsilon = \frac{u_p}{c_0}$$

$$\dot{\varepsilon} = \frac{d\varepsilon}{dt} = \frac{1}{2c_0}\frac{du_{FS}}{dt} \approx \frac{1}{2c_0}\frac{\Delta u}{\Delta t} \tag{6.95}$$

where Δt is the time difference between the time measurements of u_{max} and u_{min}.

Up to about a strain rate of 10^6 s^{-1}, the strain creation method uses gun drivers of plates or high explosives, while the higher strain rates are achieved with high-power pulsed laser drivers. The minimum pressure, in the pressure-specific volume curve, gives the EOS theoretical spall strength. As already mentioned above, in reality the theoretical value of spall is not achieved due to material defects. However, for extremely high strain rates, the theoretical values from the EOS are expected, since the material defects do not have enough time to combine and induce a spall, so the interatomic forces dominate the strength of the material in this case. The spall strength depends also on the temperature of the shocked target.

Last, but not least, from the free surface measurements the SW pressure is known if the EOS is available (Equation 6.39)

$$P_H = \rho_0 u_s u_p \approx \rho_0 \left(c_0 + \alpha \frac{u_{max}}{2} \right) \frac{u_{max}}{2} \tag{6.96}$$

6.2 Diagnostics and EOS

Two independent parameters must be measured in SW experiments to obtain an EOS point. For example, measuring the shock velocity u_s and the particle velocity u_p, the final pressure and density P,ρ are calculated by the Hugoniot relations:

$$P - P_0 = \rho_0 u_s u_p$$

$$\frac{\rho}{\rho_0} = \frac{u_s}{u_s - u_p}$$

The temperature is a fundamental thermodynamic parameter of the EOS, but it is not a part of the Hugoniot relations.

To determine the above EOS parameters a variety of diagnostics were developed. Two types of diagnostics may be distinguished: passive diagnostics, which are based on detecting the emission of light in the visible range from the shocked heated material. The second type of diagnostics uses a probe beam, which may be a laser, x-ray, or particle beam. Shock velocity and rear surface temperature are measured by using the target self-illumination. Particle velocity, density, and electronic properties are determined using active diagnostics, such as VISAR, laser probe beams, x-ray and proton radiography, and x-ray diffraction.

In order to obtain reliable data on EOS high-accuracy measurements are required for target sizes of about 1 mm within a time of 10^{-9} s, therefore implying a spatial resolution needed of 10 μm and a temporal resolution of a few 10^{-12} s.

6.2.1 VISAR

Velocity interferometry, and in particular the VISAR, developed originally by Barker and Hollenbach in 1972 (Barker and Hollenbach, 1972; Asay and Barker, 1974; Hemsing, 1979) for gas gun and explosively driven experiments, has become a standard diagnostic technique in SW research for measuring particle, surfaces velocities, and ionizing shock fronts in transparent media. The VISAR technique was implemented for laser-induced SWs experiments by Moshe et al. (1996) and used in measurements of dynamic spall strengths in aluminum at high strain rates. The motion of a surface is probed by a pulsed laser beam focused on the surface and reflected from it, undergoing a Doppler shift proportional to the surface velocity. The reflected frequency shifted light is collected, collimated, and entered in a Michelson interferometer. The light exciting the interferometer produces an interference pattern of parallel fringes. The interference pattern is imaged by a cylindrical lens, to a set of bright points on the entrance slit of a streak camera. The velocity history of the moving surface is determined from the change in the interference pattern. The fringes move in a direction perpendicular to their orientation. The velocity of the moving surface $V(t)$ is related to the vertical position $y(t)$ of the interference pattern by

$$V(t) = \frac{\lambda c}{4L_e\left(n - \frac{1}{n}\right)} \frac{y(t)}{(1+\delta)d}$$

where
  λ is the probe laser wavelength
  d is the fringe spacing
  L_e is the length of the etalon

n is its index of refraction

$(1 + \delta)$ is a correction term due to the wavelength dependence of the refractive index

The VISAR technique was further improved by Cellier et al. (1998). In their system the probe laser is brought to the studied surface through a multimode fiber. This illumination differs from previous systems, which used spatially coherent point illumination on the surface, combined with diffuse reflecting surfaces. The spatially incoherent beam illuminates the target on a well-defined spot with a field of rays incident from all possible angles allowed within the numerical aperture. This configuration is less sensitive to vignette of the returned light caused by local tilt or curvature of a specular reflecting surface.

An additional upgrade of the VISAR technique is line-VISAR, which provides spatial resolution (Baumung et al., 1996; Cellier et al., 2004a).

Advanced implementations of line-VISAR systems are currently operating at Omega, Phebus, Luli, and Vulcan laser facilities (Cellier et al., 2004a). These systems include generally two interferometers, allowing two fringe sensitivities, a variety of etalon thicknesses, corresponding to velocity sensitivities ranging between few tens of km/s/fringe to several hundreds of meter/s/fringe, Michelson or Mach-Zender interferometer configurations, fiber-coupled illumination, dual wavelength operation, and optical recording channel to record thermal luminescence data simultaneous with the VISAR data. These line-VISAR systems serve as routine diagnostics in various types of experiments.

Shock breakout measurements in stepped opaque metallic targets at pressures of multi-Mbar are performed with a temporal resolution around 20 ps. In general these experiments are done in an impedance matching configuration for EOS determination. In fact for breakout measurements self-illumination detection can provide the required arrival time. However, the temporally resolved VISAR diagnostic enables the detection of shock planarity and preheat as well.

Motion of free surfaces as a function of time is measured for spallation research. The velocity history of the free surface is a signature of the spallation formation mechanism, revealing the shock, release, spallation, and pullback. In the acoustic approximation, the spall strength and the strain rate are calculated from the free surface velocity profile and given by Equations 6.94 and 6.95. With short pulse laser-induced loading, very high strain rates can be obtained, enabling the measurement of the strength of materials approaching the theoretical limit predicted by the EOS (Moshe et al., 2000). This approach to the ultimate strength was measured (Moshe et al., 2000) for aluminum and copper irradiated by subnanosecond laser pulses, at strain rates in the range $(1.5–5) \times 10^8$ s^{-1}. The pressures obtained from the VISAR measurements were up to 430 kbar in aluminum and 1300 kbar in copper. The corresponding temperatures on the release isentrope, 300–660 K in aluminum and 310–1340 K in copper, and the theoretical spall strengths corresponding to these temperatures, 101–108 kbar in aluminum and 172–211kbar in copper, were calculated using a wide range semiempirical EOS for metals. The highest values of the spall strengths deduced from the measured free surface velocity profile were 80 ± 10 kbar in aluminum and 156 ± 20 kbar in copper, therefore very close to the ultimate strengths given by the EOS.

The Hugoniot EOS of transparent dielectric materials using the impedance matching method is currently measured with line-VISAR diagnostics.

Laser-induced SW is produced in a reference material and transmitted in the sample material. A stepped aluminum base-plate is usually used as the reference material (Cellier et al., 2004a,b). The shock must be strong enough to ionize the sample into a reflecting state. The EOS of several materials, such as LiF, sapphire, diamond, polystyrene, fused silica, and water, was measured using this method.

Absolute EOS measurements of iron in the pressure range of 1–8 Mbar were conducted (Benuzzi-Mounaix et al., 2002) at the LULI facility. The shock velocity and the rear surface velocity were measured simultaneously using two VISAR diagnostics with two different velocity sensitivities and step targets. An additional interesting finding of this work was an experimental verification of the doubling rule $V_{fs} = 2U_p$ at high pressures. V_{fs} is the free surface (rear target) velocity measured with the VISAR and the particle velocity U_p is calculated from

$$V_{fs} = U_p + \int_0^{P_H} \frac{dP}{\rho c_S}; \quad c_S = \left(\frac{\partial P}{\partial \rho}\right)_S^{1/2}$$

using the SESAME EOS. The above doubling rule holds for weak shocks. For pressure up to 3 Mbar the discrepancy between the two estimations for U_p was less than 3%. As pointed by Benuzzi-Mounaix et al. this indicates that pressures up to 3 Mbar correspond to the weak shock regime. For these pressures the free surface remained highly reflective after the shock breakout, suggesting that it did not vaporize. For pressures in the range 4–8 Mbar, at times larger than 200 ps after the shock breakout, the VISAR probe beam was completely absorbed by the unloading plasma. Moreover, the measured rear surface velocity was higher by about 10% than twice the calculated velocity U_p. In this case it was claimed that the VISAR measured the velocity of the critical density surface. This possibility was consistent with simulations, showing a difference of 10%–15% between the free surface velocity and the critical density surface velocity, for the above experimental conditions.

6.2.2 X-Ray Diffraction

X-ray Bragg diffraction provides a time-resolved measurement of the lattice spacing of crystal structure. The x-ray scatter from the atoms in the lattice and a constructive interference peak in the scattering signal appears at the Bragg condition:

$$n\lambda = 2d\sin(\theta)$$

where
  d is the lattice spacing
  θ is the incident angle, equal to the diffractive angle
  λ is the x-ray wavelength

SW compression of the lattice will induce a change in the lattice spacing, which may be measured as a change in the angle satisfying the Bragg condition. The shift in the angle of diffraction may be recorded on x-ray time-integrating film or a time-resolving x-ray streak camera. Using in situ transient x-ray diffraction, the compression of the unit cell may be measured while crystal is being shocked (Wark et al., 1989). Furthermore, a follow-up of the motion and dislocations in the crystal structure, which induce stresses and consequently plastic deformation, may be observed. In addition, x-ray diffraction is a diagnostic to measure solid–solid and solid–liquid phase transitions.

In situ x-ray diffraction experiments require a large laser system to generate the shock pressure and simultaneously generate the x-ray back lighter for x-ray diffraction. Using multi-keV for diffraction, depths of several tens of microns in the crystal may be probed. The x-ray diffraction technique was demonstrated with a shocked silicon crystal (Kalantar et al., 1999; Kalantar et al., 2000; Loveridge et al., 2001). The SW was generated by a nearly Planckian radiation created with eight Nova laser beams, inside a cylindrical gold hohlraum. The x-ray back lighter for diffraction was a 1.85 Å He-like Fe spectral emission, induced by two beams of Nova. At a peak pressure ramp of 200 kbar, reaching approximately 120 kbar at the free surface, before shock breakout, it was measured that the lattice spacing was compressed by approximately 6% parallel to the shock propagation, while no compression was observed in the orthogonal direction. Recent experiments conducted with the Vulcan laser measured hydrostatic-like compression in copper, i.e., the sequence of lattice modifications leading to plastic deformation (Hawreliak et al., 2003; Kalantar et al., 2003). This sequence included the crystal compression along the direction of the shock propagation, following the generation of a large shear stress normal to the direction of shock propagation and the compression in the lateral directions, leaving the crystal with the same structure as initially, now compressed in all three directions.

Melting in gallium was detected by x-ray diffraction in experiments with the Trident laser (Luo et al., 2003).

The x-ray diffraction diagnostic was further extended for investigation of polycrystalline samples. A superposition of uniaxial strain in the elastic wave and isotropic strain in the plastic wave was measured in beryllium (Swift et al., 2005). This experiment needed to be performed over a wide range of angles, since in addition to plasticity a possible solid–solid phase transition and melting are expected to occur at the studied pressure range.

6.2.3 Pyrometry

The temperature of shocked and released states may be measured using the pyrometry diagnostics. The temperature is fundamental to the EOS and it is not part of the Hugoniot relations and therefore must be measured separately from the pressure and the density, providing important and independent details on the state of the material. The most important EOS issues that are explored by using pyrometric measurements are phase transitions, such as solid–liquid, solid–solid, and metal–insulator. Usually temperature measurements are conducted simultaneously with other measurements, including shock or particle velocity and reflectivity, along the Hugoniot curve. In many cases a phase transition is recorded as a change of slope along the Hugoniot curve.

Multifrequency pyrometry is based on an absolute measurement of the emission $I(\lambda)$ from the shock front. The temperature T is inferred by fitting to a gray body Planck spectrum:

$$I(\lambda) = \varepsilon(\lambda)\frac{2\pi hc^2}{\lambda^5}\frac{1}{\exp(hc/\lambda k_B T) - 1}$$
$$\varepsilon(\lambda) = 1 - R(\lambda)$$

where
  $\varepsilon(\lambda)$ is the wavelength-dependent emissivity
  $R(\lambda)$ is the reflectivity
  h, c, k_B are the Planck constant, the light velocity, and Boltzmann constant, respectively

Direct temperature measurements can be performed for materials that are transparent in their initial state (Zeldovich and Raizer, 1966). In the case of shock-compressed metals, temperature measurement could only be made at the time the shock arrives at the solid–vacuum interface, before the plasma expands into the vacuum. The finite temporal resolution of instruments makes such direct measurements of the shock temperature impossible and leads to the measurement of the temperature of the expanding plasma instead. Moreover, the radiation emerging from the hot target regions is screened by cooled vapors of the material released by the RW.

The emission from the shock may be observed through a transparent window, such as sapphire or LiF. In this case the self-emission of the interface between the sample and the window are recorded. Some issues have to be addressed in order to obtain reliable measurements, such as the quality of the sample–window interface, the properties of the window material and the thermal conduction across the interface.

Measurements of the rear surface temperature in laser-compressed targets have been made since 1980 (McLean et al., 1980; Ng et al., 1985) to study mechanisms of energy transport in the material, especially preheating by x-rays and fast electrons. This preheating has an undesirable effect since in that case the shock propagates in a medium whose initial state is unknown, making the determination of the EOS impossible. The first temperature experiments (McLean et al., 1980) measured rear surface continuum emission with monochromators coupled to photomultipliers and using blackbody emission assumptions. Further measurements (Ng et al., 1985) were temporally resolved using a streak camera and measured brightness and spectral rear surface temperatures. These experiments required an absolute calibration of the complete optical system.

Later experiments (Hall et al., 1997; Batani et al., 1999) measured color temperature by recording the space time-resolved rear surface emissivity in two different spectral regions on the same streak camera. The temperature was calculated using a model for opacity for the absorption of visible light in metallic vapors.

An important theoretical issue, which was addressed using temperature measurements, is the transition of hydrogen from a molecular insulator to a liquid metal. Temperature measurement along the principal Hugoniot up to pressures of 230 GPa was performed with the Nova laser (Cellier et al., 2000; Collins et al., 2001). The absolute spectral radiance $I(\lambda)$ of the shock front was measured at several wavelengths using a fiber-coupled pyrometer. The emissivity was either obtained from the measured reflectivity or from the calculated reflectivity, based on Drude model for the complex index of refraction of a metal in the free electron approximation:

$$n^2 = 1 - \frac{\omega_p^2 \tau_e^2}{1 + \omega^2 \tau_e^2} + i \frac{\omega_p^2 \tau_e^2}{\omega \tau_e \left(1 + \omega^2 \tau_e^2\right)}$$

$$R(\lambda) = \left| \frac{n-1}{n+1} \right|^2$$

$$\tau_e = \frac{R_0}{v_F}$$

The electron relaxation time τ_e is related to the interatomic distance R_0 and the Fermi velocity v_F. ω_p is the plasma frequency and $\omega = 2\pi/\lambda$, where λ is the measured wavelength.

The experimental results showed that at pressures in the range of 30–50 GPa, the deuterium becomes highly compressive and reflective and the corresponding Hugoniot temperatures are around 0.45 eV lower than those predicted by most models. This behavior indicates the metallization of the deuterium.

Recent experiments performed with the Luli laser used pyrometry and VISAR diagnostics simultaneously to measure EOS and phase transitions in two planet inner core materials, water (Koening et al., 2004) and iron (Koening et al., 2004; Huser et al., 2005). Knowledge of the melting curve of iron is an important issue in geophysics, especially near 3.3 Mbar, the inner-core outer-core boundary pressure. The experiments were done using the impedance mismatch method, where the shock velocity was simultaneously measured in two materials, the sample material, and aluminum, the reference material, since its EOS is known up to 40 Mbar. Two VISAR interferometers with different sensitivities and wavelength were used simultaneously to resolve 2π fringe shift ambiguities and to permit reflectivity measures at two wavelengths.

In the experiments aimed to study the EOS of water the target consisted of an aluminum step and a cell filled with water, imbedded in a diamond anvil cell. This configuration enabled to vary the initial conditions of the Hugoniot and reach the pressure range of 0.5–1 Mbar at a temperature range of 0.25–0.35 eV, interesting for planetary studies. When water is metallized the VISAR probe beam is reflected from the shock front and the shock velocity is determined directly from the fringe shift $FS = n_0 u_s$, where S is the sensitivity of the VISAR and n_0 is the index of refraction of water. When water remains partially transparent, the shock velocity, the particle velocity, and the index of refraction of the compressed water are related by

$$FS = n_0 u_s - n(u_s - u_p).$$

The shock velocity in aluminum is determined from the transit time in the step. The EOS measurements were in agreement with the SESAME EOS and experiments conducted with the other lasers. The measured index of refraction as a function of density indicates that for densities up to 2.1 g/cm³, water remains transparent. As the density increases water becomes absorbing and the index of refraction increases reaching a value 3.8 at a density 2.4 g/cm³, corresponding to a pressure of about 1 Mbar on the principal Hugoniot. This indicates metallization of water. At that pressure a change in the slope of the Hugoniot curve was measured as well.

In the iron experiments the interface velocity and the temperature of the shock released iron into an LiF window were recorded. As the shock was transmitted in the window, the index of refraction of the compressed LiF was considered to calculate the interface velocity, following the model of Kormer:

$$n(\sigma) = n_0 + \frac{dn}{d\sigma}(\sigma - 1)$$

where $\sigma = \rho/\rho_0$ is the compression of the LiF window.

The shock temperature in iron was calculated from

$$T_{shock} = T_{release} \exp \int_{V_{shock}}^{V_{release}} \frac{\gamma}{V} dV$$

where
    V is the volume
    γ is the Gruneisen parameter

As the interface moves at the released particle velocity in iron, which decreases with time, a range of pressures could be measured on the same laser shot.

The temperature versus pressure upon partial release was obtained from the experiments in the pressure range of 0.5–1.5 Mbar in which the temperature varied in the range of 4000–5000 K. As the pressure decreased the temperature dropped sharply and then reached a plateau-like phase. This plateau was attributed to the coexistence of solid and liquid phases as the iron passes the melting line during the release.

By extrapolating the experimental results along the Lindemann melting curve, the shock temperature at a shock pressure of 3.3 Mbar was 7800 ± 1200 K, higher than that obtained with static methods. This behavior was attributed to superheating (Luo and Ahrens, 2003) occurring during shock compression. Another possibility that was suggested to explain this discrepancy is the existence of a solid–solid phase transition around 2 Mbar (Luo and Ahrens, 2004).

6.2.4 Reflectivity and Chirped Pulse Reflectometry

The reflectivity of shocked compressed water and the Hugoniot EOS were measured up to pressures of 790 GPa using line-imaging VISAR operating at 1064 and 532 nm (Cellier et al., 2004b). The experiments were performed with several large lasers, including Phebus and Luli facility in France, Omega laser in Rochester, NY, and Vulcan laser in the United Kingdom. It was found that the reflectivity increased continuously with a pressure above 100 GPa, and saturates at 45% above 250 GPa. This value is well above an increase in reflectivity of about 4%, resulting from a compression-driven increase in the index of refraction. Therefore, the increase in the reflectivity of 45% was attributed to free carriers generated by thermal activation across a mobility gap. This electronic conduction is important for understanding magnetic field generation in giant planets, such as Neptune and Uranus, and may explain the large magnetic fields measured in these planets by the Voyager 2 probe.

Additional reflectivity experiments conducted with the Vulcan laser with precompressed water (Collins et al., 2000) revealed information regarding the origin of the above insulator–conductor transition. The reflectivity measured at ~4 Mbar in precompressed water was significantly less than the value found on the principal Hugoniot at the same pressure, indicating that precompression reduced the electronic conductivity. If the metal–insulator transition were due to a Mott transition, the higher density in the precompressed targets should increase the electronic conductivity rather than decrease it. Therefore, it was suggested that the insulator–conductor transition on the principal Hugoniot in water was thermally induced.

Chirped pulse reflectometry (CPR) is a frequency-domain interferometry (FDI) technique, which can measure shock breakout by recording the reflectivity change at breakout as a modulation in the spectrum of a chirped probe pulse. A chirped pulse is a high bandwidth pulse that has been stretched in time by forcing the different frequencies to travel different path lengths using gratings. If the frequencies of the chirped pulse change linearly with time, event timing is directly proportional to the frequency at which the reflectivity changes sharply. Spectral oscillations may appear due to a near discontinuous change in reflectivity in the time domain. Spectral oscillations act as precision timing fiducials, which are reproducible, since they depend on the probe chirp rate and the shock breakout dynamics. The probed target region is imaged on a spectrometer slit, allowing simultaneous measurement of the probed target regions. Simultaneously recording breakout times in a target with steps with different accurately known thicknesses allows shock velocity measurements. The temporal resolution of the CPR technique is larger by more than an order of magnitude than the temporal resolution of streak cameras, and in addition, avoids the sensitivity, dynamic range, noise, and nonlinearity problems associated with streak-camera-based measurements.

The CPR technique was proposed and demonstrated (Gold et al., 2000) by measuring the change in reflectivity of an aluminum target irradiated by a 130 fs pulse, with intensity of the order of 10^{14} W/cm^2. The plasma temperature in this case is of the order of several eV, similar to shocked rear surface temperatures. The plasma reflectivity was measured with a resolution of 3 μm. The temporal resolution, determined by the input pulse width, chirp rate, spectrometer spectral dispersion, and the detection camera, was less than 1% over 20 ps.

The CPR technique as originally suggested was not further implemented for measuring SW velocities, but was combined with FDI technique to measure shock parameters as a function of time with high temporal resolution over many picoseconds.

6.2.5   Frequency Domain Interferometry

The FDI technique permits the measurement of both the amplitude and the phase shift difference induced by a change in the optical properties of a perturbed material between a pair of femtosecond probe pulses, with simultaneously high spatial and temporal resolution. This method allows on the same laser shot the very precise determination of the shock breakout, due to the high temporal resolution of the CPR. In addition, it permits a precise measurement of the particle velocity and the shock velocity (depending on the experimental configuration) as a function of time. In addition, with FDI it is possible to measure the quality of the shock front and observe microstructures not observable with a streak camera.

Like in the CPR technique, the probe pulse is a chirped femtosecond pulse that is further divided by the two arms of a Michelson interferometer into two collinear pulses, which are separated temporally by a time Δt by adjusting one of the arm lengths. The first-reference probe pulse hits the target surface before the pump pulse/shock breakout, and the second twin pulse probes the surface after the pump. After being reflected from the surface the second probe pulse undergoes a phase change $\Delta\Phi$ from the first pulse and its intensity is decreased by a factor R. The focal spot size of the probe is larger than the focal spot size of the pump to permit uniform illumination of the studied surface. A lens upon the entrance slit of a spectrometer images the region illuminated by the twin probe pulses. The imaging lens provides spatial resolution along a diameter of the surface under test. The typical spatial resolution of the system is a few microns. The temporally separated twin pulses can interfere due to the dispersion of the spectrometer grating, which broadens the pulses and makes them overlap in time. The power spectrum of the interference pattern is

$$I(\omega) = I_0(\omega)[1 + R + 2\sqrt{R}\cos(\omega\Delta t + \Delta\Phi)]$$

where $I_0(\omega)$ is the spectrum of the reference pulse. The reflection coefficient and the phase difference can be obtained from the fringe shift and the fringe contrast. The induced phase shift can be measured as a function of time by changing the delay between the pump pulses and the probe pulse.

The FDI technique was demonstrated (Geindre et al., 1994) by measuring the reflection of a twin 40 fs probe pulse from the front surface of aluminum plasma generated by a 77 fs laser pulse with an irradiance of 3×10^{15} W/cm^2. In this case the phase shift of the probe beam is induced by different processes at different time delays: the propagation of the probe pulse in underdense plasma, the reflection and the absorption at the critical surface, and the expansion of the plasma (Doppler effect).

The FDI technique was demonstrated in EOS measurement (Evans et al., 1996a,b). Simultaneous measurements of the shock and particle velocity in the range of megabar pressures were conducted in aluminum, revealing results in agreement with SESAME EOS. The targets in those experiments consisted of aluminum layers of thicknesses in the range of 10–400 nm coated on 2 mm thick fused silica substrates. The aluminum layers were shocked by a 120 fs laser at intensities around 10^{14} W/cm^2. The reflection of a pair of 120 fs probe pulses separated by 18 ps, from the interface between the aluminum and the fused silica, was measured as a function of the time after the pump laser, implying the particle velocity. The shock velocity was determined from breakout times measured for the same laser intensities and two aluminum thicknesses.

Benuzzi-Mounaix et al. (1999) did the simultaneous measurement of the shock and particle velocity in aluminum at multi-Mbar pressures using the FDI techniques. Their experiment was preformed with the 100 TW Luli laser. The duration of the pump pulse was 550 ps and the thickness of the Al layers was 10 μm, more suitable to EOS experiments. A streak camera for measuring shock breakout was used as well for comparison with the FDI technique. In addition to EOS measurement, the FDI technique allowed evidence of the presence of hot spots in the laser beam and their effects on the local breakout and preheating.

6.2.6   Ellipsometry

Ellipsometry measures the polarization-dependent reflectivity of a probe beam from a surface and can distinguish the real and imaginary parts of the time-dependent dielectric function $\varepsilon(\omega)$ or the index of refraction $\tilde{n}(\omega) = \sqrt{\varepsilon(\omega)}$, giving access to fundamental properties of the material, such as phase transitions and electrical conductivity. ω is the probe laser frequency.

The polarization state of the light incident upon the sample may be decomposed into an S and a P component (the S component is oscillating perpendicular to the plane of incidence and parallel to the sample surface, and the P component is oscillating parallel to the plane of incidence). The amplitudes of the S and P components, after reflection and normalized to their initial value, are denoted by r_s and r_p, respectively. Ellipsometry measures the ratio of r_s and r_p: $\eta = (r_p/r_s)e^{i\delta}$, where δ is the phase difference between r_p and r_s. The reflected amplitudes r_p and r_s and the phase difference δ are determined by four detector photopolarimeters as described by Yoneda et al. (2003).

Using the boundary conditions of the electromagnetic field and Snell's law one gets the real and imaginary complex index of refraction, and the dielectric function can be calculated from the knowledge of η.

$$\tilde{n}_2 = \tilde{n}_1 \tan\phi \left[1 - \frac{4\eta}{(1+\eta)^2} \sin^2\phi \right]^{1/2}$$

where the laser probe enters from medium 1 (complex index of refraction $\tilde{n}_1$; for vacuum $\tilde{n}_1 = 1$) into medium 2 with a complex index of refraction $\tilde{n}_2$. The angle of laser incidence is φ.

The index of refraction is related with the free carrier density, effective mass, and interband absorption via the Drude model. In cases where the studied surface is expanding, the complex index of refraction is determined by solving the Maxwell equation for the probe beam propagating in the expanding plasma.

The probe laser beam may be monochromatic in standard ellipsometry or broadband in spectroscopic ellipsometry, covering a spectral range in the infrared, visible, and ultraviolet. In this case, the complex index of refraction can be obtained in certain spectral regions, enabling probing of lattice-phonons and free carriers-plasmon properties, transparency, and interband transitions.

Since ellipsometry measures the ratio (or difference) of two values (rather than the absolute value of either), it is very robust, accurate, and reproducible. For instance, it is relatively insensitive to scatter and fluctuations and requires no standard sample or reference beam.

In laser-induced SWs studies the ellipsometry is combined with other diagnostics, such as VISAR, pyrometry, or x-ray diffraction, providing a precise characterization of the dynamic properties of the material. Distinct changes in reflectivity are expected to occur at pressures corresponding to phase transitions. Solid–solid and solid–liquid transitions were detected from time-resolved ellipsometry measurements on silicon and tin, irradiated with the Trident laser at Los Alamos National Laboratory (Luo, 2004; Tierney et al., 2004). A probe-pulsed laser beam at a wavelength of 660 nm was reflected from the sample surface at a glancing angle, through a release window. The intensities of the S and P polarized light containing the encoded information on the optical properties were recorded using a streak camera with a temporal resolution of tens of picoseconds. The experiments were performed at pressures below and above the transitions pressures. The reflectivity of both S and P components changed (S, increased and P, decreased) at shock breakout in silicon shocked at pressures below the diamond bct transition. At 14 GPa, corresponding to the phase transition the reflectivity of both S and P components decreased dramatically, indicating the occurrence of the phase transition. In contrast, for measurements on tin, no measurable changes in the reflectivity were observed between different pressures below the pressure at which melt occurred on release. At about 40 GPa, above the melt pressure predicted by the authors, an increase in the S-polarization and a decrease in the P-polarization reflectivity were observed.

Time-resolved laser ellipsometry can be used to probe the expanding front surface of intense laser-irradiated targets as well. The optical properties of expanding plasma are related to the density and the temperature via the collision frequency and the ionization state, and the density and the temperature during the expansion are dependent on the EOS of the material. Therefore, polarization-dependent reflectivity measurements may shed light on EOS models. Ultrashort laser pump–probe ellipsometry experiments were performed with gold targets, irradiated by femtosecond pulses at intensities in the range of 2×10^{12}–5×10^{13} W/cm^2 (Yoneda et al., 2003). The reflectivity of the S and P components were measured for different delays between the pump and the probe, allowing monitoring of the expansion as a function of time. The experiments were modeled (Yoneda et al., 2003) using a nonideal EOS calculated from the Saha equation and a rarefaction flow given by a self-similar Riemann solution. This modeling revealed that an interesting thermodynamic regime was measured in these experiments. When the low-density gold plasma cooled to temperatures below 1 eV, recombination to mainly neutral atomic vapor dominated. The authors attribute this behavior to the high ionization potential (9.2 eV) and strong electron affinity (2.3 eV) of neutral gold. Many previous short pulse pump–probe experiments have been conducted mainly with aluminum targets and were probed at shorter time delays.

6.2.7 X-Ray and Proton Radiography

The impedance matching method relies on the knowledge of the EOS of a reference material, usually aluminum. This method is widely used in SW experiments at pressures up to several Mbars. However, at higher pressures and higher compressions, the EOS of the reference material might not be accurate, and small errors in the velocity measurements may induce errors in the pressure and density determination. A direct measurement of the density of the compressed material is desirable for the range of multi-Mbar pressures. The method of simultaneous measurement of the shock or particle velocity using VISAR and the density using multi-keV x-rays or proton beams is emerging.

Proton beams were first used for probing compressed matter in a laser-driven implosion experiment performed with the 100 TW Vulcan laser (Mackinnon et al., 2002). Plastic microballoons were compressed by 6 ns laser beams, with a maximum energy of 900 J on the target. A 100 J, 1 ps laser beam was used to generate a diagnostic proton beam of about 7 MeV or a diagnostic x-ray beam of about 4.5 keV. By varying the delay between the implosion beams and the diagnostic beam, the history of the implosion process was monitored with a ps temporal resolution. Proton radiography revealed high temporal and spatial resolution throughout all stages of the implosion. X-ray radiography resulted in good resolution as well, but high-contrast images were obtained only when the density was high.

6.2.8   Isochoric Heating

For EOS studies plasmas with uniform densities and single temperatures are required. However, when a material is heated at a temperature of several eV, pressures of more than Mbar are induced, causing the plasma to expand. In this case measurements of the plasma properties are extremely difficult. One way to overcome this problem is by isochoric heating with subpicosecond laser pulses or proton beams at temperatures of eV up to few tens of eV and solid densities. These values cannot be obtained in SWs experiments and therefore a characterization of these plasma states is very important for EOS research. Lasers are limited in their penetration of solid matter to a few microns skin depth. Therefore, only a thin surface laser is directly heated, while the inner part of the sample is heated on a longer timescale by heat conduction. MeV protons have a penetration depth of a few microns and are more attractive for volumetric heating.

With the advent of multi-TW ultrashort laser pulses, proton beams are produced by irradiating thin solid targets with 100 fs, 5×10^{20} W/cm^2. Up to 10^{11}–10^{13} protons per laser pulse, originating from hydrocarbons found in surface contaminations on the target back surface, are emitted with energies up to 20–50 MeV, from a spot of 200 μm. The duration of the proton beam is a few times larger than the duration of the laser pulse. The conversion efficiency of the laser energy to protons ranges between 2% and 7%. The protons are emitted in a cone with a half angle of 15°–20°. These energetic proton beams were discovered recently (Clark et al., 2000; Snavely et al., 2000) and are now produced at many TW laser facilities. It was further demonstrated that these proton beams can be focused by generating them with hemispherical targets (Patel et al., 2005).

The volumetric heating of solid density aluminum to a 23 eV plasma state was demonstrated (Patel et al., 2005) using a laser delivering 10 J in 100 fs. The proton beam was focused on a 10 μm thick Al foil and the temperature of the heated foil was determined by measuring the thermal emission at the wavelength of 570 nm from the hot rear surface with a fast optical streak camera. The temporal and spatial resolutions were 70 ps and 5 μm, respectively. In addition, the proton flux required to heat the aluminum foil at the temperature estimated in the experiment was calculated using a Monte Carlo simulation of the proton energy deposition as a function of distance in the target. The results of this simulation were found in agreement with measured values of the conversion of the laser energy into protons.

In 2006, the largest laser facilities delivered hundreds of Joules in subpicosecond pulses, allowing the generation of proton beams of few tens of Joules, and therefore enabling isochoric heating at temperatures of the order of keV and pressures in the gigabar range, widely extending the thermodynamic range for EOS measurements. Moreover, it was suggested to use isochoric heating of precompressed targets with nanosecond pulses in large-scale laser facilities combining nanosecond and subpicosecond laser beams, accessing EOS points that cannot be reached with other experiments.

6.3 Summary and Discussions

EOS of materials at high pressures and high temperatures are fundamental to numerous fields of science such as astrophysics, geophysics, plasma physics, inertial confinement physics, and so on. EOS at pressures up to a few Mbar can be measured in static experiments using diamond anvil techniques. For higher pressures the SW technique is used. Laser-induced SWs techniques enable the study of EOSs and related properties, such as phase transitions, conductivity, and material properties at pressures up to Gbar, expanding the thermodynamic range reached by conventional gas gun SWs and static loading experiments.

At very high pressures the statistical models, TF and TFD (Thomas-Fermi-Dirac) become accurate, but the exact validity of this limit is not known. In the intermediate regime, at pressures smaller than 1 Gbar, experimental measurements are needed. Two basic techniques are used in laser-induced SWs research, direct drive and indirect drive. In direct drive, one or more beams irradiate the target. In indirect drive, thermal x-rays generated in laser-heated cavities create the SW. Both direct and indirect drive can be used to accelerate a small foil-flyer and collide it with the studied sample, creating a shock in the sample, similar to gas-gun accelerated plates experiments.

By using lasers with various energy, time duration, and focal spot laser, the shock pressure can be varied in a controllable way from less than 10 kbar to more than several hundreds of Mbar. The pulse duration of laser-produced SWs may range from hundreds of femtoseconds to 100 ns, enabling material properties studies on a microscale to a mesoscale—subgrain size, single crystal, to pseudobulk. The strain rates achievable in laser-produced SWs vary from about 10^6 to 10^9 s^{-1}. At this very high strain rate one can measure the strength of materials predicted by the EOS (Moshe et al., 2000). Tailoring the temporal shape of the laser pulse allows achieving both quasi-isentropic ramp wave and shock loading. In an isentropic compression experiment one uses a smoothly increasing drive pressure applied to the sample and the evolution of the compression wave is measured as it propagates through different thicknesses of the sample. A complete determination of an isentrope up to the peak pressure can be obtained in a single ramp wave experiment, while only a single Hugoniot point is measured in an SW experiment. This quasi-isentropic compression may explore states relatively far from the Hugoniot. An additional important advantage of laser-induced SWs is the possibility to synchronize shocks with a variety of optical, x-ray, and electronic diagnostics.

Most of the laser-induced SWs experiments in the last decade used impedance matching. Recently, absolute EOS measurements were performed, measuring two shock quantities simultaneously on the same shock.

We summarize some of the major achievements according to the irradiated material.

6.3.1   Metals

EOS points on the principal Hugoniot of copper up to 20 Mbar and gold and lead up to 10 Mbar have been made with an accuracy of 1% in shock velocity, using the HELEN laser (Rothman et al., 2002). The experiments were performed in the indirect drive configuration and used the impedance match method. Shock breakout from base and steps was detected by monitoring light emission from the rear surface of the target with optical streak cameras, and shock velocities were derived from the transit times across known height steps. The experimental results were in agreement with the SESAME EOS and previous experimental results.

Absolute measurements of the EOS of iron at pressures in the range of 1–8 Mbar, relevant to planetary physics, were performed with step targets at Luli Laser (Benuzzi-Mounaix et al., 2002). The shock velocity and the free surface velocity have been simultaneously measured by self-emission and VISAR diagnostics.

The Hugoniot of tantalum up to pressures of 40 Mbar was measured with the Gekko/Hyper laser (Ozaki et al., 2004). Tantalum is a material typically used in dynamic high-pressure studies to study the reflected shock for a material or projectile. EOS measurements of tantalum are limited up to 10 Mbar by conventional techniques such as gas-gun. The laser-induced SW measurements were based on the impedance match method, and the shock breakout was detected from the self-emission and the reflection of a probe laser from the rear surface. A radiation pyrometer based on a color temperature measurement was used as well.

6.3.2 Plastics

Plastics play very important roles as shell materials in ICF and their EOS data are needed for target design and analyzing the experimental data. Plastics are important in laser-induced SWs experiments, since they are constituents of diverse targets. Unlike metals they are largely transparent to high-energy x-rays, i.e., x-rays can be used to back light relatively thick samples of plastic and provide information on the sample as a function of time.

The Hugoniot curves of Parylene-C and brominated CH at pressures up to 8 Mbar have been made using the HELEN laser (Rothman et al., 2002).

EOS of dielectric materials, sapphire (Al_2O_3) and lithium fluoride (LiF) up to 20 Mbar was measured using the Omega laser and two line-imaging VISAR (Hicks et al., 2003). The measured Hugoniot data indicated that the SESAME EOS provides a good description of the EOS of both sapphire and lithium fluoride.

Transition of transparent wide gap insulators above 5 Mbar into partially degenerate liquid semiconductors was measured for sapphire and lithium fluoride (Hicks et al., 2003). The transformation was observed as an increase in the reflectivity with pressure up to 45% in sapphire at 20 Mbar and 20% in LiF at 10 Mbar.

6.3.3 Foams

Foams, which are low-density porous materials, have many applications in the physics of high pressures, in particular related to ICF and astrophysics. In laser-irradiated foam buffered targets, an efficient thermal smoothing of laser energy is achieved. In indirect drive, low-density foam placed inside the hohlraum prevents cavity closure due to the inward motion of the high-Z plasma from the wall. In EOS experiments the use of foams enables reaching states of matter with higher temperatures at lower than solid densities. Moreover, foams may be used to increase pressures due to impedance mismatch on foam–solid interface. Temperature and shock velocities in $800\,mg/cm^3$ foams shocked to pressures of a few Mbar were performed with the Luli laser (Koenig et al., 2005b). The experiments were based on the impedance mismatch method and VISAR and pyrometry diagnostics were used.

In recent experiments performed with the PALS laser, the EOS of lower-density foams in the range of 60–$130\,mg/cm^3$ up to pressures of 3.6 Mbar was measured (Dezulian et al., 2006). The EOS data were obtained using aluminum as reference material and the shock breakout from double layer Al–foam targets. Samples with different values of initial density were used, enabling the study of a wide region of the phase diagram. Shock acceleration when the shock crossed the Al–foam interface was measured as well. The experimental results showed that Hugoniot of low-density foams at high pressures is close to that of a perfect gas with the same density.

To conclude, the EOS research with lasers has become an important tool at very high pressures, densities, and temperatures. Although many new diagnostics have been developed, laser–EOS research still lacks the accuracy needed for EOS study.

References

Altshuler, L. V. 1965. Use of shock waves in high pressure physics, *Sov. Phys. Usp.* 8, 52.

Asay, R. A. and Barker, L. M. 1974 Interferometric measurement of shock induced internal particle velocity and spatial variations of particle velocity, *J. Appl. Phys.* 45, 2540.

Asay, R. A. and Shahinpoor, M. (eds). 1993. *High-Pressure Shock Compression of Solids.* Springer-Verlag, New York.

Babuel-Peyrissac, J. P., Fauquignon, C., and Floux, F. 1969. Effect of powerful laser pulse on low Z solid material. *Phys. Lett. A* 30, 290.

Barker, L. M. and Hollenbach, R. E. 1972. Laser interferometer for measuring high velocity of any reflecting surface, *J. Appl. Phys.* 43, 4669.

Batani, D., Bossi, S., Benuzzi, A., Koenig, M., Faral, B., Boudenne, J. M., Grandjouan, N., Atzeni, S., and Temporal, M. 1996. Optical smoothing for shock-wave generation: Applications to the measurement of equations of state. *Laser Part. Beams* 14, 211.

Batani, D., Keonig, M., Benuzzi, A., Krasyuk, I. K., Pashinin, P. P., Semenov, A. Yu., Lomonosov, I. V., and Fortov, V. E. 1999. Problems of measurement of dense plasma heating in laser shock wave compression, *Laser Part. Beams* 17, 265.

Baumung, K., Singer, J., Razorenov, S., and Utkin, A. 1996. *Shock Compression of Condensed Matter—1995*, S. Schmidt and W. Tao (eds.), AIP, Woodbury, NY, 1996, p. 1015.

Benuzzi-Mounaix, A., Koenig, M., Boudenne, J. M., Hall, T. A., Batani, D., Scianitti, F., Masini, A., and Di Santo, D. 1999. Chirped pulse reflectivity and frequency domain interferometry in laser driven shock experiments, *Phys. Rev. E* 60, R2488.

Benuzzi-Mounaix, A., Koenig, M., Huser, G., Faral, B., Batani, D., Henry, E., Tomasini, M., Marchet, B., Hall, T. A., Boustie, M., DeResse'guier, T. H., Hallouin, M., Guyot, F., Andrault, D., and Charpin, Th. 2002. Absolute equation of state measurements of iron using laser driven shocks, *Phys. Plasmas* 9, 2466.

Boustie, M. and Cottet, F. 1991. Experimental and numerical study of laser induced spallation into aluminum and copper, *J. Appl. Phys.* 69, 7533.

Bushman, A. V., Kanel, G. I., Ni, A. L., and Fortov, V. E. 1993. *Intense Dynamic Loading of Condensed Matter*, Taylor & Francis, Washington, D.C.

Cauble, R., Phillion, D. W., Hoover, T. J., Holmes, N. C., Kilkenny, J. D., and Lee, R. W., 1993. Demonstration of 0.75 Gbar planar shocks in x-ray driven colliding foils, *Phys. Rev. Lett.* 70, 2102.

Cauble, R., Phillion, D. W., Lee, R. W., and Hoover, T. J. 1994. X-ray driven flyer foil experiments near 1.0 Gbar, *J. Quant. Spectrosc. Radiat. Transfer* 51, 433.

Celliers, P. M., Collins, G. W., Da Silva, L. B., Gold, D. M., and Cauble, R. 1998. Accurate measurements of laser-driven shock trajectories with velocity interferometry, *Appl. Phys. Lett.* 73, 1320.

Cellier, P. M., Da Silva, L. B., Gold, D. M., Cauble, R., Wallace, R. J., Foord, M. E., and Hammel, B. A. 2000. Shock induced transformation of liquid deuterium into a metallic fluid, *Phys. Rev. Lett.* 84, 5564.

Cellier, P. M., Collins, G. W., Hicks, D. G., Koenig, M., Henry, E., Benuzzi-Mounaix, A., Batani, D., Bradley, D. K., Da Silva, L. B., Wallace, R. J., Moon, S. J., Eggert, J. H., Lee, K. K., Benedetti, L. R., Jeanloz, R., Masclet, I., Dague, N., Marchet, B., Rabec Le Gloahec, M., Reverdin Ch, Pasley, J., Willi, O., Neely, D., and Dans, C. 2004a. Electronic conduction in shocked compressed water, *Phys. Plasmas* 11, L41.

Celliers, P. M., Bradly, D. K., Collins, G. W., Hicks, D. G., Boehly, T. R., and Amstrong, W. J. 2004b. Line-imagining velocimeter for shock diagnostics at the OMEGA laser facility, *Rev. Scientific Instrum.* 75, 4916.

Clark, E. L., Krushelnick, K., Davies, J. R., Zepf, M., Tatarakis, M., Beg, F. N., Machcek, A., Norreys, P. A., Santala, M. I. K., Watts, I., and Dangor, A. E. 2000. Measurements of energetic proton transport through magnetized plasmas from intense laser interaction with solids, *Phys. Rev. Lett.* 84, 670.

Collins, G. W., Celliers, P. M., Hicks, D. G., Mackinnon, A. J., Moon, S. J., Cauble, R., DaSilva, L. B., Koenig, M., Benuzzi-Mounaix, A., Huser, G., Jeanloz, R., Lee, K. M., Benedetti, L. R., Henry, E., Batani, D., Loubeyre, P., Willi, O., Pasley, J., Gessner, H., Neely, D., Notly, M., and Danson, C. 2000. Using Vulcan to recreate planetary cores, CLF-RAL Annual Report 2000/2001, 37.

Collins, G. W., Celliers, P. M., Da Silva, L. B., Cauble, R., Gold, D. M., Ford, M. E., Holmes, N. C., Hammel, B. A., and Wallace, R. J. 2001. Temperature measurements of shock compressed deuterium up to 230 GPa, *Phys. Rev. Lett.* 87, 165504.

Cottet, F., Hallouin, M., Romain, J. P., Fabbro, R., Faral, B., and Pepin, H. 1985. Enhancement of a laser driven shock wave up to 10 TPa by the impedance match technique, *Appl. Phys. Lett.* 47, 678.

Courant, R. and Friedrichs, K. O. 1948. *Supersonic Flow and Shock Waves*, Interscience Publishers, New York.

Dekel, E., Eliezer, S., Henis, Z., Moshe, E., Ludmirsky, A., and Goldberg, I. B. 1998. Spallation model for the high strain rate range, *J. Appl. Phys.* 84, 4851.

De Resseguier, T. and Cottet, F. 1995. Experimental and numerical study of laser induced spallation in glass, *J. Appl. Phys.* 77, 3756.

Dezulian, R., Canova, F., Barbanotti, S., Orsenigo, F., Redaelli, R., Vinci, T., Lucchini, G., Batani, D., Rus, B., Polan, J., Kozlova, M., Stupka, M., Praeg, A. R., Homer, P., Havlicek, T., Soukup, M., Frousky, E., Skala, J., Dudzak, R., Pfeifer, M., Kilpio, A., Shashkov, E., Stuchebrukhov, I., Vovchenko, V., Chernomyrdin, V., and Krasuyk, I. 2006. Hugoniot data of plastic foams obtained from laser-driven shocks, *Phys. Rev. E* 73, 047401.

Eliaz, N., Moshe, E., Eliezer, S., and Eliezer, D. 2000. Hydrogen effects on the spall strength and fracture characteristics of amorphous Fe-Si-B alloy at very high strain rates, *Metal. Mater. Trans. A* 31, 1085.

Eliezer, S. 2002. *The Interaction of High-Power Lasers with Plasmas*, Institute of Physics Publishing, Bristol.

Eliezer, S. and Ricci, R. A. (eds). 1991. *High Pressure Equations of State: Theory and Applications*, North Holland, Amsterdam.

Eliezer, S., Ghatak, A., and Hora, H. with a forward by E. Teller 1986. An Introduction to Equations of State: Theory and Applications, Cambridge University Press, Cambridge; (2002) *Fundamentals of Equations of State*, World Scientific, Singapore.

Eliezer, S., Gilath, I., and Bar-Noy, T. 1990. Laser induced spall in metals: Experiment and simulation, *J. Appl. Phys.* 67, 715.

Evans, R., Badger, A. D., Fallies, F., Mahdieh, M., Hall, T. A., Audebert, P., Geindre, J. P., Gauthier, J. C., Mysyrowicz, A., Grillon, G., and Antonetti, A. 1996a. Time-and-space-resolved optical probing of femtosecond laser driven shock waves in aluminum, *Phys. Rev. Lett.* 77, 3359.

Evans, A. M., Freeman, N. J., Graham, P., Horsfield, C. J., Rothman, S. D., Thomas, B. R., and Tyrrell, A. J. 1996b. Hugoniot EOS measurements at Mbar pressures, *Laser Part. Beams* 14, 113.

Fabbro, R., Faral, B., Virmont, J., Pepin, H., Cottet, F., and Romain, J. P. 1986. Experimental evidence of the generation of multi-hundred megabar pressures in 0.26 μm wavelength laser experiments, *Laser Part. Beams* 4, 413.

Fortov, V. E., Kostin, V. V., and Eliezer, S. 1991. Spallation of metals under laser irradiation, *J. Appl. Phys.* 70, 4524.

Geindre, J. P., Audebert, P., Rousse, A., Fallies, F., Gauthier, J. C., Mysyrowicz, A., Dos Santos, A., Hamoniaux, G., and Antonetti, A. 1994. Frequency-domain interferometer for measuring the phase and amplitude of a femtosecond pulse probing a laser-produced plasma, *Opt. Lett.* 19, 1997.

Gilath, I., Eliezer, S., Dariel, M. P., and Kornblith, L. 1988. Brittle-to-ductile transition in laser induced spall at ultra high strain rate in the 6061-T6 aluminum alloy, *Appl. Phys. Lett.* 52, 1207.

Gold, M. D., Cellier, P. M., Collins, G. W., Budil, K. S., Cauble, R., DaSilva, L. B., Foord, E., Stewart, R. E., Wallace, R. J., and Young, D. 2000. Interferometric and chirped optical probe techniques for high-pressure equation of state measurements. *Astrophys. J. Suppl. Ser.* 127, 333.

Grady, D. E. 1988. The spall strength of condensed matter, *J. Mech. Phys. Solids* 36, 353.

Hall, T. A., Benuzzi, A., Batani, D., Beretta, D., Bossi, S., Faral, B., Koenig, M., Krishnan, J., Lower, Th., and Mahdieh, M. 1997. Color temperature measurement in laser-driven shock waves, *Phys. Rev. E* 55, R6356.

Hawreliak, J., Rosolankova, K., Sheppard, Wark J. S., and Kalantar, D. 2003. Observation of hydrostatic-like shock compression in copper using in-situ X-ray diffraction from multiple crystal planes, CLF Annual Report 2002/2003, RAL-TR-2003-018, 49.

Hemsing, W. F. 1979. Velocity sensing interferometer (VISAR) modification, *Rev. Sci. Instrum.* 50, 73.

Hicks, D. G., Celliers, P. M., Collins, G. W., Eggert, J. H., Moon, S. J. 2003. Shock-induced transformation of Al_2O_3 and LiF into semiconducting liquids, *Phys. Rev. Lett.* 91, 035502.

Huser, G., Koenig, M., Benuzzi-Mounaix, A., Henry, E., Vinci, T., Faral, B., Tomasini, M., Telaro, B., and Batani, D. 2005. Temperature and melting of laser iron on LiF window, *Phys. Plasmas*, 12, 060701.

Kalantar, D. H., Chandler, E. A., Colvin, J. D., Lee, R., Remington, B. A., Weber, S. B., Wiley, L. G., Hauer, A., Wark, J. S., Loveridge, A., Failor, B. H., Meyers, M. A., and Ravichandan, G. 1999. Transient x-ray diffraction used to diagnose Si crystals on the Nova laser, *Rev. Sci. Instrum.* 70, 629.

Kalantar, D. H., Remington, B. A., Colvin, J. D., Mikaelian, O., Weber, S. V., Wiley, L. G., Wark, J. S., Loverdge, A., Allen, M., Hauer, A. A., and Meyers, M. A. 2000. Solid-state experiments at high pressure and strain rate, *Phys. Plasmas* 7, 1999.

Kalantar, D. H., Bringa, E., Caturla, M., Colvin, J., Lorenz, K. T., Kurmar, M., Stolken, J., Allen, A. M., Rosolankova, K., Wark, J. S., Meyers, M. A., Scneider, M., and Boehly, T. R. 2003. Multiple film plane diagnostics for shocked lattice measurements, *Rev. Sci. Instrum.* 74, 1929.

Kato, Y., Mima, K., Miyanaga, N., Arinaga, S., Kitagawa, Y., Nakatsuka, M., and Yamanaka, C. 1984. Random phasing of high-power lasers for uniform target acceleration and plasma instability suppression, *Phys. Rev. Lett.* 53, 1057.

Koenig, M., Faral, B., Boudenne, J. M., Batani, D., Benuzzi, A., and Bossi, S. 1994. Optical smoothing techniques for shock wave generation in laser-produced plasmas, *Phys. Rev. E* 50, R3314.

Koenig, M., Henry, E., Huser, G., Benuzzi-Mounaix, A., Faral, B., Martinolli, E., Lepape, S., Vinci, T., Batani, D., Tomasini, M., DaSilva, L., Caubles, R., Hicks, D., Bradley, D., MacKinnon, A., Patel, P., Eggerts, J., Pasley, J., Willi, O., Neely, D., Notley, M., Danson, C., Borghesi, M., Romagnanis, L., Boehly, T., and Lee, K. 2004. High pressure generated by laser driven shocks: Applications to planetary physics, *Nucl. Fusion* 44, S208.

Koenig, M., Benuzzi-Mounaix, A., Ravasio, A., Vinci, T., Ozaki, N., Lepape, S., Batani, D., Huser, G., Hall, T., Hicks, D., MacKinnon, A., Patel, P., Parks, H. S., Boehly, T., Borghesi, M., Kar, S., and Romagnani, L. 2005a. Progress in the study of warm dense matter, *Plasma Phys. Control Fusion* 47, B441.

Koenig, M., Benuzzi-Mounaix, A., Batani, D., Hall, T., and Nazarov, W. 2005b. Shock velocity and temperature measurements of plastic foams compressed by smoothed laser beams, *Phys. Plasmas* 12, 012706.

Landau, L. D. and Lifshitz, E. M. 1987. *Fluid Mechanics*, Pergamon Press, Oxford.

Lehmberg, R. and Obenschain, S. 1983. Use of induced spatial incoherence for uniform illumination, *Opt. Commun.* 46, 1983.

Loveridge, A., Allen, A., Belak, J., Boehly, T., Hauer, A., Holian, B., Kalantar, D., Kyrala, G., Lee, R. W., Lomdahl, P., Meyers, M., Paisley, D., Pollaine, S., Remington, B., Swift, D. C., Weber, S., and Wark, J. S. 2001. Anomalous elastic response of silicon to uniaxial shock compression on nanosecond time scales, *Phys. Rev. Lett.* 86, 2349.

Luo, S. N. and Ahrens, T. J. 2003. Superheating systematics of crystalline solids, *Appl. Phys. Lett.* 82, 1836.

Luo, S. N. and Ahrens, T. J. 2004. Shock induced superheating and melting curves of geophysical important minerals, *Phys. Earth Planet. Inter.*, 134–144, 369.

Luo, S. N., Swift, D. C., Tierny, T., Xia, K., Tschauner, and Asimow P. D. 2003. Time resolved x-ray diffraction investigation of superheating-melting behavior of crystals under ultrafast heating, in *Shock Compression of Condensed Matter—2003*, Furnish, M. D., Gupta, Y. M., and Forbes, J. W. (eds.), AIP Conference Proceeding, CP706, Portland, Oregon, p. 95.

Mackinnon, A. J., Patel, P. K., Hatchett, S., Koch, J., Phillips, T. H., Town, R., Key, M. H., Romagnani, L., Borghesi, M., King, J. A., Snavely, R., Freeman, R. R., Clarke, R. J., Habara, H., Heathcote, R., Lancaster, K. L., Norreys, P. A., Stephens, R., Cowan, T. C. 2002. Picosecond radiography of a laser implosion, *CLF Annual report* 2002/2003 RAL, p. 40.

McLean, E., Gold, S. H., Stamper, J. A., Whitlock, R. R., Griem, H. R., Obenschain, S. P., Gitomer, S. J., and Matzen, M. K. 1980. Preheat studies of foils accelerated by radiation due to laser irradiation, *Phys. Rev. Lett.* 45, 1246.

McQueen, R. G. 1991. *High Pressure Equations of State: Theory and Applications*, Eliezer, S. and Ricci, R. A. (eds.), North Holland, Amsterdam.

Moshe, E., Dekel, E., Henis, Z., and Eliezer, S. 1996. Development of an optically recording interferometer system (ORVIS) for the laser induced shock waves measurements, *Appl. Phys. Lett.* 69, 1379.

Moshe, E., Eliezer, S., Henis, Z., Werdiger, M., Dekel, E., Horovitz, Y., Maman, S., Goldberg, I. B., and Eliezer, D. 2000. Experimental measurements of the strength of metals approaching the theoretical limit predicted by the equation of state, *Appl. Phys. Lett.* 76, 1555.

Ng, A., Parfneiuk, D., and DaSilva, L. 1985. Hugoniot measurements for laser-generated shock waves in aluminium, *Phys. Rev. Lett.* 54, 2604.

Ng, A., Celliers, P., and Parfeniuk, D. 1987. Dynamics of laser-driven shock waves in fused silica, *Phys. Rev. Lett.* 58, 214.

Obenschain, S. P., Whitlock, R. R., Lean, E. A., and Ripin, B. H. 1983. Uniform ablative acceleration of targets by laser irradiation at 10^{14} W/cm^2, *Phys. Rev. Lett.* 50, 44; Ozaki, N., Tanaka, A., Ono, T., Shigemori, K., Nakai, M., Azechi, H., Yamanaka, T., Wakabayashi, K., Yoshida, M., Nagao, H., and Kondo, K. 2004. GEKKO/HYPER-driven shock waves and equation of state measurements at ultrahigh pressures, *Phys. Plasmas* 11, 1600.

Patel, P. K., Mackinnon, A. J., Key, M. H., Cowan, T. E., Foord, M. E., Allen, M., Price, D. F., Ruhl, H., Springer, P. T., and Stephens, R. 2005. Isochoric heating of solid-density matter with an ultrafast proton beam, *Phys. Rev. Lett.* 91, 125004-1.

Rosenberg, Z., Luttwak, G., Yeshurun, Y., and Partom, Y. 1983. Spall studies of differently treated 2024A1 specimens, *J. Appl. Phys.* 54, 2147.

Rothman, S. D., Evans, A. M., Horsfield, C. J., Graham, P., and Thomas, B. R. 2002. Impedance match equation of state experiments using indirectly laser-driven multimegabar shocks, *Phys. Plasmas* 9, 1721.

Skupsky, S., Short, R. W., Kessler, T., Craxton, R. S., Letzring, S., and Soures, J. M. 1989. Improved laser-beam uniformity using the angular dispersion of frequency modulated light, *J. Appl. Phys.* 66, 3456.

Snavely, R. A., Key, M. H., Hatchett, S. P., Cowan, T. E., Roth, M., Phillips, T. W., Stoyer, M. A., Henry, E. A., Sangster, T. C., Singh, M. S., Wilks, S. C., Mackinnon, A., Offenberger, A., Pennington, D. M., Yasuike, K., Langdon, A. B., Lasinski, B. F., Johnson, J., Perry, M. D., and Cambell, E. M. 2000. Intense high-energy proton beams from petawatt-laser irradiation of solids, *Phys. Rev. Lett.* 85, 2945.

Steinberg, D. J. 1996. *Equation of State and Strength Properties of Selected Materials*, UCRL-MA-106439, Livermore, CA, USA.

Swift, D. C., Tierney, I. V., Kopp, R. A., and Gammel, J. T. 2004. Shock pressure induced in condensed matter by laser ablation, *Phys. Rev. E* 69, 36406.

Swift, D. C., Tierney, T. E., Luo, S. N., Paisley, D. L., Kyrala, G. A., Hauer, A., Greenfield, S. R., Koskelo, A. C., McClellan, K. J., Lorenzana, H. E., Kalantar, D., Remington, B. A., Peralta, P., and Loomis, E. 2005. Dynamic response of materials on nanosecond time scales, and beryllium for inertial confinement fusion, *Phys. Plasmas* 12, 056308.

Tierney, T. E., Swift, D. C., and Johnson, R. P. 2004. Novel techniques for laser-irradiation driven, dynamic materials experiments, in *Shock Compression of Condensed Matter—2003*, Furnish, M. D., Gupta, Y. M., and Forbes, J. W. (eds.), AIP, Woodbury, NY, 2004, p. 1413.

van Kessel, C. G. M. and Sigel, R. 1974. Observation of laser-driven shock waves in solid hydrogen, *Phys. Rev. Lett.* 33, 1020.

Veeser, L. R., Solem, J. C., and Liebar, A. J. 1979. Impedance-match experiments using laser-driven shock waves, *Appl. Phys. Lett.* 35, 761.

Wark, J. S., Whitlock, R. R., Hauer, A. A., Swain, J. E., and Solone, P. J. 1989. Sub-nanosecond X-ray diffraction from laser shocked crystals, *Phys. Rev. B* 40, 5705.

Yoneda, H., Morikami, H., Ueda, K., and More, R. M. 2003. Ultrashort pulse laser ellipsometric pump-probe experiments on gold targets, *Phys. Rev. Lett.* 91, 075004-1.

Zeldovich, Ya. B. and Raizer, Yu. P. 1966. *Physics of Shock Waves and High-Temperature Hydrodynamic Phenomena*, Hayes, W. D. and Probstein, R. F. (eds.), Academic Press, New York.

7

Material Processing with Femtosecond Lasers

Masayuki Fujita

CONTENTS

7.1 Introduction

Since the advent of chirped pulse amplification technology, ultrashort pulse lasers have had significant progress and have given scientists opportunities to explore new physics and allowed industrial people to pioneer innovative applications. Nowadays, femtosecond laser pulses are easily available from commercial lasers and even attosecond laser pulses are realized in laboratories. In this chapter, we focus our discussion on the ultrashort pulse laser process in the order of 100 fs, which is the most popular pulsewidth in laboratories and industrial research.

The features of the femtosecond laser pulses are ultrafastness and ultrahigh electric field. The length of a 100 fs laser pulse is only 30 μm, which is almost a thin pancake. In the 100 fs, one can concentrate electromagnetic energy as much as possible, only limited by engineering capability. For example, 1 kJ energy in 100 fs gives us 10 PW, and only 100 mJ in 100 fs gives us 1 TW, which is comparable to the instant power consumed on the earth. If one focuses the 1 TW peak power in a spot of 10 μm, focused intensity becomes 10^{18} W/cm², which is equivalent to the electric field of 2×10^{12} V/m. As the electric field binding an electron in a hydrogen atom is 5×10^{11} V/m, 1 TW light pulse can be used to directly strip the electron from the nuclear.

7.2 Laser–Matter Interactions in Femtosecond Timescale

7.2.1 Interaction with Electrons

As the laser beam is a coherent electromagnetic wave, electrons put in the laser beam get energy from the oscillating electric field. Especially in high field, say more than 10^{18} W/cm², the motion of the electron becomes relativistic. The oscillatory energy of electrons in the electromagnetic field of laser pulses is given by a parameter, ponderomotive potential U_p [1]

$$U_p = \left(e^2 E_0^2 / 4m\omega^2\right)\left(1 + \alpha^2\right) = 9.33 I_{14} \lambda_{\mu m}^2 \, [\text{eV}] \tag{7.1}$$

where
  e and m are electron charge and mass
  E_0 and ω are electric field and angular frequency of the laser
  α is 0 for linear polarization and 1 for circular polarization
  I_{14} is a laser intensity in the unit of 10^{14} W/cm²
  $\lambda_{\mu m}$ is the laser wavelength in the unit of μm

For the wavelength of 1 μm and the intensity of 10^{18} W/cm², U_p becomes 100 keV. The electrons with 100 keV energy are almost relativistic (close enough to the electron rest mass energy $m_0 c^2 = 511$ keV). They are oscillating with velocity close to the light velocity. In a field more than 10^{18} W/cm², the electron mass gets heavier as it gets more energy from the field.

In the high field of laser pulses, electrons behave in a relativistic manner. Hence a parameter a_0, electron momentum divided by the rest mass, is defined as follows:

$$a_0 = p/m_0 c = eE_0/m_0 c\omega \tag{7.2}$$

where p is the momentum that an electron receives from the laser field. In a practical form, Equation 7.2 can be expressed by

$$a_0 = 0.857 \times 10^{-9} \lambda(\mu m)\sqrt{I(\text{W/cm}^2)} \tag{7.3}$$

For more than 10^{18} W/cm², $a_0 > 1$ and the relativistic electrons are generated. With relativistic factor γ, a_0 is expressed by $a_0 = (\gamma^2 - 1)^{1/2}$, where the electron mass is given by $m = m_0\gamma$ and $\gamma = (1 - \beta)^{-1/2}$, $\beta = v/c$, and v is the electron velocity.

7.2.2 Interaction with Atoms and Molecules (Optical Field Ionization)

In the high field of laser pulses, atoms and molecules are easily ionized. Keldysh parameter, which is a ratio of the ionization potential and electron ponderamotive potential, is given by [2]

$$\Gamma = \sqrt{I_p/2U_p} \tag{7.4}$$

where I_0 is the ionization potential. In the case of $\Gamma > 1$ or $I_p > U_p > hv_0$, where hv_0 is the photon energy, multiphoton absorption becomes a dominant process and ionization, called above threshold ionization, happens. On the contrary, in the case of $\Gamma < 1$, ionization with distorted nuclear potential occurs. If the distortion of the potential is small, electrons try to escape from atoms through the tunneling effect, which is called tunnel ionization. For more distortion, the potential will be suppressed even lower than the electron energy level and barrier-suppression ionization comes off. Figure 7.1 shows schematics of the potential distortion and resulting ionization. When such a high field is applied to molecules in femtosecond timescale, one can ionize them by keeping their original structure, which is useful for detecting environmentally hazardous molecules like dioxins [3].

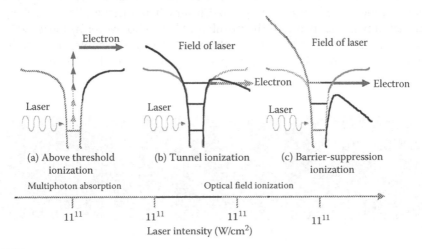

FIGURE 7.1
Schematics of the potential distortion induced by a high field of intense laser beams and the resulting ionization mechanism.

7.3   Material Processing

One of the most important applications of the femtosecond laser pulses is the industrial application such as micromachining. Even with a small pulse energy, for example, 1 mJ, the femtosecond laser pulses give us high peak intensity. Depending on the intensity (W/cm²) or fluence (J/cm²), distinctive phenomena occur. Recent studies of femtosecond laser processing cover a wide variation of material like metal, semiconductor, and dielectric materials. For a transparent material such as glass, quartz, and wideband gap semiconductor, internal micromachining using multiphoton absorption is well known. However, we describe surface processing mainly for metals and semiconductors here.

7.3.1   Above Ablation Threshold

The most popular industrial application of the femtosecond laser is micromachining, such as drilling or cutting, with a negligible heat affected zone. Such a process uses laser ablation and hence requires pulse energy enough to give us laser fluence above ablation threshold. The ablation threshold depends on the laser pulsewidth, even for the same material. It is well known that the ablation threshold is lower for shorter pulsewidth [4,5]. The ablation threshold decreases with $\tau^{1/2}$ scaling down to the pulsewidth τ in the order of 10 ps, which corresponds to the timescale of the thermal diffusion. When the laser pulsewidth becomes shorter than the timescale of the thermal diffusion, sharp ablation with negligible thermal effects can be realized. Figure 7.2 shows scanning electron microscopy (SEM) images of ablated holes on Si targets by pulsed lasers. The pulsewidth, wavelength, irradiated peak fluence, and the number of laser shot were: (a) 100 fs, 800 nm, 4.2 J/cm², 600 shots; (b) 200 ps, 800 nm, 6.1 J/cm², 600 shots; and (c) 10 ns, 1064 nm, 67.5 J/cm², 300 shots, respectively. The laser focal spot size was

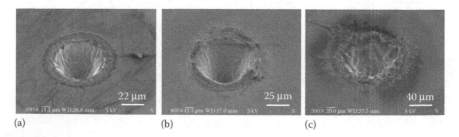

(a)   (b)   (c)

FIGURE 7.2
Comparison of SEM images of an ablated hole. Pulsewidth, wavelength, irradiated peak fluence, number of laser shot were (a): 100 fs, 800 nm, 4.2 J/cm², 600 shots, (b): 200 ps, 800 nm, 6.1 J/cm², 600 shots, (c): 10 ns, 1064 nm, 67.5 J/cm², 300 shots, respectively. The target material was Si.

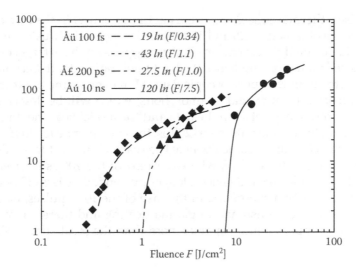

FIGURE 7.3
Pulsewidth dependences of ablation rate as a function of irradiated laser fluence. The target material was Si.

$30 \pm 5\,\mu m$ ($1/e$). For longer pulsewidth, a melting zone becomes noticeable. In the case of 100 fs (Figure 7.3c), one can see two regions: one is a drilled hole and the other is a weakly ablated region surrounding the hole, where irradiated laser fluence is lower than the center of the spot. The femtosecond laser process has two process thresholds. As the influence of the thermal diffusion is negligible, the laser intensity profile is directly reflected on the ablated profile.

More quantitatively, dependences of ablation rate on the laser pulsewidth are shown in Figure 7.3. Solid diamonds, solid triangles, and solid circles correspond to the pulsewidth (wavelength) of 100 fs (800 nm), 200 ps (800 nm), and 10 ns (1064 nm), respectively. Curves in Figure 7.3 represent model fittings, which is given by [6]

$$L = l\left[\ln\left(F_0/F_{th}\right)\right] \tag{7.5}$$

where
  L is the ablation rate
  F_0 and F_{th} are the peak laser fluence and threshold fluence
  l is a characteristic length, which can be the light penetration depth
    (skin depth) or the thermal diffusion length

By fitting Equation 7.5 to the experimental data, one can decide F_{th} and l. In the case of 100 fs, where two process thresholds were seen, two functions were used to fit the data. The light penetration depth and the thermal diffusion length, $l = 19$ and 43 nm, were obtained. Even with the femtosecond laser pulses, thermal processing occurs at high fluence. In the case of 200 ps and 10 ns, we obtained $l = 28$ and 120 nm as the thermal diffusion length.

It should be noted that carrier excitation happens on the Si target even with low fluence irradiation and hence the light penetration depth becomes effectively shorter than the spectroscopic data. Moreover, as the Si target shows phase transition from crystal to amorphous, which will be described later, there is a discrepancy between the data and the model near the threshold.

Comparing the data for 100 fs and 200 ps, the ablation rate for 100 fs is larger than that for 200 ps at the same fluence. The reason is that a 100 fs pulse interacts only with a solid surface, whereas a part of the 200 ps laser energy is absorbed by the plasma and hence a longer pulsewidth is less efficient in the viewpoint of ablation energetics. In the case of the 10 ns pulses, most of the laser energy will be consumed in plasma heating and therefore melting of the surface is apparent and expands more than the focal spot as shown in Figure 7.2c.

7.3.2   Near Ablation Threshold

Under the femtosecond irradiation, an interesting phenomenon called laser-induced periodic surface structure (LIPSS) [7] happens near the ablation threshold. Inside the laser spot after the laser irradiation, we will find periodical surface structure; the period is in the order of a laser wavelength, say submicron. Without laser interference, one can generate grooves like a grating. Figure 7.4 shows SEM images of LIPSS on a copper target generated by 100 fs laser pulses. Although it looks like a somewhat irregular structure, one can see a rainbow-colored reflection of a white light from the surface as if it is the diffraction from a grating. It is a regular periodic structure on average. The direction of the groove is perpendicular to the polarization of the laser light. The structure period is in the order of the irradiated laser wavelength, for example, 600 nm period with 800 nm laser wavelength in the case of Figure 7.4. It has also been reported that the period varies with the irradiated

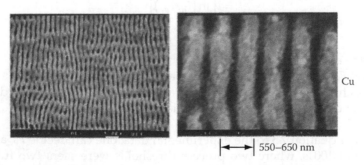

FIGURE 7.4
LIPSS on copper target irradiated by 100 fs laser pulses near the ablation threshold.

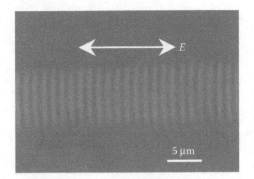

FIGURE 7.5
LIPSS on Si target without ablation. Irradiated laser wavelength, pulsewidth, peak fluence were 800 nm, 100 fs, and 0.2 J/cm², respectively. Pulse repetition rate was 1 kHz and the scanning speed was 800 μm/s.

laser fluence. This kind of submicron surface structure has the effect of reducing surface friction with lubrication and improving adhesion of surface coating such as diamond-like coating.

Figure 7.5 shows an SEM image of LIPSS on a Si target without ablation. Irradiated laser wavelength and peak fluence were 800 nm and 0.2 J/cm², respectively. The laser beam was scanned at 0.8 mm/s at a pulse repetition rate of 1 kHz. The arrow in the figure denotes the polarization of the irradiated laser. The bright stripes were swelled to 1.5 nm, where amorphization occurred. Ablation would occur with higher fluence and one can generate LIPSS like that in Figure 7.4.

LIPSS can be generated on the dielectric material, which is usually transparent for infrared to visible light and hence multiphoton absorption must take place. Figure 7.6a shows an example of SEM images of LIPSS on a quartz (SiO_2) plate. A LIPSS with 600–800 nm periods was produced for the laser fluence more than 4 J/cm². The laser wavelength was 800 nm. Compared to the metal and the semiconductor, more than 10 times higher fluence was required, as the target was transparent. However, one can reduce the required laser fluence with a thin metal coating on it. Figure 7.6b shows SEM images of LIPSS on a copper-coated quartz (SiO_2) plate. The coating thickness was 100 nm. After laser ablation of the coating, about 150 nm period LIPSS was produced for a laser fluence of 0.3–3 J/cm². In spite of this, there existed 600–800 nm LIPSS on the metal layer before complete ablation of the coating and 150 nm LIPSS was produced on the quartz plate. The reason for this is unknown as yet and is still under investigation. Nevertheless, applying this kind of submicron structure can be used to reduce the Fresnel reflection of optics.

7.3.3 Below Ablation Threshold

Even below the ablation threshold, the femtosecond laser pulses leave traces like footprints on the target as they have a high peak intensity. When the femtosecond laser pulses are irradiated on a monocrystalline Si wafer,

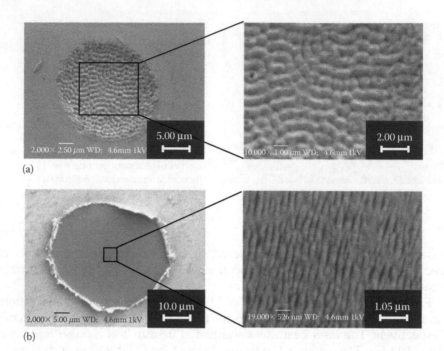

FIGURE 7.6
LIPSS on (a) SiO$_2$ and (b) Cu-coated SiO$_2$ (Cu thickness was 100 nm). Irradiated laser wavelength and pulsewidth were 800 nm and 100 fs. The peak fluence and number of laser shots were (a) 7.3 J/cm^2, 5 shots and (b) 2.7 J/cm^2, 40 shots.

amorphization occurs without ablation. Figure 7.7 shows laser microscope images of a Si wafer irradiated by femtosecond laser pulses with a peak fluence of 0.33 J/cm^2 and laser wavelength of 800 nm. The number of the laser shot was 6 for Figure 7.7a and 15 for Figure 7.7b. The ring-shaped bright region corresponds to the amorphized region. As the intensity profile of the laser spot was Gaussian, the amorphization was limited to a certain range of the laser fluence, and also a multiple laser shot resulted in an expansion of the amorphized area. It is noted that ablation would happen with more laser shots at the same spot (it was 18 shots in the case of Figure 7.7). Figure 7.8 shows

FIGURE 7.7
Ring-shaped amorphized area on Si target after the irradiation of femtosecond laser pulses at fluence of 0.33 J/cm^2. The number of laser pulses are (a) 6 shots and (b) 15 shots.

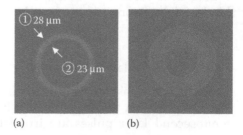

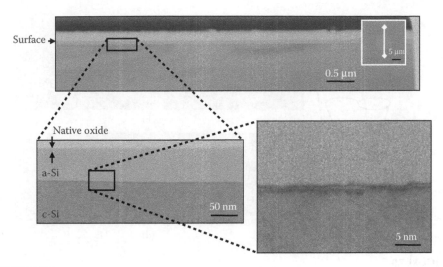

FIGURE 7.8
Cross-sectional TEM images of the amorphized Si sample.

cross-sectional transmission electron microscopy (TEM) images of the amorphized area, where the laser fluence was adjusted to cause filled circle amorphization. The amorphized layer was quite uniform in spite of the Gaussian intensity profile and the boundary was sharp and smooth. The thickness of the amorphized layer is independent of the irradiated fluence and the number of the laser shots, whereas it depends on the laser wavelength [8].

Even below the threshold for amorphization, it has been reported that amorphous Si can be crystallized with the femtosecond laser irradiation [9]. Experiments have been done with a monocrystalline Si (c-Si) wafer on which 8 nm amorphous Si (a-Si) was sputtered (see left schematics in Figure 7.9). As the experiments were conducted in air, a native oxide layer of 3 nm resided on the surface. Figure 7.9 shows laser microscope images of the target irradiated by 100 laser pulses with a wavelength of 800 nm and pulsewidth of 100 fs for various peak fluences. The laser microscope had an optical source of 488 nm wavelength for which reflectivity of a-Si and c-Si was 46% and 39%, respectively. The dim regions in Figure 7.9, which shows lower reflectivity, might correspond to recrystallized regions. Figure 7.10 shows cross-sectional TEM images of the target, which correspond to the irradiation conditions of Figure 7.9b [10]. Again considering the intensity profile was Gaussian, the central part with higher fluence showed amorphization, whereas lower fluence caused recrystallization in the ring-shaped dim region. The crystallization proceeded from the boundary with the monocrystalline substrate, which meant the crystallization was the epitaxial growth.

In summary, ultrashort pulse lasers, here namely femtosecond lasers, certainly give us new subjects for investigation both in the scientific and industrial field. Their high field enables us to explore direct interactions of light with

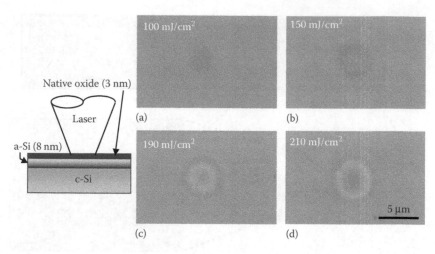

FIGURE 7.9
Laser microscope images of the target surface irradiated by 100 fs laser pulses with various laser fluences. Left schematics show cross section of the target.

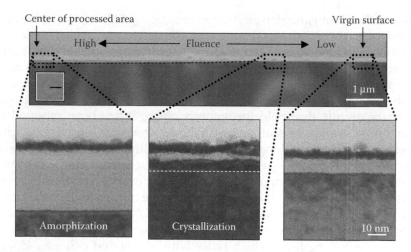

FIGURE 7.10
Cross-sectional TEM images of the crystallized Si sample. (From Fujita, et al., *Rev. Laser Eng.*, May 2008. With permission.)

TABLE 7.1

Dependences of Threshold Fluence for a Si Target
in Femtosecond Laser Processing

	Crystallization	Amorphization	LIPSS	Ablation
F_{th} (mJ/cm^2)	100	190	200	340

nuclear. Relativistic physics can be investigated in small-scale laboratories. Moreover, even in a medium- or low-intensity regime, exotic industrial applications are expected to come out. Table 7.1 summarizes threshold fluences for the above-mentioned processing phenomena; crystallization, amorphization, LIPSS, and ablation. Just as in a range of fluence varied by a factor of 3–4, one can see drastic change in the material.

References

1. S. Augst et al., *J. Opt. Soc. Am. B* 8, 858, 1991.
2. L. V. Keldysh, *Sov. Phys. JEPT.* 20, 1307, 1965.
3. H. Harada et al., *Chem. Phys. Lett.* 342, 563, 2001.
4. D. Du, X. Liu, G. Kom, J. Squier, and G. Mourou. *Appl. Phys. Lett.* 64, 3071, 1994.
5. P. P. Pronko, S. K. Dutta, D. Du, and R. K. Singh. *J. Appl. Phys.* 78, 6233, 1995.
6. S. Nolte, C. Momma, H. Jacobes, A. Tunnermann, B. N. Chichikov, B. Wellegehausen, and H. Welling. *J. Soc. Am. B.* 14, 2716, 1997.
7. M. Birnbaum. *J. Appl. Phys.* 36, 3688, 1965.
8. Y. Izawa, Y. Setuhara, M. Fujita, R. Sasaki, H. Nagai, M. Yoshida, and Y. Izawa. *Appl. Phys. Lett.* 90, 044107, 2007.
9. Y. Izawa, M. Fujita, Y. Setuhara, M. Hashida, and Y. Izawa. *Proceedings of the 4th International Congress on Laser Advanced Materials Processing*, Kyoto, #06-75, 2006.
10. Fujita, et al., Current status and prospects of the femtosecond laser processing, *Rev. Laser Eng.*, 35, 91, May 2008.

8

Nanoparticles Induced by Femtosecond Lasers

Shalom Eliezer

CONTENTS

8.1   Introduction

A particle with a dimension larger than about 1 nm and smaller than 100 nm is arbitrarily defined as a nanoparticle (NP). Sometimes particles less than 10 nm are referred to as NPs and the larger particles up to about 1 µm are called mesoparticles, where the material's properties are closer to those in its bulk. NPs are of great interest because of the physics associated with them, namely, NPs possess a large surface-to-volume ratio and they are macroscopic objects described by quantum physics.

NPs can exhibit very different physical and chemical properties in comparison to macroscale particles. Opaque matter (such as copper) becomes transparent to visible light, insulators transform into conductors, inert materials (like platinum or gold) behave as catalysts, melting temperature of material changes significantly, etc. The fascinating nanotechnology follows from these unique quantum physics and the large surface-to-volume ratio of the nanosystem. These NPs may also prove very important and useful to many industrial and biological applications. Thus, the synthesis and study of

NPs of various elements and compounds is of great interest both in techno-logical applications and for fundamental research (Mitura, 2000; Cao, 2004; Gogotsi, 2006; Wolf, 2006).

Traditionally, NPs are produced by techniques such as arc discharge, vapor deposition, electrochemical deposition, and laser ablation through a long pulse duration (about a few nanoseconds) in an appropriate gas atmosphere. In this chapter we describe NPs (Eliezer et al., 2004, 2005) and nanotubes (Eliezer et al., 2005) created by femtosecond laser–solid target interaction. These ultrashort laser pulses allow the completion of the energy deposition into the target well before the target expansion begins, thus effectively decoupling these two stages of the process. The laser energy is initially deposited in the electron subsystem, within a target surface layer of few tens of nanometer thickness (Fisher et al., 2001). This stage continues during the laser pulse duration. After the pulse terminates and before the expansion sets in, the electrons in the surface layer undergo cooling by heat diffusion and by energy transfer to the ions. This stage continues for several picoseconds followed by a cooling process that lasts about 1 ns. If during this cooling a mixture of liquid plus vapor is obtained, then the particles condense into NPs or nanotubes.

The timescales and appropriate dimensions of this scenario can be described by the following:

$$\tau_1 \equiv \text{energy deposition} \sim 10^{-13}\,[\text{s}] \Rightarrow c_s\tau_1 \sim 5 \times 10^{-8}\,[\text{cm}]$$
$$\tau_2 \equiv \text{thermal relaxation}\,(T_e = T_i) \sim 10^{-12}\,[\text{s}] \Rightarrow c_s\tau_2 \sim 5 \times 10^{-7}\,[\text{cm}] \qquad (8.1)$$
$$\tau_3 \equiv \text{nanoparticle production} \sim 10^{-9}\,[\text{s}] \Rightarrow c_s\tau_3 \sim 5 \times 10^{-4}\,[\text{cm}]$$

where c_s is of the order of the appropriate speed of sound taken in the above equation as 5×10^5 [cm/s]. Note that at room temperature, the speed of sound for aluminum is 5.4×10^5 [cm/s], for graphite carbon it is 2.63×10^5 [cm/s], while for diamond it is 1.96×10^5 [cm/s]. It is interesting to see that when a femto-second laser interacts with a solid target, NPs can be created in a few nano-seconds during which time the material expands into the vacuum at a distance of a few microns. For adiabatic expansion into the vacuum one has to replace c_s by a velocity of about 10 times greater, thus increasing the estimations in Equation 8.1 by a factor of 10. Even in this case the NPs and nanotubes are created during the plasma expansion in less than 100 μm.

It has also been suggested (Gamaly et al., 2000) that by using the high repetition rate of ultrafast laser pulses one can obtain carbon films. NPs from a high repetition rate laser were obtained (Wang et al., 2007) with a XeCl excimer laser (wavelength 308 nm, pulse duration 15 ns, laser fluence 4 J/cm²) focused at the surface of a single crystalline Si wafer. Nanocrystal films of magnetic materials were also synthesized with femtosecond laser ablation in vacuum (Ausanio et al., 2004, 2006; Amoruso et al., 2006). Laser pulses of 300 fs were used to produce a plume of NPs, which were deposited on a substrate to produce a film with a thickness of about 1 μm.

8.2 Why Femtosecond Lasers?

The femtosecond laser heats the material without changing its density to a good approximation. In this case, the initial thermodynamic conditions are calculated (Fisher et al., 2001; Eliezer, 2002; Colombier et al., 2006). The laser can heat any material at a solid density plasma state; in particular, critical point temperatures are easily achieved. The use of femtosecond lasers decouples the plasma formation of the target from the subsequent dynamics of the removed material, contrary to the interaction of nanosecond (or more) laser pulses.

The laser pulse ends before the expansion starts, therefore the laser does not interact with the ejected material. In this case the adiabatic expansion is a very good approximation. The laser absorption time duration is shorter than the electron lattice thermalization (~few ps), implying a rapid electron cooling creating fast atoms as well as NPs.

Furthermore, a very important advantage of the femtosecond lasers is the accuracy in material removal due to the ultrashort pulse duration (Du et al., 1994; Nolte et al., 1997; Gamaly et al., 1999). Using molecular dynamics simulation (Allen and Tildesley, 1987; Vidal et al., 2001; Lorazo et al., 2003; Cheng and Xu, 2005) or thermodynamic calculations (Eliezer et al., 1986, 2004; Gamaly et al., 2004), the phase transition occurs near the critical point. The femtosecond laser pulses are much shorter than the heat conduction timescale; therefore, the irradiated metal can reach a state of superheated liquid (Henis and Eliezer, 1993) with a temperature higher than its critical temperature T_c. During adiabatic cooling the liquid decomposes into tiny droplets and vapor. The size of the droplet is determined by the correlation length λ_c, given by (Stanley, 1971; Liu et al., 2007)

$$\lambda_c = r_0 \left(\frac{T - T_c}{T_c} \right)^{-\nu} \tag{8.2}$$

where ν is the scaling exponent and r_0 is the distance of the nearest interacting particles. λ_c describes the average dimension of density fluctuation in the fragmented liquid, thus giving the average magnitude of the droplet. The typical physical values of $r_0 \sim 1\,nm$, $\nu = 0.6$, and $(T - T_c)/T_c \sim 0.01$ imply $\lambda_c \sim 16\,nm$. This predicts the tendency of NP formation induced by femtosecond laser interaction with metals.

8.3 Laser–Plasma Interaction

In our experiments the NPs are collected on a silicon crystal at a distance of about 3 cm from the target, therefore allowing free expansion of NPs into the

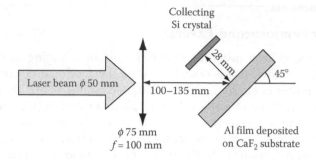

FIGURE 8.1
A schematic view with experimental description of the experiment.

vacuum (Eliezer, 2002). These experiments, described schematically in Figure 8.1, are carried out with a titanium sapphire laser with 0.8 μm wavelength, 50 fs pulse duration, and 0.5 J of energy with intensities varying between 3×10^{12} and 5×10^{15} W/cm². The irradiated targets are composed of a 100 nm thick foil either of aluminum or of carbon, deposited on a transparent heat-insulating glass substrate (e.g., CaF_2). The NP debris are collected on a silicone wafer.

For this class of lasers the solid target is heated faster than the timescale for hydrodynamic expansion. The plasma medium in this case has a density equal to its initial solid density during the laser pulse duration. The typical laser pulse duration is of the order of τ_L ~100 fs. During this time the laser is absorbed and heats the target along a skin depth δ ~200 Å, a value much larger than the surface expansion ~10 Å. For a laser spot much larger than the skin depth one can assume a one-dimensional (1D) problem.

We assume that the laser is radiated in the z-direction and is normally incident on a solid target (positioned in the x–y plane) at z = 0. The wave equation of the electric field for a linearly polarized monochromatic wave at normal incidence is

$$\frac{\partial^2 E(z)}{\partial z^2} + \frac{\varepsilon(\omega, z)\omega^2 E(z)}{c^2} = 0 \tag{8.3}$$

This equation is also known as the Helmholtz equation (Milchberg and Freeman, 1989), where ε is the dielectric permittivity of the target related to the electrical conductivity σ_E by

$$\varepsilon = 1 + \frac{i4\pi\sigma_E}{\omega} \tag{8.4}$$

For a constant ε and a vanishing electric field at infinity, the solution of Equation 8.3 is

$$E(z) = E(0)\exp\left(-\frac{z}{\delta}\right) \tag{8.5}$$

$E(z)$ is the electric field transmitted into the solid target. Denoting by E_i and E_r the incident and reflected electric field accordingly, one can write

$$E_i = E(0)\exp[i(kz - \omega t)]$$
$$E_r = \sqrt{R}E(0)\exp[i(-kz - \omega t)]$$

(8.6)

where R is the reflection coefficient. The absorption coefficient A is given by (Landau and Lifshitz, 1975; Born and Wolf, 1980)

$$A = 1 - R$$
$$R = \left|\frac{1-\sqrt{\varepsilon}}{1+\sqrt{\varepsilon}}\right|^2 = \frac{(n_R - 1)^2 + n_I^2}{(n_R + 1)^2 + n_I^2}$$

(8.7)

where the complex index of refraction is defined as

$$\sqrt{\varepsilon} = n_R + in_I$$

(8.8)

Substituting the solution of Equation 8.5 into Equation 8.3 yields the skin depth δ

$$\delta = \frac{i}{\sqrt{\varepsilon}}\frac{c}{\omega}$$

(8.9)

Substituting for the dielectric function

$$\varepsilon = 1 - \frac{\omega_{pe}^2}{\omega(\omega + i\nu)}$$

(8.10)

where ω_{pe} is the plasma frequency and ν is the electron effective collision frequency (e.g., $\nu = \nu_{ei}$), one gets

$$\delta = \frac{c}{\omega_{pe}}\left(\frac{1+i(\nu/\omega)}{1-(\omega^2/\omega_{pe}^2)-i(\nu\omega/\omega_{pe}^2)}\right)^{1/2} \approx \frac{c}{\omega_{pe}}\left(1+i\frac{\nu}{\omega}\right)^{1/2} \quad \text{for } \omega_{pe} \gg \omega \quad (8.11)$$

In our plasma medium (solid target) $\omega_{pe} \gg \omega$ and therefore, the last approximation is justified. The skin depth is therefore given by

$$\delta = \begin{cases} \dfrac{c}{\omega_{pe}}, & \text{for } \dfrac{\nu}{\omega} \ll 1 \text{ and } \dfrac{\omega}{\omega_{pe}} \ll 1 \\[4mm] \dfrac{c}{\omega_{pe}}\left(\dfrac{\nu}{2\omega}\right)^{1/2}, & \text{for } \dfrac{\nu}{\omega} \gg 1 \text{ and } \dfrac{\omega}{\omega_{pe}} \ll 1 \end{cases}$$

(8.12)

where $c/\omega_{pe} \sim 10\,nm$ for solid density plasma. For the last stage the following algebraic relation was used

$$\sqrt{a+ib} = \frac{1}{\sqrt{2}} \left(a^2+b^2\right)^{1/4} \left[\left(1+\frac{a}{\sqrt{a^2+b^2}}\right)^{1/2} + i\left(1-\frac{a}{\sqrt{a^2+b^2}}\right)^{1/2} \right] \tag{8.13}$$

Using the dielectric function from Equation 8.10 in the absorption coefficient A in Equation 8.7 one gets

$$A = \begin{cases} \dfrac{2v}{\omega_{pe}}, & \text{for } \dfrac{v}{\omega} \ll 1 \text{ and } \dfrac{\omega}{\omega_{pe}} \ll 1 \\[2em] \dfrac{2v}{\omega_{pe}}\sqrt{\dfrac{\omega}{v}}, & \text{for } \dfrac{v}{\omega} \gg 1 \text{ and } \dfrac{\omega}{\omega_{pe}} \ll 1 \end{cases} \tag{8.14}$$

For the above analysis it was assumed that the dielectric function ε is a constant in space and time. However, in general this is a very crude approximation, since during the laser pulse duration the dielectric function ε is not constant due to the change in electron and ion temperatures as well as possible change in the electron number density. The electron collision frequency v and the electrical conductivity σ_E are functions of T_e, T_i, and n_e, which are functions of space and time. In this case the following approach is more suitable.

Joule heating gives the laser power deposition into the target per unit volume

$$Q(z,t) = \frac{1}{2}\text{Re}\{\sigma_E(z,t)\}|E|^2 \tag{8.15}$$

Note that $\sigma_E(\omega,z)$ and $\varepsilon(\omega,z)$ vary with time t on a timescale much longer than $2\pi/\omega$. Both σ_E and ε depend on z and t through $T_e(z,t)$ and $T_i(z,t)$, respectively, while Q and $E(z)$ depend on t also via the laser pulse temporal intensity profile, $I_L(t)$. A good approximation for the intensity profile is

$$I_L \approx I_0 \sin^2\left(\frac{\pi t}{2\tau_L}\right) \quad \text{for } 0 \le t \le 2\tau_L$$

$$I_L \approx 0 \quad \text{for } t \le 0, t \ge 2\tau_L \tag{8.16}$$

where τ_L is the FWHM (full width half maximum) pulse duration. The absorption coefficient A measured at a given location (x,y) on the target surface $(z=0)$ is averaged over the temporal pulse profile:

$$A = \frac{\displaystyle\int_0^{2\tau_L} dt \int_0^{\infty} Q(z,t)dz}{\displaystyle\int_0^{2\tau_L} I_L(t)dt} \tag{8.17}$$

In order to calculate the absorption coefficient A in Equation 8.17 one has to use the heat deposition $Q(z,t)$ given in Equation 8.15, which requires a knowledge of the electric field E and electric conductivity σ_E inside the target.

Since the interaction of the femtosecond laser is with a solid state of matter during the laser pulse duration, a knowledge of quantum solid state physics is required (Ashcroft and Mermin, 1976; Gantmakher and Levinson, 1987; Ziman, 1995; Peierls, 2001). In the following we shall outline the ingredients and some approximated formulas for the calculation of the absorption of femtosecond lasers by solid targets.

The dielectric permittivity ε, or equivalently the electrical conductivity σ_E, contains the contribution from both the interband and intraband absorption mechanism:

$$\sigma_E = \sigma_{bb} + \sigma_D \qquad (8.18)$$

The subscript bb and D stand for interband (band–band transition) absorption and intraband absorption, known as the Drude contribution (the inverse bremsstrahlung absorption) given by

$$\sigma_D = \frac{n_e e^2}{m_{eff}} \left(\frac{\nu + i\omega}{\nu^2 + \omega^2} \right) \qquad (8.19)$$

Regarding the band–band transitions there are two distinct possibilities: transition can occur between the Bloch electron bands (between the lower [e.g., first band] and the upper [e.g., second band] electron bands in the first Brillouin zone), like in aluminum metal (Ashcroft and Sturm, 1971). The other possibility is the transition between the occupied atomic states and the Fermi surface. The bb transitions will not be further discussed here, although they can contribute significantly to the absorption of femtosecond lasers (Fisher et al., 2001).

The electric conductivity in Equation 8.19 is a function of the electron collision frequency ν, which in our case is the total momentum relaxation rate for electrons in the solid. This collision frequency has two contributions

$$\nu = \nu_{e,ph}(T_e, T_i) + \nu_{e,e}(T_e) \qquad (8.20)$$

where $\nu_{e,ph}$ and $\nu_{e,e}$ are the electron momentum relaxation rates due to electron–phonon and electron–electron collisions, respectively. Since ν is a function of T_e and T_i it is necessary to solve the energy equations

$$C_e(T_e)\frac{\partial T_e}{\partial t} = \frac{\partial}{\partial z}\left(\kappa(T_e)\frac{\partial T_e}{\partial z} \right) - U(T_e, T_i) + Q(z,t)$$

$$C_i(T_i)\frac{\partial T_i}{\partial t} = U(T_e, T_i) \qquad (8.21)$$

where
  C_e and C_i are the electron and ion heat capacities at constant volume
    (dimension [erg/(cm^3 K)])
  κ is the electron heat conductivity (dimension [erg/(cm s)])
  U is the energy transfer rate from electrons to ions and can also be
    written by

$$U = \gamma(T_e - T_i). \tag{8.22}$$

To calculate κ, U, and σ_E as functions of T_e and T_i a knowledge of the phonon
spectrum (i.e., the dispersion relation) and the electron–phonon interaction
matrix element is required (Kaganov et al., 1957).

The phonon spectrum $\omega_{ph}(q)$ for longitudinal phonons, where q is the
phonon wave vector, can be approximated by (Dederichs et al., 1981)

$$\omega_{ph}(q) = qs \quad \text{for } q \leq q_b$$
$$\omega_{ph}(q) = q_b s \quad \text{for } q \geq q_b \tag{8.23}$$

where
  s is the longitudinal sound velocity (e.g., $s = 6.4 \times 10^5$ cm/s for aluminum)
  q_b is a parameter determined by fitting the DC electrical conductivity
    data

Since electron–phonon scattering in metals occurs primarily via a single
longitudinal phonon absorption or emission (by an electron), the following
matrix element square is assumed (Gantmakher and Levinson, 1987):

$$\left| M(\mathbf{k} \rightarrow \mathbf{k}' \pm \mathbf{q}) \right|^2 = \begin{cases} \dfrac{\hbar w_0^2 q}{2V\rho s} \delta_{k'-k,\mp q} & \text{for } q \leq q_b \\[2ex] \dfrac{\hbar w_0^2 q_b}{2V\rho s} \delta_{k'-k,\mp q} & \text{for } q \geq q_b \end{cases} \tag{8.24}$$

where
  w_0 is the deformation potential constant
  ρ and V are the medium density and volume, respectively
  $\hbar \mathbf{q}$ is the quasimomentum of the absorbed (or emitted) phonon as the
    electron wave vector has changed from $\mathbf{k}$ to $\mathbf{k}'$

The δ in Equation 8.24 takes care of the momentum conservation in these
processes. The transition probability is given by

$$W(\mathbf{k} \rightarrow \mathbf{k}' \pm \mathbf{q}) = \frac{2\pi}{\hbar} \left| M(\mathbf{k} \rightarrow \mathbf{k}' \pm \mathbf{q}) \right|^2 \left(N_q + \frac{1}{2} \pm \frac{1}{2} \right) \delta \left(E_k - E_{k'} \mp \hbar \omega_{ph}(q) \right) \tag{8.25}$$

$E_k = E_F + \hbar^2\{k_F[|k|-|k_F|]/2m_{eff}$ and the δ describe the energy conservation. N_q is the population of longitudinal phonon mode in the first Brillouin zone with the reduced momentum $q_{red} = q - K$, where K is a corresponding vector of the reciprocal lattice. q may lie outside the first Brillouin zone (Umklapp process) but q_b is inside the first Brillouin zone. N_q is the Bose–Einstein distribution function given by

$$N_q = \left\{ \exp\left[\frac{\hbar \omega_{ph}(q_{red})}{k_B T_i} \right] - 1 \right\}^{-1} \tag{8.26}$$

The rate of electron–phonon (i.e., electron–ion) energy exchange U is calculated by the method described by Kaganov et al. (Kaganov et al., 1957; Allen, 1987; Fisher et al., 2001)

$$U = \frac{1}{(2\pi)^3} \int \left(\frac{dN_q}{dt} \right) \hbar \omega_q 4\pi q^2 dq \tag{8.27}$$

Using the phonon dispersion relation (Equation 8.23) and the matrix element (Equation 8.24), we get Equation 8.27. Note that $U(T_e = T_i) = 0$ as required from physical consideration. For aluminum with ion temperatures larger than 300 K, Equation 8.27 yields a γ value in the domain $(3.7–3.9)\times 10^{18}$ erg/ (cm^3 s K). This value is in very good agreement with Wang and Downer (1992) and is larger than the empirical values of Eidman et al. by a factor of 1.5–3 (2000).

$$U = \gamma(T_e - T_i) = \frac{s w_0^2 m_{eff}^2}{4\pi^3 \hbar^2 \rho} \left\{ \left(\frac{\hbar s}{k_B T_e} \right)^{-5} \int_0^{\hbar q_b s/k_B T_e} \frac{x^4 dx}{e^x - 1} - \left(\frac{\hbar s}{k_B T_i} \right)^{-5} \right.$$

$$\times \left. \int_0^{\hbar q_b s/k_B T_i} \frac{x^4 dx}{e^x - 1} + \frac{q_b^3}{2} (4k_F^2 - q_b^2) \left[\exp\left(\frac{\hbar q_b s}{k_B T_e} \right) - 1 \right] \right\} \tag{8.28}$$

The electron heat conductivity is given by

$$\kappa(T_e, T_i) \approx \frac{C_e}{3\nu} \left(v_F^2 + \frac{3k_B T_e}{m_e} \right) \tag{8.29}$$

$v_F = \hbar k_F/m_{eff}$ is the Fermi velocity, m_{eff} is the effective mass of the electron in the solid and $\hbar = 2\pi h$, where h is the Planck constant. The Fermi momentum k_F is related to the Fermi energy $E_F = (\hbar k_F)^2/2m_e$, where m_e is the (free) electron mass. ν is given by Equation 8.20 and it has two contributions, one from the electron–electron collisions and the second from the electron–phonon collisions defined by

$$v_{e,ph} = \sum_{k'} W(\mathbf{k} \to \mathbf{k}' \pm \mathbf{q}) \left[\frac{(\mathbf{k} - \mathbf{k}') \cdot \mathbf{k}}{k^2} \right] \left[\frac{1 - f(k', T_e)}{1 - f(k, T_e)} \right] \qquad (8.30)$$

$f(k,T_e)$, the Fermi–Dirac distribution, describes the electron population

$$f(k, T_e) = \left\{ \exp\left[\frac{E_k - \mu(T_e)}{k_B T_e} \right] + 1 \right\}^{-1} \qquad (8.31)$$

where μ is the free electron chemical potential. One can see from Equation 8.30 that v is the electron momentum relaxation rate due to collisions with phonons. For temperatures higher than the Fermi temperature the $v_{e,ph}$ has to be changed to the plasma frequency collisions v_{ei}.

We do not give here a detailed calculation of $v_{e,ph}$, but point out that for temperatures smaller than the Fermi temperature the electron–phonon collision frequency is independent of T_e and the following approximation is valid:

$$v_{e,ph} \approx B_1 T_i \quad \text{for } T_i \leq T_F = \frac{E_F}{k_B} \qquad (8.32)$$

where B_1 is a constant. If solid–liquid phase transition takes place then the approximation (Equation 8.32) is valid for $T_e = T_i < T_m$ (T_m is the melting temperature) and for $T_F > T_e = T_i > T_m$, and $v_{e,ph}$ is expected to have a jump at T_m. However, on the femtosecond timescale T_e is not equal to T_i and it is conceivable that the solid–liquid transition time is longer than the laser pulse duration, so that in this case Equation 8.32 might be a good approximation. For $T_e > T_F$ this dependence on T_i is no longer satisfied; however, in this domain $v_{e,ph}$ is significantly smaller than the momentum relaxation rate in electron–electron collisions $v_{e,e}$, given by

$$v_{e,e} \approx \begin{cases} \left| \frac{E_F}{\hbar} \left(\frac{T_e}{T_F} \right)^2 \right| \equiv B_2 T_e^2 & \text{for } T_e < T_F \\ \\ \frac{E_F}{\hbar} \left(\frac{T_e}{T_F} \right)^{-3/2} & \text{for } T_e > T_F \end{cases} \qquad (8.33)$$

In the second approximation of Equation 8.33 the logarithmic factor in T_e was ignored.

The values of w_0 and q_b in the above equations are determined by fitting the DC electrical conductivity with the experimental data (Zinovev, 1996) as a function of temperature. Using Equations 8.19, 8.32, and 8.33 one has to fit the following expression to obtain the values of w_0 and q_b.

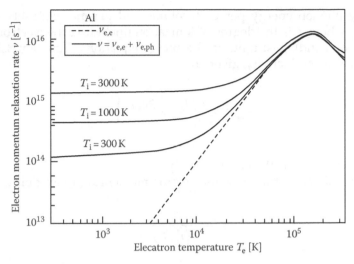

FIGURE 8.2
Calculated collision frequency $v(T_e, T_i)$ for several values of ion temperature T_i.

$$\sigma_E(DC) \equiv \sigma_E(\omega = 0) = \frac{n_e e^2}{m_{eff} v}$$

$$v(T_e = T_i = T < T_m) = B_1 T + B_2 T^2 \tag{8.34}$$

For example, for aluminum with an electron effective mass $m_{eff} = 1.20 m_e$ one gets $w_0 = 5.16\,eV$ and $q_b = 4.77 \times 10^7\,cm^{-1}$. Figure 8.2 shows the calculated values of the collision frequency $v(T_e, T_i)$ as a function of the electron temperature for several values of ion temperatures.

In order to solve the energy equations (Equation 8.21), one has to know also the electron and ion heat capacities at constant volume C_e and C_i. C_e can be approximated by

$$
\begin{aligned}
C_e(low) &= \frac{\pi^2 n_e k_B^2 T_e}{2 E_F} \quad \text{for } k_B T_e \ll E_F \\
C_e(high) &= 1.5 k_B n_e \quad \text{for } k_B T_e \gg E_F \\
C_e &= \frac{C_e(low) C_e(high)}{\sqrt{C_e(low)^2 + C_e(high)^2}} \quad \text{for } k_B T_e \sim E_F
\end{aligned}
\tag{8.35}
$$

The ion heat capacity at constant volume is given by

$$
C_i = \left(\frac{\partial E_i}{\partial T_i} \right)_V \approx
\begin{cases}
3 k_B n_i & \text{for } T_i \geq 300\,K \\
\sim T_i^3 & \text{for } T_i \ll 300\,K
\end{cases}
\tag{8.36}
$$

$$E_i = \sum_{phonon\ modes} \int_0^{q_{max}} \frac{dq\, 4\pi q^2 \hbar \omega(q)}{(2\pi)^3 \left[\exp\left(\hbar \omega(q)/k_B T_i\right) - 1 \right]}$$

E_i is the phonon energy per unit volume and plasma neutrality requires $n_e = Zn_i$, where Z is the degree of ionization (including the electrons in the conduction band). Last input to the energy equations (Equation 8.21) is the electron heat conductivity, given by

$$\kappa(T_e, T_i) \approx \frac{C_e}{3v}\left(v_F^2 + \frac{3k_B T_e}{m_e} \right) \tag{8.37}$$

where $v_F = \hbar k_F / m_{eff}$ is the Fermi velocity.

The characteristic timescale for electron momentum relaxation τ_{ke} is

$$\tau_{ke} = \frac{1}{v(T_e, T_i)} \tag{8.38}$$

For aluminum at room temperature this characteristic timescale is dominated by electron–phonon collision time, $\tau_k = 8\,\text{fs}$, while at the Fermi temperature ($T_F = 0.135 \times 10^6\,\text{K}$) the electron–electron collisions dominates and $\tau_k = 0.1\,\text{fs}$.

The characteristic times of electron energy relaxation (cooling) τ_{Ee} and ion (i.e., lattice) energy relaxation (heating) τ_{Ei} are defined by

$$\tau_{Ee} = \frac{C_e(T_e - T_i)}{U} = \frac{C_e}{\gamma}$$
$$\tau_{Ei} = \frac{C_i}{\gamma} \tag{8.39}$$

For aluminum the electron energy relaxation time increases from $\tau_{Ee}(T_e = 300\,\text{K})$ ~100 fs to $\tau_{Ee}(T_e = T_F)$ ~ 10 ps mainly due to the increase in $C_e(T_e)$, while $\tau_{Ei}(T_e > 300\,\text{K})$ ~6 ps since both C_i and γ are nearly constant in T_e and T_i at room temperature and above.

Experimental data and theoretical calculations of femtosecond laser absorption by an aluminum target is given in the literature (Fisher et al, 2001; Amoruso et al, 2005a,b). Figure 8.3 shows the detailed results of the above calculations for a titanium–sapphire laser ($\tau_L = 50\,\text{fs}$, $\lambda_L = 800\,\text{nm}$) interacting with a copper target. The experimental data indicate the domain where the above approximations are relevant.

It is interesting to point out that the laser absorption can increase with surface roughness. In particular for a gold target (Vorobyev and Guo, 2005) with absorption of about 25% in one 60 fs laser pulse (of Ti:sapphire with a wavelength of 800 nm) at a fluence ~1 J/cm², one gets with a large number of applied laser pulses an absorption close to 100%.

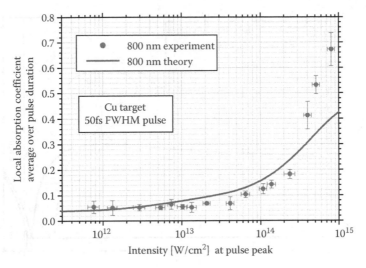

FIGURE 8.3

Absorption of a 50 fs titanium–sapphire laser ($\lambda_L = 0.800\,\mu m$) by Cu target.

8.4 Aluminum Nanoparticles

The experimental setup and conditions are described in Figure 8.1. The pressure in the vacuum chamber is 10^{-4} Torr. The laser absorption coefficient changes between 16% and 40% for laser irradiances of 5×10^{12} and $2 \times 10^{14}\,W/cm^2$, respectively.

The density–temperature (ρ,T) diagram of aluminum is presented in Figure 8.4. Here we are using the accurate equation of state (EOS) similar to the one we developed in analyzing the ultimate strength of materials (Eliezer et al., 2002). The initial conditions of the laser-created plasma are given by the solid density ρ_0 and temperature T_0 determined by the laser–plasma interaction. The area between the spinodal and binodal on the right side in Figure 8.4 is the superheated liquid domain. If during the adiabatic expansion from the initial condition ($\rho_0\,T_0$) the plasma enters the superheated liquid domain, then the liquid phase is achieved followed by a fast crystallization of the aluminum. Adiabatic paths starting at T_0 in the domain of 0.2 to 4 eV are the most promising in the context of NP productions; since they reach the spinodal lower than the critical density where a large fraction of the material transforms into a liquid state. For initial temperatures T_0 is larger than 10 eV, the plasma expands, cools, recombines into gas, and is eventually deposited on the vacuum vessel walls as individual atoms.

For the appropriate initial conditions, the NPs are created during a period of order of 1 ns. These NPs expand into the vacuum perpendicular to the

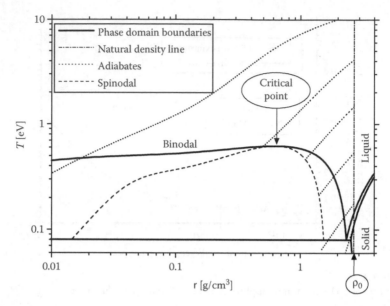

FIGURE 8.4
Phase diagram of aluminum.

aluminum target and are collected on a silicon wafer for x-ray diffraction (XRD), for scanning electron microscopy (SEM), and for atomic force microscopy (AFM). For transmission electron microscopy (TEM), the debris are caught on a copper grid covered on one side with a carbon membrane.

A typical AFM image (out of many images of different scales) of the NPs deposited onto a silicone substrate is given in Figure 8.5. In this particular

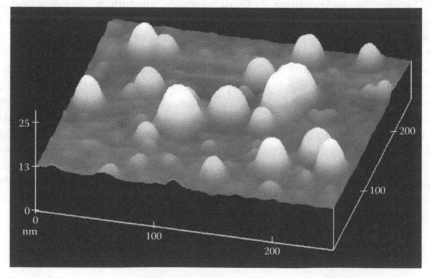

FIGURE 8.5
An AFM image for aluminum NPs.

FIGURE 8.6
An AFM image for large aluminum plate.

figure, the scan size in the x–y is 250×250 nm and 25 nm in the z-direction. This image reveals particles ranging from 10 to 50 nm in diameter and from 2 to 10 nm in height.

Figures 8.6 through 8.8 show AFM images (Grossman et al., 2007) of a very large aluminum particle, nickel NPs, and iron NPs, respectively. The large

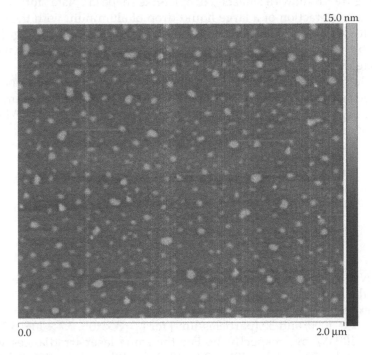

FIGURE 8.7
An AFM image for nickel NPs.

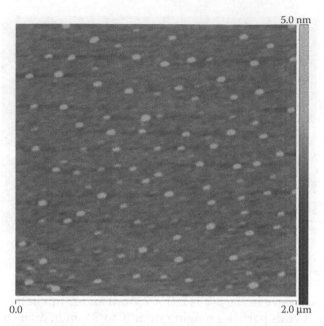

5.0 nm

0.0   2.0 μm

FIGURE 8.8
An AFM image for iron NPs.

NP in Figure 8.6 shows a smashed droplet-like shape (a "plate" form) created during the interaction of a large liquid drop of aluminum with the silicone substrate. While the NPs cool very fast and solidify before hitting the silicone substrate, the large particles of the order of one micron collide with the substrate while being in a liquid state of matter.

The aluminum size distribution of the NPs is determined with the aid of an image analysis software package based on the AFM data. The cumulative number density $f(S)$ of NPs with an area from ∞ to S (in units of nm^2) fits well the following formula:

$$f(S)\left[\text{particles}/\mu m^2\right] = \alpha_1 \exp\left(-\frac{S}{\mu_1}\right) + \alpha_2 \exp\left(-\frac{S}{\mu_2}\right)$$

$$N\left[\text{particles}/\mu m^2\right] = f(0) = \alpha_1 + \alpha_2 \qquad (8.40)$$

$$<S> = \frac{\alpha_1\mu_1 + \alpha_2\mu_2}{\alpha_1 + \alpha_2}$$

where N is the total number of particles per unit area and $<S>$ is the average area of the NPs.

Our experiments show the following data:

N = 4, 300, 1000, and 30 [particles/μm^2] for I_L = 3×10^{12}, 3.7×10^{13}, 5.3×10^{13}, and 5×10^{14} W/cm^2, respectively. For the same laser irradiances we get appropriately $<S>$ = 1.2×10^4, 4.2×10^2, 1.2×10^2, and 5×10^3 nm^2. These experimental values are in agreement with the qualitative model described above.

The TEM data characterizes the structure and the composition of the NPs at a subnanometer resolution. Line arrays of atoms are measured indicating the crystalline nature of the aluminum NPs. The distance between the atomic layers of the crystal is found to be 2.354 ± 0.118 Å, which is in good agreement with the strongest reflection from the cubic Al phase (2.338 Å for planes {111}). It is important to point out that these crystals contain less than 10 atoms in each direction.

The dynamics of the laser ablation plume expansion of aluminum was recently analyzed by using space and time resolved soft x-ray absorption spectroscopy (Okano et al., 2006). For a laser intensity of 10^{14} W/cm^2, the ions expanded with a velocity of 10^6 cm/s while the NPs had a velocity of about 10^5 cm/s.

8.5 Nickel NPs

The generation of nickel NPs with a 500 fs pulsed laser was investigated (Liu et al., 2007). The laser, with a wavelength of 1 μm, is focused to about 30 μm on a nickel target in vacuum (pressure ~1 × 10^{-5} Pa). The repetition rate of the laser is up to 200 kHz with a maximum pulse energy of 100 μJ. A silicon substrate collects the NPs for AFM and a copper grid collection is used separately for high-resolution TEM (HRTEM) analysis.

The sizes of the NPs were measured for the laser fluence between 0.1 and 10 J/cm^2 (for the 500 fs pulses). In particular, for a fluence of 0.4 J/cm^2 the particles have a very narrow size distribution peaked at 8 nm, while at 8 J/cm^2 the NPs have a wider distribution with a tail of very large sizes. The large particles (~100 nm and more) show a smashed droplet-like shape. In addition, by supplying different background gases, various crystalline structures are obtained (e.g., NiO cubes and Ni/NiO core/shell spheres).

In analyzing the laser ablation of nickel in vacuum (<10^{-5} Pa) the NPs of nickel were detected (Amoruso et al., 2007). These experiments were done with a 527 nm laser wavelength and a pulse duration of 300 fs, about 1.3 mJ per pulse with a repetition rate of 33 Hz, within the fluence domain of 0.1 to 1 J/cm^2. Time-gated optical imaging and optical emission spectroscopy are used to characterize the expansion dynamics. It is found that an atomic fast cloud of atoms leaves the target surface (on ns timescale) significantly before the NPs cloud does on a μs timescale. The velocity of the atomic cloud is of the order of 10 km/s and it is larger by a factor of 10 or more than the flow of the NP cloud. The dimensions of the NPs are between 10 and 100 nm. The final ablation depth in these experiments is of the order of 100 nm. The experimental analysis is supported by hydrodynamic modeling and molecular dynamics simulations (Perez and Lewis, 2003; Amoruso et al., 2005a,b, 2007; Cheng and Xu, 2005; Lorazo et al., 2006).

8.6   Carbon Targets

For carbon the EOS is more complicated and we do not have all the necessary data and knowledge to make a rigorous analysis like the one for aluminum in Section 8.4. Therefore, a simplified analysis is done here for the carbon case.

The energy equation used in this simplified analysis equates the absorbed laser energy to the integrated heat capacity:

$$(1-R)(F-F_m)\alpha = nc_P(T-T_b) \tag{8.41}$$

The nanotubes are obtained with a laser fluence of $F = 7.5$ [J/cm^2], implying a laser irradiance of $I = 1.5 \times 10^{14}$ W/cm^2. The fluence of the femtosecond laser that melts (actually sublimates) the graphite is F_m [J/cm^2] = 0.13 (Reitze et al., 1992). From the reflection experiments one gets for our case $R = 0.2$, that is, 80% of the laser energy is absorbed. The $\alpha^{-1} = 10^{-5}$ cm is the carbon layer thickness, assumed to have an initial constant temperature T, $T_b = 4500$ K is the boiling temperature, $c_P = 3k_B$ (k_B is the Boltzmann constant) is the Dulong–Petit value of the heat capacity, and $n = 1.12 \times 10^{23}$ [cm^{-3}] is the initial density of the graphite plasma (for $\rho_0 = 2.25$ g/cm^3). Equation 8.41 yields an initial temperature of $T = 111{,}000$ K.

Our laser focal spot is large compared with the plasma expansion before it solidifies. Therefore, it is conceivable to assume a 1D adiabatic expansion described by the following equations:

$$\frac{T}{T_b} = \left(\frac{\rho}{\rho_0}\right)^{\gamma-1} = \left(\frac{x_0}{x_0+ut}\right)^{\gamma-1}$$

$$u = \frac{2}{\gamma-1}\left[\frac{Zk_BT}{M}\right] \approx 2\times 10^6\,[T(\text{eV})]^{1/2}\left[\frac{\text{cm}}{\text{s}}\right] \tag{8.42}$$

where
   $x_0 = \alpha^{-1}$ is the graphite foil thickness
   $\gamma = 1.6$ is the adiabatic constant
   u is the plasma expansion velocity
   $Z = 4$ is the appropriate ionization of carbon
   M is the carbon mass

For the temperature derived in Equation 8.41, one gets $u = 6 \times 10^6$ cm/s. This value yields from Equation 8.42 a time $t = 0.35$ ns, that is, the plasma has cooled to the boiling temperature T_b in a time of 0.35 ns while expanding only 60 μm. We do not know the time required for the nanotube production; however, it is estimated to be in the order of one to a few nanoseconds.

Figure 8.9 was taken with a TEM. In Figure 8.9, about 10 concentric tubes are seen when the internal tube has a diameter of about 5 nm. (Dresselhaus et al., 1996).

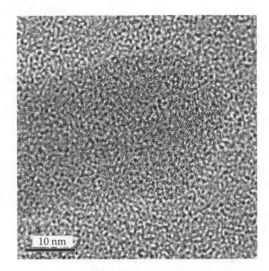

FIGURE 8.9
Nanotubes of carbon as measured by TEM.

8.7 Conclusions

The main methods used to synthesize nanotubes are carbon arc discharge, chemical vapor deposition, and ion bombardment. NPs have also been achieved with these techniques and also by laser ablative deposition in an appropriate gas atmosphere. In this chapter it has been suggested to use femtosecond lasers to create nanocrystals and nanotubes in a vacuum medium. One of the advantages of this scheme is that the interaction of the external field (the laser in our case) with the material under consideration during the NP or nanotube formation is nonexistent. The femtosecond laser is used only to achieve appropriate initial conditions. Therefore, in these experiments it might be easier to understand the mechanism of the production of nanotubes and nanocrystals. Furthermore, with the laser, there are no limits on the density and temperature of the initial conditions. Very high temperatures and solid densities can be easily achieved. More work and experiments are required to comprehend the new ideas.

References

Allen P. B. 1987 Theory of thermal relaxation of electrons in metals, *Phys. Rev. Lett.* 59, 1460.

Allen M. P. and Tildesley D. J. 1987 *Computer Simulations of Liquids*, Clarendon, Oxford.

Amoruso S., Ausanio G., Bruzzese R., Vitiello M., and Wang X. 2005a Femtosecond laser pulse irradiation of solid targets as a general route to nanoparticle formation in a vacuum, *Phys. Rev. B* 71, 033406.

Amoruso S., Bruzzese R., Vitiello M., Nedialkov N. N., and Atanasov P. A. 2005b Experimental and theoretical investigations of femtosecond laser ablation of aluminum in vacuum, *J. Appl. Phys.* 98, 044907.

Amoruso S., Ausanio G., Bruzzese R., Lanotte L., Scardi P., Vitiello M., and Wang X. 2006 Synthesis of nanocrystal films via femtosecond laser ablation in vacuum, *J. Phys.: Condens. Matter* 18, L49.

Amoruso S., Bruzzese R., Wang X., Nedialkov N. N., and Atanasov P. A. 2007 Femtosecond laser ablation of nickel in vacuum, *J. Phys. D: Appl. Phys.* 40, 331.

Ashcroft N. W. and Mermin N. D. 1976 *Solid State Physics*, Harcourt Brace College Publishers, Fort Worth, TX.

Ashcroft N. W. and Sturm K. 1971 Interband absorption and the optical properties of polyvalent metals, *Phys. Rev. B* 3, 1898.

Ausanio G., Barone A. C., Iannotti V., Lanotte L., Amoruso S., Brúcese R., and Vitiello M. 2004 Magnetic and morphological characteristics of nickel nanoparticles films produced by femtosecond laser ablation, *Appl. Phys. Lett.* 85, 4103.

Ausanio G., Barone A. C., Iannotti V., Scardi P., D'Incau M., Amoruso S., Vitiello M., and Lanotte L. 2006 Morphology, structure and magnetic properties of $(Tb_{0.3}Dy_{0.7}Fe_2)_{100-x}Fe_x$ nanogranular films produced by ultrashort pulsed laser deposition, *Nanotechnology* 17, 536.

Born M. and Wolf E. 1980 *Principles of Optics*, 6th edn., Pergamon, Oxford.

Cao G. 2004 *Nanostructures and Nanomaterials. Synthesis, Properties and Applications*, Imperial College Press, London.

Cheng C. R. and Xu X. F. 2005 Mechanisms of decomposition of metal during femtosecond laser ablation, *Phys. Rev. B* 72, 165415.

Colombier J. P., Combis P., Rosenfeld A., Hertel I. V., Audouard E., and Stoian R. 2006 Optimized energy coupling at ultrafast laser-irradiated metal surfaces by tailoring intensity envelopes: Consequences for material removal from Al samples, *Phys. Rev. B* 74, 224106.

Dederichs P. H., Schober H. R., and Sellmyer D. J. 1981 *Landolt-Bornstein*, 3rd edn., Vol. 13a, Springer-Verlag, Berlin, p. 11.

Dresselhaus M. S., Dresselhaus G., and Eklund P. C. 1996 *Science of Fullerenes and Carbon Nanotubes*, Academic Press, San Diego.

Du D., Liu X., Korn G., Squier J., and Mourou G. 1994 Laser-induced breakdown by impact ionization in SiO_2 with pulse widths from 7 ns to 150 fs, *Appl. Phys. Lett.* 64, 3071.

Eidman K., Meyer-ter-Vehn J., Schlegel T., and Huller S. 2000 Hydrodynamic simulation of subpicosecond laser interaction with solid-density matter, *Phys. Rev. E* 62, 1202.

Eliezer S. 2002 *The Interaction of High Power Lasers with Plasmas*, Institute of Physics, Bristol, UK.

Eliezer S., Ghatak A., Hora H., and Teller E. 1986 *An Introduction to Equations of State: Theory and Applications*, Cambridge University Press, Cambridge; 2002 *Fundamentals of Equations of State*, 2nd edn., World Scientific, Singapore.

Eliezer S., Moshe E., and Eliezer D. 2002 Laser-induced tension to measure the ultimate strength of metals related to the equation of state, *Laser Part Beams* 20, 87.

Eliezer S., Eliaz N., Grossman E., Fisher D., Gouzman I., Henis Z., Pecker S., Horovitz Y., Fraenkel M., Maman S. Ezersky V., and Eliezer D. 2005 Nanoparticles and nanotubes induced by femtosecond lasers, *Laser Part. Beams* 23, 15.

Eliezer S., Eliaz N., Grossman E., Fisher D., Gouzman I., Henis Z., Pecker S., Horovitz Y., Fraenkel M., Maman S., and Lereah Y. 2004 Synthesis of nanoparticles with femtosecond laser pulses, *Phys. Rev. B* 69, 144119.

Fisher D., Fraenkel M., Henis Z., Moshe E., and Eliezer S. 2001 Interband and intraband (Drude) contributions to femtosecond laser absorption in aluminum, *Phys. Rev. E* 65, 016409.

Gamaly E. G., Rode A. V., and Luther-Davies B. 1999 Ultrafast ablation with high-pulse-rate lasers. Part I: Theoretical considerations, *J. Appl. Phys.* 85, 4213.

Gamaly E. G., Rode A. V., and Luther-Davies B. 2000 Formation of diamond-like carbon films and carbon foam by ultrafast laser ablation, *Laser Part. Beams* 18, 245.

Gamaly E. G., Rode A. V., Uteza O., Kolev V., Luther-Davies, B., Bauer T., Koch J., Korte F., and Chichkov B. N. 2004 Control over a phase state of the laser plume ablated by femtosecond laser: Spatial pulse shaping, *J. Appl. Phys.* 95, 2250.

Gantmakher V. F. and Levinson Y. B. 1987 *Carrier Scattering in Metals and Semiconductors*, North Holland, Amsterdam.

Gogotsi Y. (ed.), 2006 *Nanomaterials Handbook*, Taylor & Francis, Boca Raton, FL.

Grossman E., Gouzman I., Louzon E., Maman S., and Eliezer S. 2007 Atomic force microscopy analysis of nanoparticles created by femtosecond laser pulses, to be published.

Henis Z. and Eliezer S. 1993 Melting phenomenon in laser induced shock waves, *Phys. Rev. E* 48, 2094.

Kaganov M. I., Lifshitz I. M., and Tanatarov L. V. 1957 Relaxation between electrons and crystalline lattice, *Sov. Phys. JETP* 4, 173.

Landau L. D. and Lifshitz E. M. 1975 *Electrodynamics of Continuous Media*, Pergamon Press, Oxford.

Liu B., Hu Z., and Che Y. 2007 Nanoparticle generation in ultra-fast pulsed laser ablation of nickel, *Appl. Phys. Lett.* 90, 044103.

Lorazo P., Lewis L. J., and Meunier M. 2003 Short-pulse laser ablation of solids: From phase explosion to fragmentation, *Phys. Rev. Lett.* 91, 225502.

Lorazo P., Lewis L. J., and Meunier M. 2006 Thermodynamic pathways to melting, ablation, and solidification in absorbing solids under pulsed laser irradiation, *Phys. Rev. B* 73, 134108.

Milchberg H. M. and Freeman R. R. 1989 Light absorption in ultrashort scale length plasmas, *J. Opt. Soc. Am. B*, 6, 1351.

Nolte S., Momma C., Jacobs H., Tunnermann A., Chichkov N., Wellegehausen B., and Welling H. 1997 Ablation of metals by ultrashort laser pulses, *J. Opt. Soc. Am. B* 14, 2716.

Okano Y., Oguri K., Nishikawa T., and Nakano H. 2006 Observation of femtosecond laser induced ablation plumes of aluminum using space and time resolved soft x-ray absorption spectroscopy, *Phys. Rev. Lett.* 89, 221502.

Peierls R. E. 2001 *Quantum Theory of Solids*, Clarendon Press, Oxford.

Perez D. and Lewis L. J. 2003 Molecular-dynamics study of ablation of solids under femtosecond laser pulses, *Phys. Rev. B* 67, 184102.

Reitze, D. H., Ahn, H., and Downer, M. C. 1992 Optical properties of liquid carbon measured by femtosecond spectroscopy, *Phys. Rev. B* 45, 2677.

Stanley H. E. 1971 *Introduction to Phase Transitions and Critical Phenomena*, Oxford University Press, New York, pp. 39–46.

Vidal F., Johnston T. W., Laville S., Barthelemy O., Chaker M., Le Drogoff B., Margot J., and Sabsabi M. 2001 Critical-point phase separation in laser ablation of conductors, *Phys. Rev. Lett.* 86, 2573.

Mitura S. (ed.), 2000 *Nanotechnology in Material Science*, Pergamon, Oxford.

Vorobyev A. Y. and Guo C. 2005 Enhanced absorption of gold following multipulse femtosecond laser ablation, *Phys. Rev. B* 72, 195422.

Wang X. W. and Downer M. C. 1992 Femtosecond time-resolved reflectivity of hydrodynamically expanding metal surfaces, *Opt. Lett.* 17, 1450.

Wang Y. L., Xu W., Zhou Y., Chu L. Z., and Fu G. S. 2007 Influence of pulse repetition rate on the average size of silicon nanoparticles deposited by laser ablation, *Laser Part. Beams* 25, 9.

Wolf E. L. 2006 *Nanophysics and Nanotechnology. An Introduction to Modern Concepts in Nanoscience*, 2nd edn., Wiley, New York.

Ziman J. M. 1995 *Principles of the Theory of Solids*, 2nd edn., Cambridge University Press, Cambridge.

Zinovev V. E. 1996 *Handbook of Thermophysical Properties of Metals at High Temperatures*, Nova Science, New York.

Index

Milton Keynes UK
Ingram Content Group UK Ltd.
UKHW040447071024
449327UK00020B/1054

9 780367 452476